Nurul Islam

Reducing Rural Poverty
in Asia
Challenges and Opportunities
for Microenterprises
and Public Employment Schemes

*Pre-publication
REVIEWS,
COMMENTARIES,
EVALUATIONS . . .*

"**T**he case studies in this book provide a wealth of information on public employment schemes and rural nonfarm enterprises in Bangladesh, India, and the Philippines, including program descriptions, analysis of problem areas, and findings from recent micro-level surveys. Lessons drawn from these case studies can make a substantial contribution to improving the design and performance of programs and projects aimed at raising real incomes of the poor in Asia."

Paul Dorosh, PhD
*Senior Economist,
South Asia Agriculture
and Rural Development Unit,
World Bank*

"**A**ny attack on world poverty must place major emphasis on rural poverty in Asia, which accounts for the majority of the world's poor. Although the long-run prospects for eliminating rural poverty depend on rapid economic growth and structural transformation, in the short-to-medium term, rural poverty needs to be tackled directly, via programs to raise incomes and productivity. The countries covered in this book—India, Bangladesh, and the Philippines—have unprecedented experience with numerous schemes of this sort.

This comprehensive, thorough, and insightful book reviews mechanisms that have been adopted to extend credit to the rural poor and employment schemes in countries throughout the region. The book provides invaluable guidance for anyone concerned with rural poverty in Asia, particularly for policymakers, for whom this should be compulsory reading. Academics and students concerned with rural development in Asia will also benefit greatly."

Frances Stewart, DPhil
*Professor of Development Economics,
University of Oxford,
United Kingdom*

More pre-publication
REVIEWS, COMMENTARIES, EVALUATIONS . . .

"Dr. Islam provides a comprehensive assessment of the policies needed to help Asian smallholder farmers escape poverty, with emphasis on policies related to the capital and labor markets in rural areas of Asia. The policy conclusions are based on in-depth empirical evidence from three Asian countries and the author draws on his extensive experience and understanding of Asian rural development in general. This book provides a very valuable set of generalizable lessons for the design of future policies for rural development in Asia and elsewhere. It also provides a large number of very specific lessons related to microcredit schemes and public employment undertakings and valuable policy recommendations to facilitate the transition from small-scale farming to expanded nonfarm rural employment.

This book is a must-read for students and scholars of rural development in Asia and elsewhere."

Per Pinstrup-Andersen, PhD
*World Food Prize Laureate
and H. E. Babcock Professor,
Cornell University*

Food Products Press®
An Imprint of The Haworth Press, Inc.
New York • London • Oxford

Reducing Rural Poverty in Asia

in Asia

Challenges and Opportunities
for Microenterprises
and Public Employment Schemes

FOOD PRODUCTS PRESS®
Global Food & Nutrition Security
Suresh Babu, PhD
Senior Editor

Economic Reforms and Food Security: The Impact of Trade and Technology in South Asia edited by Suresh Chandra Babu and Ashok Gulati

Reducing Rural Poverty in Asia: Challenges and Opportunities for Microenterprises and Public Employment Schemes by Nurul Islam

Reducing Rural Poverty in Asia

Challenges and Opportunities for Microenterprises and Public Employment Schemes

Nurul Islam

Food Products Press®
An Imprint of The Haworth Press, Inc.
New York • London • Oxford

For more information on this book or to order, visit
http://www.haworthpress.com/store/product.asp?sku=5368

or call 1-800-HAWORTH (800-429-6784) in the United States and Canada
or (607) 722-5857 outside the United States and Canada

or contact orders@HaworthPress.com

Published by

Food Products Press®, an imprint of The Haworth Press, Inc., 10 Alice Street, Binghamton, NY 13904-1580.

PUBLISHER'S NOTE
The development, preparation, and publication of this work has been undertaken with great care. However, the Publisher, employees, editors, and agents of The Haworth Press are not responsible for any errors contained herein or for consequences that may ensue from use of materials or information contained in this work. The Haworth Press is committed to the dissemination of ideas and information according to the highest standards of intellectual freedom and the free exchange of ideas. Statements made and opinions expressed in this publication do not necessarily reflect the views of the Publisher, Directors, management, or staff of The Haworth Press, Inc., or an endorsement by them.

Cover design by Jennifer M. Gaska.

Library of Congress Cataloging-in-Publication Data

Islam, Nurul, 1929-
    Reducing Rural Poverty in Asia : challenges and opportunities for microenterprises and public employment schemes / Nurul Islam.
        p. cm.
    Includes bibliographical references and index.
    ISBN-13: 978-1-56022-300-9 (hc. : alk. paper)
    ISBN-10: 1-56022-300-6 (hc. : alk. paper)
    ISBN-13: 978-1-56022-301-6 (pbk. : alk. paper)
    ISBN-10: 1-56022-301-4 (pbk. : alk. paper)
    1. Poverty—Government policy—Asia. 2. Rural poor—Government policy—Asia. 3. Economic assistance, Domestic—Asia. 4. Small business—Government policy—Asia. 5. Public service employment—Asia. 6. Rural development—Government policy—Asia. I. Title.

HC415.P6I85 2005
339.4'6'095091734—dc22
                                                                                2005010168

DEDICATED TO THE POOR AND THE HUNGRY

CONTENTS

ABOUT THE AUTHOR

Nurul Islam, PhD, received his doctorate in economics at Harvard University. He has been a professorial fellow at the Yale University Economic Growth Center and a fellow of St. Anthony's College (Oxford University, United Kingdom). He has served as director of the Pakistan Institute of Development Economics and founding director of the Bangladesh Institute of Developmental Studies. He was deputy chairman (minister) of the Planning Commission in Bangladesh, assistant director general of the Economic and Social Policy Department of the Food and Agriculture Organization of the United Nations (FAO), and policy advisor of the International Food Policy Research Institute (IFPRI), where he is currently an emeritus fellow. Dr. Islam also served as chair of the UN Committee for Development Planning (later Policy for Development)—a committee composed of independent development experts from all over the world which undertook annual review and analysis of salient developments in the world, including developments in poor countries, and recommended relevant policies to the UN Economic and Social Council for national and international action. In addition, he has served as a consultant to the United Nations, the United Nations Development Programme, the International Labour Office, UNCTAD, the World Bank, and the Asian Development Bank.

CASE CONTRIBUTORS

Roehlano M. Briones is currently a fellow at the World Fish Center in Malaysia and has taught at several Philippine universities. He was educated at the University of the Philippines, where he received his PhD in economics. He served at various times as consultant to the World Bank, government of the Philippines, the UNDP (United Nations Development Program), and the U.N. Children's Fund (UNICEF). His research interests and publications include macroeconomics, trade, poverty, agrarian reform, rural credit, employment, and research priorities in aquatic resources, including world demand and supply potentials of world fish resources. His recent publications include *Agricultural Impacts and the Pace of Land Reform, Property Rights Reform in the Philippines,* and *Modeling the Asian Fish Sector.*

Raghav Gaiha is currently a professor of public policy, Faculty of Management Studies, Delhi University, India. He was educated at universities in India and obtained his PhD in economics at the University of Manchester in the United Kingdom. He was a visiting fellow for various periods at Cambridge, Harvard, Stanford, MIT, and Pennsylvania universities, and at the World Bank. He has served as consultant/advisor in a wide range of national and international institutions, including the World Bank, IFAD, UN, and IFPRI. His research interests and publications include income distribution, rural poverty and employment, institutions, and decentralization. Among his most recent publications are *Agricultural Technology and Rural Poverty* and *Rural Public Works and Millennium Development Goals and Decentralization.*

Rushidan Islam Rahman is a research director at the Bangladesh Institute of Development Studies (BIDS). She was educated at Dhaka and Sussex universities, and she obtained a PhD in economics at Canberra University. She has worked as consultant to UN/ESCAP, ILO, ADB, and the World Bank, and served on various national governments and nongovernmental advisory committees. Her research interests and publications include poverty, unemployment and labor market, microfinance, agricultural development, and rural nonfarm activities—in each of these areas she has written extensively. Her important publications include *Poverty Alleviation and Empowerment through Microfinance: Two Decades of Experience, Savings, and Farm Investment in Bangladesh* (co-author), and *Agriculture and Rural Development in Bangladesh* (co-author).

Sajjad Zohir was until recently a senior research fellow at the Bangladesh Institute of Development Studies and is currently the executive director of the Economic Research Group in Dhaka. He was educated at Dhaka and Dalhousie universities, and he obtained his PhD degree at the University of Toronto. He served at various times as visiting fellow at IFPRI, IDS, and Manchester University as well as consultant to the World Bank, UNDP, ADB, and UN/ESCAP. His research interests and publications include agricultural diversification, social accounting, trade, and delivery of social services. Among his recent publications are *Analysis of Markets As Conduits of Macro and Micro Transmission in Bangladesh, Agricultural Growth through Crop Diversification in Bangladesh* (co-author), and *NGOs in Bangladesh: An Overview of NGO Sector* (unpublished draft).

List of Acronyms

ADB	Asian Development Bank
AGOA	African Growth and Opportunity Act
AIDS	acquired immunodeficiency syndrome
APQLI	Adjusted Physical Quality of Life Index
ARCDP	Agrarian Reform Communities Development Project
ARP	Access Road Project
ASA	Association for Social Advancement
ASEAN	Association of South East Asian Nations
ATDP	Agriculture Technology Development Project
BDO	block development officer
BIPOOL	big PO (partner organization) operating over a large area
BIS	Bureau of Indian Standards
BKB	Bangladesh Krishi Bank
BOM	Bank of Maharashtra
BPL	below poverty line
BRAC	Bangladesh Rural Advancement Committee
BRDB	Bangladesh Rural Development Board
BSCIC	Bangladesh Small and Cottage Industries Corporation
CAR	capacity assessment rating
CARE	Cooperative for American Relief Everywhere
CB	commercial bank
CDF	credit and development form
CDP	Committee of Development Policy
CEDP	Community Employment and Development Program
CIF	cost insurance freight
CIP	Communal Irrigation Project
CPD	Centre for Policy Dialogue
CPI	Corruption Perception Index
DA	Department of Agriculture
DAR	Department of Agrarian Reform
DCCB	district cooperative credit bank
DHRUVA	Dharampur Utthan Vahini
DILG	Department of Interior and Local Government
DOLE	Department of Labor and Employment
DPC	district project coordinator

DPCC	District Project Coordination Committee
DPWH	Department of Public Works and Highways
DRDA	District Rural Development Authority
EAS	Employment Assurance Scheme
EBA	Everything But Arms
EDI	Economic Development Institute
EDII	Entrepreneurship Development Institute of India
EDT	entrepreneurship development training
EGS	Employment Guarantee Scheme
EIU	economic intelligence unit
EPZ	export processing zones
ESCAP	Economic and Social Commission for Asia and Pacific
EU	European Union
EVI	Economic Vulnerability Index
FDI	foreign direct investment
FFW	food-for-work
FGD	focused group discussion
FO	field officer
FTA	Free Trade Agreement
FWWB	Friends of Women's World Banking
GATT	General Agreement on Trade and Tariff
GB	Grameen Bank
GBF	Grameen Bank Family
GDP	gross domestic product
GNI	gross national income
GOB	government of Bangladesh
GOI	government of India
GOM	government of Maharashtra
GSK	Gano Sasthya Kendra
GSP	generalized scheme of preferences
GU	Grameen Uddog
HAI	Human Assets Index
HYVs	high-yielding varieties
IA	Irrigation Association
IAS	impact assessment studies
ICCDRB	International Center for Cholera and Diarrhearial Research in Bangladesh
ICED	International Centre for Entrepreneurship Development
ICRISAT	International Crops Research Institute for the Semi-Arid Tropics
IDA	International Development Association (Affiliate of World Bank)

IDO	irrigation development officer
IDS	international development strategy
IFAD	International Fund for Agricultural Development
IFPRI	International Food Policy Research Institute
IGA	income generating activities
IGVGD	Income Generation for Vulnerable Group Development Program
IIED	Indian Institute for Entrepreneurship Development
ILO	International Labour Organization
IMF	International Monetary Fund
IP	intellectual property
IPR	intellectual property rights
IRDP	Integrated Rural Development Program
JGSY	Jawahar Gram Samriddhi Yojana
JRY	Jawahar Rozgar Yojana
KST	Kishoreganj Sadar Thana
LBES	labor-based equipment-supported methods
LDC	least developed countries
LFS	Labor Force Survey
LGU	local government unit
LIDP	Local Infrastructure Development Program
MAVIM	Mahila Arthik Vikas Mahamandal
MCA	Millennium Challenge Account
MCED	Maharashtra Centre for Entrepreneurship Development
MDC	Municipal Development Council
MDG	millennium development goals
MECRO	microenterprise customer relations officer
MEDU	Microenterprise Development Unit
MERCOSUR	Mercado Comüne del Sur (South American Common Market)
MF	microfinance
MFI	microfinance institution
MIS	management information system
MITCON	Maharashtra Industrial and Technical Consultancy Organization
MNC	multinational corporation
MRCP	Maharashtra Rural Credit Project
MRP	maximum retail price
MTR	midterm review
NABARD	National Bank for Agriculture and Rural Development
NAFTA	North America Free Trade Agreement
NBFC	Nonbanking Financial Corporation

NCAER	National Council for Applied Economic Research
NCB	nationalized commercial bank
NEDA	National Economic Development Authority
NFA	nonfarm activities
NFS	nonfarm sector
NGO	nongovernment organization
NIA	National Irrigation Administration
NORAD	Norwegian Development Aid Agency
NSS	National Sample Survey
OBC	other backward classes
OECD	Organisation for Economic Co-Operation and Development
OOSA	organizations operating over a small area
PAP	People's Action Plan
PBR	plant breeders' rights
PDC	Provincial Development Council
PIC	project implementation committee
PKSF	Palli Karma Sahayak Foundation
PO	partner organization
PSC	project steering committee
PVP	plant variety protection
PWE	public works employment
R&D	research and development
RBI	Reserve Bank of India
RD–12	Rural Development–12 (project)
RDC	Regional Development Council / Rural Development Council
RDD	Rural Development Department
RDRS	Rangpur-Dinajpur Rural Services
RIP	Rural Industries Project
RIS	Research Information System for the Nonaligned and Other Developing Countries
RMP	Road Maintenance Program
RNA	rural nonfarm activity
RNF	rural nonfarm
RPW	Rural Public Works
RRB	regional rural bank
RSME	rural small and microenterprises
RWP	Rural Works Program
SAARC	South Asian Association for Regional Cooperation
SAPAP	South Asian Poverty Alleviation Programme
SC/ST	scheduled caste/scheduled tribe

SDT	skill development training
SEDP	Small Enterprises Development Project
SFMC	SIDBI Foundation for Micro Credit
SGSY	Swarnjayanti Gram Swarozgar Yojana
SHG	self-help group
SIDBI	Small Industries Development Bank of India
SO	social organizer
SPS	sanitary and phytosanitary
SRO	self-regulatory organization
SRRIP	Second Rural Roads Improvement Project
SSGV	Shram Shaktidware Gram Vikas
SSS	Society for Social Services
SYG	Sahyogini/animator/field worker
TI	Transparency International
TRIPS	trade-related aspects of intellectual property rights
UK	United Kingdom
UN	United Nations
UNDP	United Nations Development Programme
UNICEF	United Nations Children's Emergency Fund
UPOV	International Convention on the Protection of New Varieties of Plants (L'union internationale pour la protection des obtentions vegetales)
USA	United States of America
USAID	United States Agency for International Development
VDA	village development assembly
VDC	village development council
VLW	village-level worker/gram sevak
VO	village organization
WFP	World Food Program
WSP	Water Supply Project
WTO	World Trade Organization

Foreword

Nonfarm rural activities are increasingly being recognized as playing an important role in alleviating poverty in developing countries. Development researchers have made significant advances toward understanding the contributions of nonfarm employment and income in this regard. In spite of improved understanding of the issues and challenges that policymakers face in promoting rural nonfarm employment, several issues remain to be addressed regarding the choice of appropriate policy interventions. What range of interventions is available? What strategies are needed to provide a better enabling environment for rural nonfarm enterprises? How crucial is the role of microfinance? How about access to productivity-improving technologies? How is efficient marketing ensured? How do we best combine microfinance as the critical lever with other supporting measures? What is the relative importance of various nonfarm activities, for example, trade, small scale and household industry, services, and construction? What are the policies needed to diversify the nonfarm sector and promote small entrepreneurs? How do we bring together a multisectoral approach to the improvement of rural nonfarm income and employment? What role can rural public employment schemes play in enhancing the income of the poor? What are the ways to design and implement them consistently with the objective of providing productive infrastructure for rural development?

Professor Nurul Islam and his collaborators have attempted to provide answers to these and other questions by conducting research and preparing a set of country case studies. The book addresses two major programs focusing on microfinance and public wage employment schemes in order to make an assault on rural poverty. A wide range of experiences from Asian countries and the lessons and policy conclusions based on the analysis of specific country experiences are brought together in this book.

It is my hope that this book will stimulate debate and encourage further research on this important subject of addressing rural poverty through rural nonfarm employment and income-generating programs, and that it will lead to improved programs and policies for addressing poverty that underlies and aggravates the problem of malnutrition and food insecurity.

Joachim von Braun
Director General
International Food Policy Research Institute
Washington, DC

Preface

Asia provides the shortest and quickest road to realizing the millennium development goals (MDG). This is because the majority of the world's poor live in Asia and most of them live in rural areas. Also, food insecurity and malnutrition, closely interrelated with poverty, are widespread in rural Asia. Making a modest dent on Asian poverty has the potential of realizing large gains in global human development.

Most of the rural poor are employed in agriculture; many are engaged in nonfarm activities as a supplementary occupation. Landless and near-landless households depend predominantly on nonfarm earnings. In the war against rural poverty and malnutrition, the rural nonfarm sector plays a crucial role. Simultaneously, with a fuller appreciation of its nature and characteristics, it is necessary to examine policies and programs needed to promote the nonfarm sector. The economywide policies such as trade, fiscal, industrial, and labor policies as well as rural development policies embracing human, social, and physical infrastructure can contribute to growth. No less important are the specific policies relating to access to finance, technology, and markets as well as institutions for the development of the nonfarm sector.

First, the government policies can be directed to self-employment in nonfarm microenterprises or wage employment in medium-scale enterprises. How best to promote them? The cost benefit analysis of the different modes of delivery and types of assistance—e.g., financial, technical—is essential for making relevant policy choices. Second, nonfarm employment can also be promoted through rural public employment schemes relating to construction and maintenance of physical infrastructure as well as labor components of development projects in different sectors. This raises a different set of policy issues and implementation problems.

The book addresses the prospects and constraints of making an assault on rural poverty by means of these two major programs in which Asia has a long and varied experience. It derives lessons and policy conclusions based on an analysis of specific country experiences. Part I provides a framework for the analysis of rural microenterprises with a focus on microfinance; it highlights the main findings of the country case studies and suggests a set of guidelines for an appropriate strategy with emphasis on targeting the poor, alleviating poverty, and achieving financial sustainability. Second, it analyzes the issues relating to public wage employment schemes and the principal findings of the case studies; it draws policy conclusions for the formulation of effective public employment schemes. The country studies conducted in three countries, India, Bangladesh, and the Philippines, constitute Part II of the book. They are included with an indication of the authorship of the various country studies.

To attain an MDG such as reducing poverty by half in 2015, which is essential for achieving global food security, an assessment of the pros and cons of the past and the present experience is crucial. The book seeks to make a contribution to this endeavor. It may provide lessons for other similarly situated countries in Asia and beyond, such as Africa, which is facing the challenge of developing a high productivity rural nonfarm sector for reducing rural poverty.

The book is expected to be useful not only for academic researchers, but also for development experts and practitioners in national governments and in the wider development community. This book is based on the research that was undertaken earlier at the request of and with financial assistance from the International Fund for Agricultural Development (IFAD), for which I am grateful. The original study has been revised and edited—the country studies greatly abridged and heavily edited—for the purpose of the present book. I am also grateful to the International Food Policy Research Institute (IFPRI) for the research facilities provided for this work and to Vicky Lee for preparing the manuscript for publication.

PART I:
SYNTHESIS AND MAJOR LESSONS

Chapter 1

Rural Poverty and the Nonfarm Sector in Rural Asia

RURAL POVERTY AND THE NONFARM EMPLOYMENT NEXUS

Poverty is multidimensional; it encompasses not only what is called income poverty, i.e., deprivation of income/consumption such as the satisfaction of minimum level of food and other basic needs, but also limited access to health, nutrition, and education services, which aggravates the impact of income poverty, resulting in child mortality, short life expectancy, and illiteracy. In addition, the poor are vulnerable to shocks and risks and lack ability to cope with or overcome shocks. These individuals suffer not merely from transient decline in income consumption and well-being but also sink into deeper long-term endemic poverty. Different aspects of poverty reinforce one another. In a wider sense, poverty basically connotes a lack of choice and of opportunities on the part of individuals to achieve an optimum exploitation or use of their potentials or capabilities; it implies lack of empowerment on their part to participate in or influence the decision-making process affecting their livelihoods and well-being.

South Asia, with about 1.5 billion people, contains some of the poorest/least developed countries with this population. In the early years of the twenty-first century, 40 percent of the population of South Asia is below the international poverty line (defined as an income of one dollar per day) (World Bank, 2000, 2001). The proportion of population below the poverty line varies between 42 and 29 percent among different countries of South Asia. Among the Southeast Asian countries, the Philippines and Indonesia have respectively 37 and 27 percent of their population below the national poverty lines.

In terms of social deprivation, South Asia has low life expectancy, adult literacy, and poor primary and secondary school enrollment; in 2000 the rates were 63, 55.6, and 53 percent, respectively. The corresponding rates for the Philippines were 69.3, 95.3, and 82 percent, and in Indonesia they were 66.2, 86.9, and 65 percent, respectively (UNDP, 2003).

Overall economic growth expands the employment and income earning prospects of the poor. It is especially so if growth is pro-poor in the sense of an emphasis on labor-intensive and employment-oriented sectors and projects reinforced by appropriate policies. For the poor to take advantage of the income-earning opportunities, they must acquire human assets such as good health, education, and skills and physical and financial assets such as land and equipment, and production inputs either owned by them or available through access to credit. In order for the poor to attain these assets the poor must have the voice and power to influence decisions affecting the poverty orientation of public policies and investment. This requires state or public institutions at the national and local levels that are accountable and responsive to their needs and problems through participation in political and decision-making processes. They also need to build up their own self-help or various civil society or nongovernmental organizations to influence policies as well as direct resources and policies to meet their ends.

At the same time, additional measures are necessary to ensure that the poor are able to escape poverty or do not sink into deeper levels of poverty. This is because of the low level of resources and capability of the poor who need protection from poverty-accentuating impacts of short-term shocks such as death of the family breadwinner, economic shocks, and natural disasters through insurance and safety net measures.

Most of the poor in Asia's developing countries live in rural areas, and the majority of the rural poor work in agriculture. To lift the rural poor out of poverty, improved productivity in agriculture, which results in higher per capita output, is needed. Also needed is expanding employment and income-earning opportunities in agriculture-related and other rural nonfarm activities. This will accommodate labor released from agriculture as labor requirements per unit of output decline with rising productivity. Also, larger numbers of the rural population need to be employed outside of agriculture in the urban sector as well as in the rural nonagricultural occupations. In fact, the rural nonfarm sector is already a significant source of rural income or employment. The challenge is how to increase the participation of the poor in such activities and to increase the productivity of the poor engaged in such activities.

The nonfarm rural economy accounts for 40 to 60 percent of total national employment, and 20 to 50 percent of total rural employment in the Asian region. The composition of the nonfarm economy also shows remarkable similarities across countries. Service activities dominate the nonfarm economy in both rural and urban areas, followed by manufacturing and then trade as seen in Tables 1.1 and 1.2. Service activities (including much of the "other" activities in Bangladesh and Sri Lanka) are more domi-

TABLE 1.1. Employment shares by activity in rural and urban areas, selected countries (percent).

Country	Rural population (as percent total)		Total employment		Nonfarm employment						
	1960	1994	Agricul-ture	Nonfarm	Manufac-turing	Transport	Trade	Services	Finance	Con-struction	Other
Bangladesh (1991)											
Rural			66.1	39.9	6.8	4.0		35.4		3.3	50.2
Urban			15.1	84.9	8.1	6.5		31.8		3.4	50.1
Total	95	82	54.6	45.4	7.3	5.1		33.8		3.3	50.4
Total in 2003					25.1	–	57.4	15.4	–	2.4	–
Sri Lanka (1981)											
Rural			55.7	44.3	19.8	8.3	16.5	25.2	1.5	6.6	22.1
Urban			7.3	96.7	16.0	9.7	23.9	28.7	2.8	3.7	15.0
Total	82	78	45.2	54.8	18.5	8.8	19.2	26.5	2.0	5.5	19.5
Pakistan (1981)											
Rural			71.6	28.4	21.8	9.2	18.3	33.8	1.1	12.3	3.5
Urban			8.0	92.0	21.5	10.1	25.6	31.0	2.5	7.5	1.8
Total	78	66	55.5	44.5	21.6	9.9	22.2	32.3	1.8	9.9	2.3
India (1993/94)											
Rural			76.9	23.1	30.7	6.9	19.4	26.8		11.6	4.6
Urban			17.7	82.3	22.2	12.7	25.9	38.3		3.1	2.3
Total	82	73	61.5	38.5	28.5	8.4	21.1	29.8		9.4	2.8
Total in 1999					33.0	10.0	21.0	21.0		15.0	

TABLE 1.1 (continued)

Country	Rural population (as percent total)		Total employment		Nonfarm employment						
	1960	1994	Agriculture	Nonfarm	Manufacturing	Transport	Trade	Services	Finance	Construction	Other
Philippines (1980)											
Rural	70	47	74.0	26.0	20.9	11.9	13.2	32.1	3.0	11.5	7.4
Urban			18.3	81.7	19.4	11.3	14.9	35.9	7.1	8.1	3.3
Total			51.4	48.6	19.9	11.5	14.3	34.7	5.8	9.2	4.6
Indonesia (1995)											
Rural	85	66	63.1	36.9	23.8	8.2	31.7	24.2	0.5	9.4	2.2
Urban			9.4	90.6	20.0	8.0	30.1	31.1	2.4	6.8	1.6
Total			45.9	54.1	21.8	8.1	30.9	27.9	1.5	8.0	1.8
Thailand (1996)											
Nonmunicipal	87	80	49.9	50.1	30.3	5.1	22.1	19.7		21.5	1.3
Municipal			1.9	98.1	22.6	7.0	29.9	28.8		9.7	2.0
Total			39.7	60.3	27.6	5.8	24.8	22.8		17.4	1.6

Source: Dev and Evenson (2003).

6

TABLE 1.2. Employment shares by activity and size of locality, India (in percent).

| | Total employment | | Nonfarm employment | | | | | | |
| | | | Manufacturing | | | | | | |
Country	Agriculture	Nonfarm	Household	Nonhousehold	Transport	Trade	Services	Construction	Other
India (1971)									
Rural	84.9	15.1	21.6	15.7	5.9	15.7	35.3	3.9	2.0
Rural towns	23.6	76.4	8.6	19.5	10.5	25.4	25.4	4.5	1.8
Urban towns	4.7	95.3	3.9	30.3	12.0	21.5	27.8	3.5	0.7

Source: Hazell and Haggblade (1991), p. 518.

Note: Rural towns are all urban areas under 100,000 in population; urban towns are urban areas with more than 100,000 population.

nant in the lower-income South Asian countries, while trade and manufacturing are almost as important as services in the East Asian countries. [1]

Nonfarm income shares are typically 5 to 10 percent larger than nonfarm employment shares in rural areas, a direct measure of the importance of seasonal and part-time nonfarm activity (Table 1.3). Nonfarm income shares frequently account for one-third to one-half of total rural household income. These shares have increased over time, up from 18.1 to 46.3 percent between 1971 and 1991 in Korea, and from 25 percent to 33 percent between 1967 to 1978 and 1981 to 1982 in India. Nonfarm sources account for

TABLE 1.3. Share of nonfarm income/employment in total household income/employment by farm size groups, selected countries.

Farm size (ha)	Nonfarm share (percent)	
	Employment	Income
India (1987-1988)		
Landless	46.1	
0.01-0.4	29.3	
0.41-1.0	19.0	
1.01-2.0	14.0	
2.01-4.0	11.8	
4.01 +	9.0	
South Korea (1986)		
0.0-0.5		73
0.5-1.0		49
1.0-1.5		35
1.5-2.0		26
2.0 +		19
Taiwan (China) (1979)		
0.0-0.5		67
0.5-1.0		58
1.0-1.5		48
1.5-2.0		40
2.0 +		33
Thailand (4 regions) (1980-1981)		
0.0-4.1		88
4.2-10.2		72
10.3-41.0		56
41.0 +		45

Source: India: Chadha (1993); South Korea: Rosegrant and Hazell (2000); Taiwan: Ho (1986); Thailand: Liedholm (1988).

30 to 40 percent of the average rural household income in South Asia (Start, 2001). The share of nonfarm income in total rural households in Bangladesh was 54 percent in 2000, having grown from 42 percent in 1987 (Hossain, 2002).

As population pressure grows on agriculture in the land-scarce Asian developing countries, growth in agricultural production and employment cannot absorb the increasing rural labor force; in fact, growth in agricultural productivity essential for growing per capita agricultural income, over time, decreases employment opportunities for an expanding rural labor force. The urban industrial sector cannot grow fast enough to absorb the surplus labor released from agriculture. This leaves the nonfarm rural sector to absorb the population released from agriculture and unable to be absorbed in the urban industries.

Manufacturing, service, and trade activities account for the largest shares of nonfarm employment for both male and female workers in rural areas (assuming that much of the activity classified as "other" in Bangladesh and Sri Lanka also really falls under these categories), though women are relatively more concentrated in these activities than men in most countries. Trade seems to be more important for women in the East Asian countries than in the South Asian ones, but transport and construction activities are much less important for women than men in all countries (Tables 1.4 and 1.5).

The rural nonfarm economy is especially important to the rural poor. Landless and near-landless households everywhere depend on nonfarm earnings; those with less than 0.5 hectare earn between 30 and 90 percent of their income from nonfarm sources. Nonfarm shares are strongly and negatively related to farm size. Low-investment intensive manufacturing and services (including weaving, pottery, gathering, food preparation and processing, domestic and personal services, and unskilled nonfarm wage labor) typically account for a greater share of income for the rural poor than for the wealthy (Hazell and Haggblade, 1993). Nonfarm income is also important to the poor as a means to help stabilize household income in drought years (Reardon et al., 1998).

With increasing urbanization and migration of rural workers, real rural wage rises and thus raises the opportunity cost of labor, thereby making many low-return nonfarm activities uneconomical. This leads to the demise of several traditional and low-paying craft and service activities, and to the growth of new types of employment in trade, commerce, and manufacturing. Hossain (1988b) provides evidence from the green revolution experience in Bangladesh. In villages with a majority of its rice area planted with high-yielding varieties (HYVs), agricultural wages and incomes as well as nonfarm per capita income are higher than that prevailing in villages still dependent on traditional varieties. The higher nonfarm income in prosper-

TABLE 1.4. Distribution of rural workers across nonfarm sectors, male and female (percent).

Country	Total nonfarm	Manufacture	Transport	Trade	Services	Finance	Construction	Other
Bangladesh (1991)								
Male	100	6.5	4.6		39.4		3.7	45.8
Female	100	8.4	0.4		12.2		0.8	78.2
Sri Lanka (1981)								
Male	100	18.8	9.8	18.1	20.6	1.6	7.7	23.4
Female	100	24.9	1.6	8.9	46.1	1.4	1.2	15.9
India (1993-1994)								
Male	100	26.8	8.4	21.1	27.2		12.6	3.8
Female	100	48.7	0.6	14.2	25.9		7.1	3.2
Philippines (1980)								
Male	100	16.5	19.0	9.6	22.9	3.1	18.6	10.4
Female	100	27.9	0.6	19.0	46.8	2.8	0.4	2.5
Indonesia (1995)								
Male	100	19.9	12.5	23.7	26.0	0.7	14.5	2.8
Female	100	30.9	0.3	46.2	21.0	0.2	0.3	1.0
Thailand (1996)								
Male	100	25.3	8.0	17.8	17.3		29.9	1.7
Female	100	37.5	1.0	28.2	23.1		9.7	0.5

Source: Bangladesh: Shilpi (2003).

TABLE 1.5. Women's share in total nonfarm employment by sector, rural and urban (percent).

Country	Total nonfarm	Manufacture	Transport	Trade	Services	Finance	Construction	Other
Bangladesh (1991)								
Rural	14.7	18.2	1.5			5.1	3.9	22.8
Urban	12.1	15.9	1.4			5.7	3.3	15.7
Sri Lanka (1981)								
Rural	17.9	22.3	3.4	9.6	32.8	16.1	3.3	12.9
Urban	18.5	24.3	5.5	7.6	33.7	22.9	5.8	11.4
India (1993-1994)								
Rural	19.4	30.3	1.8	14.0		18.6	11.9	16.9
Urban								
Philippines (1980)								
Rural	38.6	51.5	2.0	55.4	56.3	36.5	1.2	13.1
Urban	38.2	34.6	5.3	39.1	56.0	38.9	1.9	15.9
Indonesia (1995)								
Rural	35.6	46.2	1.1	51.9	30.8	16.9	1.3	15.0
Urban	33.5	35.7	3.0	44.8	37.0	29.0	3.7	11.9
Thailand (1996)								
Rural	41.3	51.0	8.1	52.7		48.5	18.5	17.7
Urban	45.4	46.1	14.3	48.4		56.7	24.5	40.0

Source: Rosegrant and Hazell (2000).

ous villages reflects a greater concentration of high-return nonfarm activities, including transportation and services, cottage industry, construction, and earth hauling.

Since most of the service and cottage industry activities are regional nontradables that are constrained by local demand, an increase in their aggregate supply could depress prices and incomes for some producers, unless the rural market simultaneously expands through increase in farm income or the increased output finds markets in urban areas. Cross-country studies show a positive relationship between agricultural income (measured as agricultural income per capita of the rural population) and the nonfarm share of total rural employment (Hazell and Haggblade, 1993). The relationship is particularly strong when rural areas are defined to include rural towns.

Also, higher population densities facilitate rapid attainment of efficient scales of nonfarm production and marketing. Heavily populated areas are able to afford and maintain denser networks of roads, transportation, and other rural infrastructure, which foster the development of the nonfarm sector.

The development of rural towns has positive effects on rural nonfarm employment growth because their higher income levels offer large enough markets to capture economies of scale and scope for a wide variety of nonfarm products and services, and also because their higher levels of infrastructural development help reduce costs and expand communications and market reach. As the economic transformation proceeds, towns become important centers of demand creating new market opportunities for both production inputs and consumption goods originating in the rural agricultural and nonfarm sectors. The subcontracting of many lower-level manufacturing processes to rural nonfarm enterprises boosts income and employment opportunities in the surrounding rural areas (Otsuka, 1998). These links to the rural hinterland take on particular importance in the case of those rural towns that become closely integrated into the urban economy and develop manufacturing and service activities that serve urban and export demands.

PUBLIC POLICIES FOR THE NONFARM SECTOR

Public policies for promoting the nonfarm sector can be divided into two categories. The first category relates to those economywide policies that affect the nonfarm sector such as trade, foreign exchange, fiscal, industrial, and labor policies. In a wider sense, they include policies which promote rural development in general, including agriculture (in a broader sense, crop and noncrop agriculture) as well as nonagricultural activities. Important among such policies are the development of rural physical infrastructure, such as transportation and communication, and electricity as well as human capital, especially literacy and primary and secondary education. In general, the macro or sectoral policy reforms that facilitate the growth of the nonfarm sector, especially small-scale or microenterprises in industry and trade, include

1. simplification and rationalization of entry and exit regulations as well as tax laws that lowers new entrants' costs;
2. liberalization and streamlining of export and import regulations to lower barriers for small-scale participants;
3. reforms of banking regulations so that banks are encouraged to compete in order to seek new markets;
4. reforms of property rights and collateral regulations that impair small firms' ability to receive loans; and
5. reforms of labor regulations that restrict or tax labor flows.

These reforms tend to make the playing field equal for both small-scale rural and urban enterprises.

The second category of support policies relates to specific policies directed toward the nonfarm-sector enterprises. This category includes programs and projects to provide financial assistance and credit facilities as well as technical services of various kinds to the nonfarm sector. These policies help reduce the discrimination and disincentives suffered by small-scale rural enterprises through lack of access to credit, technology, and markets. Given the urban bias in policies, these enterprises also suffer from underdevelopment of social, human, and physical infrastructure in the rural areas.

In the past, rural financial institutions failed to recognize that rural households required access to deposit or savings accounts that yielded reasonable returns. As financial services improve in rural areas, it is possible that a larger share of rural savings will remain in rural areas (including local towns). This will facilitate the growth of the rural nonfarm economy if small-scale (including part-time nonfarm) businesses, especially in the service sector, gain access to credit.

Financial assistance or credit facilities to medium, small-scale, and microenterprises are usually channeled through government-owned commercial banks or specialized financial institutions or by requiring private commercial banks to allocate a certain percentage of their loans to such enterprises; frequently these loans are backed by refinancing from central banks or specialized refinancing institutions. Small enterprises often receive loans at subsidized interest rates that necessitate credit rationing. In view of the high costs of acquiring information regarding small borrowers' creditworthiness and the quality of their loan-financed activities combined with an inadequate supervision by banks, the rate of repayment is often low. This results in a large proportion of overdue or nonperforming loans and losses for credit institutions.

Fiscal assistance to microenterprises has increasingly been channeled through nongovernment organizations (NGOs). The way in which these groups are organized, their borrowers are identified, their lending portfolios are chosen, and their projects are monitored vary widely.

Technical assistance to the nonfarm sector includes management training, training in accounting and finance, and marketing. It has been provided either directly by the government agency established for the purpose or through an NGO. Technical assistance is sometimes combined with the provision of finance. It is possible to provide technical assistance through private consultant firms with resources obtained from the government. The micro- or small-scale enterprises, mainly in manufacturing, have been the major recipients of such assistance.

One popular way of providing services to small microenterprises is to establish industrial estates that are fully endowed with infrastructure, roads, communications, electricity, and financial services in small towns or semi-rural areas. The experience with such direct assistance programs for nonfarm enterprises has been mixed and has led to generally disappointing results (see Haggblade and Mead, 1998). Industrial estates for rural areas are mostly viewed as expensive failures.

The preceding discussion has concentrated on household and microenterprises in the nonfarm sector with a focus on self-employment of the poor. Another way to promote employment in the rural nonfarm sector is to employ the rural poor through public works or employment programs devoted to a wide range of infrastructure and development projects. The public provision of employment raises a different set of policy issues and implementation problems than the development of microenterprises. They include such issues as the appropriate agencies for implementation, e.g., government, private, or NGOs, criteria for the selection of projects, efficiency of resource use, and possible ways such employment programs can be dovetailed or integrated with overall rural development projects and programs. A few additional issues such as problems of targeting the poor, the leakage to the nonpoor, and impact on poverty reduction, etc., are common to the both sets of programs.

Chapters 2 through 6 provide a framework for analysis of issues relating to the development of rural microenterprises in Asia. They highlight the main findings of a few country case studies and suggest a set of policy conclusions. Chapters 7 through 9 deal with the issues relating to the provision of wage employment through public employment schemes; they analyze the main findings of the country case studies and attempt to present a set of policy conclusions for the formulation of effective, policy-oriented public employment schemes.

Chapter 2

Rural Nonfarm Microenterprise:
Salient Features and Issues

INTRODUCTION

Most of the rural population engaged in the nonfarm sector are either "self-employed," i.e., working for themselves on a wide variety of activities located in the household or elsewhere in the village, or are engaged as wage laborers in microenterprises employing no more than ten workers. Thus, the rural nonfarm sector enterprises are basically split into two size categories: the first focuses on small-scale household enterprises; the second focuses on small- or medium-size microenterprises in rural areas and small towns.

The first enterprise category varies widely in size, location, gender, and sector of activity. Most are single-person, owner-operating units, or small units engaging family members. Most of these enterprises are subsistence or livelihood enterprises, providing employment opportunities in the absence of more profitable alternatives and often are one of several secondary sources of income, many of them intermittent, part-time, and seasonal. They are engaged in a wide range of activities with heavy concentration in trade and service activities, predominantly related either to agricultural inputs, outputs, or to consumption needs of farmers. They earn a small surplus which is frequently not reinvested but devoted to household expenditures, reflecting the poverty of the workers/owners. They require low-level skills and capital; hence there are low barriers to entry and they are overcrowded. Thus, in general, they have small potential for growth. A small number may grow over time and may start employing wage laborers

The second category of nonfarm sector enterprises mostly employs wage laborers in addition to family members, relatives, or children, and uses higher skills and capital intensity. These enterprises employ anywhere between two and ten employees. Many of them hold prospects for growth in scale, capital intensity, and market size over the long run. The major differences between the two categories of enterprises may be seen in Table 2.1. The distinction is not very rigid; in many cases, they may overlap (ADB, 1997).

TABLE 2.1. Major differences between livelihood enterprises and microenterprises.

Factor	Livelihood enterprises	Microenterprises
Capitalization	Relatively low	Higher, but initial capitalization often similar
Education	Little formal education	Usually at least secondary schooling
Skills and experience	Relatively low, except for skills acquired traditionally, as in handicrafts; trading often a fertile training ground for later manufacturing of same product	Higher, more often acquired through vocational training and/or previous wage employment
Gender	High (often majority) participation of women	Lower participation of women, but still high in many cultures
Sector	Higher proportion in livestock, backyard poultry, food processing, and petty trading	Higher proportion in manufacturing and services requiring skills
Competition	Usually function in perfectly competitive markets with low barriers to entry and little scope for cutting costs by intensive use of family labor and even by offering credit	Often occupy "niche" markets with more scope for specialization and product differentiation
Seasonality	Often seasonal, tied to crop cycle, school year, major festivals	Less affected by seasonality and function throughout the year, even if at varying levels
Contribution to household income	Usually a secondary source (although vital)	Often primary
Whether only enterprise	Usually one of several "multiple" enterprises (to compensate for seasonality and low returns)	Usually the only enterprise
Use of hired labor	Infrequent, mostly use family labor	More common, often relatives or children
Surpluses and reinvestment	Surpluses limited and often plowed back into household expenditure	Reinvestment of surpluses the norm
Use of credit	Trading activities often started on a consignment basis, livestock acquired on a profit-sharing basis, boats and rickshaws on lease; however, in order to compete, often become net lenders, especially in trading and restaurants	Credit available from a wider range (informal and semiformal) and a greater two-way flow of credit so that microenterprises are more often net lenders than livelihood enterprises

TABLE 2.1 *(continued)*

Factor	Livelihood enterprises	Microenterprises
Potential for growth	Limited in terms of new employment generation, but offer scope for increases in sale, productivity, profitability, and income; growth blocked often by demand constraints, resource constraints (artisanal fishing), and physical constraints (space in home and yard)	Have growth potential; number of workers higher, with more paid employees; employment usually of "higher quality"

An illustrative list of activities by sectoral classification is shown in Table 2.2.

However, broadly speaking, policies to promote both livelihood enterprises and microenterprises cover provisions for credit as well as noncredit inputs such as technology, skill, training, and marketing. Many livelihood enterprises require access to credit and financial resources, especially working capital, more than any other input. Others, especially as they grow in scale and those that are more in the nature of growth-oriented enterprises, need fixed capital as well as significant noncredit inputs.

As agricultural transformation takes place and urbanization proceeds, consumption as well as input/output links with urban areas are strengthened. Many small-scale livelihood enterprises including cottage industries are unable to stand competition from urban goods. Rural households spend a larger share of income on urban goods than on low-quality but low-priced products of home-based rural industries. The rural trade and transport services rapidly expand to distribute urban goods, both consumption and production inputs, in the rural areas. The repair and maintenance services for the construction and transport sector as well as for agricultural implements expand in the rural sector.

MICROFINANCE

The most important obstacles in developing rural microenterprises for the poor has been credit constraint. Rural borrowers have no collateral to offer against loans. Without any credit history or savings of their own, the credit agencies find it too risky to lend. Administration involving the appraisal, monitoring, and supervision of the use and recovery of small loans for microenterprises is too costly for the lenders. While the delivery of other microcredit inputs or services to the microenterprises has also engaged the

TABLE 2.2. A suggested sectoral classification of microenterprises.

Sector	Livelihood enterprises	Microenterprises
Noncrop agriculture-related activities	Pig, goat raising, dairy-cattle, backyard poultry, vegetable growing on leased land, fruit trees on homestead, sericulture	Stall-fed mini-dairy of high-yielding cows, "scientific" poultry or duck raising, fish fry, and fingerling raising
Trading	Small kiosk, ambulant vegetable vending, buy-and-sell fish	Larger grocery store, vegetable wholesaling
Food and refreshments	Juice stand, small tea shop, candy and snack stall	Restaurant or tea shop at busier location with hired employees
Food and agro-processing	Rice cakes	Packaged candies, processed meat products, perhaps with labels
Manufacturing:		
Textiles	Pit loom, seamstress	Power loom, garment maker, bulok orders for uniforms
Wood, rattan, bamboo, and grassproducts	Mat and basket weaving, rattan furniture	Larger furniture-making unit
Footwear and leather	Wayside shoemaker	Larger ready-made footwear-making unit, leather bags
Bricks, tiles, and pottery	Village potter	Larger brick-making unit
Fabricated metal products and repair services	Hats for local market	More sophisticated handicrafts for the export market
Other manufacturing	Simple agricultural implement maker, roadside bicycle repair	Small engineering workshop doing job work, repairs to heavier agricultural implements
Fishing	Single-person fishing boat	Larger mechanized boat with crew members
Transport	Cycle rickshaw, mule, bullock-cart	Mechanized three-wheeler, hand-tractor rental services
Other services	Wayside hairdresser	Beauty parlor

attention of the policymakers, the highest priority has been assigned to devising ways and means of removing the credit constraint for the microenterprises.

Therefore, microfinance institutions (MFIs) have been a principal means of promoting nonfarm self-employment and rural microenterprises in the various Asian countries. Given the large variety and number of rural micro-finance institutions in the South Asian region, various studies have been undertaken in recent years to analyze and appraise quite a few of them, especially in Bangladesh and India.[1] This section seeks to provide, on the basis of a survey of findings of past studies, a review of some aspects and characteristics of microfinance in the nonfarm sector. The focus of the review is on the problem of poverty targeting and the impact of microcredit on the socioeconomic welfare of the poor and poverty alleviation, including disadvantaged women; it also discusses the long-run sustainability of microcredit institutions.

Poverty Targeting

In devising cost-effective poverty microcredit programs, the most widely debated issue is the pros and cons of appropriate methods of targeting. The targeting of the poor in any antipoverty credit program can be based either on self-targeting or on indicator targeting. Indicator targeting is based on income or income proxy that is closely related to income or poverty and can be observed more easily or more cheaply than income. These indicators include income, assets, geographical targeting, landholding, gender, nutritional status, household size, and kind of dwellings and access to electricity, etc. Under the system of self-targeting, in lieu of the administrators or bureaucrats choosing participants, beneficiaries select themselves in response to incentives or disincentives that are provided to induce only the poor and discourage the nonpoor to be included in the credit system. The objective in both types of targeting is (1) to bypass or exclude the nonpoor and (2) to increase the participation of the poor, i.e., to avoid what are called "type one" and "type two" errors.

The first type of error occurs when there is a leakage to the nontarget population. The second type of error involves exclusion of some of the poor in the target group. Attention must be paid to both types of targeting errors. A targeting mechanism can scarcely be devised so that some poor will not be excluded. Frequently, the focus seems to be to minimize the leakage to the nontarget population or to the nonpoor rather than to maximize the inclusion of the poor. In the interest of extending the coverage of the poor, some leakage to the nonpoor may be accepted. It is a matter of reaching an appropriate balance between the two considerations.

When an attempt to target the poor imposes high costs of (1) identification and determination of eligibility, (2) communication with the eligible, and (3) monitoring to exclude the ineligible, cost-effective strategies must be found. Time-consuming and expensive household surveys to determine, for example, prospective participants' income and consumption levels may have to be avoided if they are not already available as part of the national statistical information. Other less expensive shortcuts may provide reasonable results. To illustrate, in one method, villages are mapped and households ranked into groups by poverty status; next, the field staff of the microcredit agency interviews those who have been identified as eligible for final selection. Another method uses the "house and compound of the house" as a crude indicator to eliminate nonpoor households from initial consideration as potential members. In the next stage, the field staff visits those who are initially chosen as poor on the basis of this criterion to verify their eligibility through a short interview that focuses on the value of their productive assets (Gibbons and Meeham, 2000).

The ownership of land or the size of the land owned is often used as an indicator of rural poverty, e.g., landless or very small landholders being poorer than those with bigger landholdings. The probability is low that the nonfarm- or nonland-based income may more than offset the poverty of the landless or smallest landholders in the rural areas, especially in the Asian developing countries.

In some cases, the size of land ownership of a household may be difficult to ascertain. For example, when sons form separate households, the land from the father may not be immediately transferred, yet they may work on the farm and receive a share of the output. Similar cases of widowed sisters who live with the brothers may be encountered. Borrowers with high dependency ratios (i.e., households with few earners to members) or households earning a major part of income from casual labor are likely to be poor.

The land- and occupation-based eligibility criteria are not perfectly correlated with poverty. Households that are marginally above the cutoff point could still fall below the poverty line. Targeting of the poor is facilitated in some cases such as the Grameen Bank in Bangladesh by the use, in addition to visible indicators such as landholding, of participatory techniques, such as local groups of population assisting in identifying the poor. Frequently, gender-based criteria (divorced, abandoned, or widowed women, female-headed households, or women relying mainly on casual labor) are used either singly or in addition to that of landlessness to identify the poor households.

Geographical targeting has also been frequently used in locating the poor households. Frequently, significant and sizable geographical effects on living standards are found after controlling for a wide range of nongeographic characteristics of households such as education, landownership, and occupational pattern, etc., characteristics that are easily observable to the policymakers (Ravallion and Wodon, 1997). Therefore, poverty targeting—its nature and extent—depends on the placement of offices or branches of the microcredit agencies, such as location in areas or regions having the largest concentration of the poor.

However, at the same time, search for financial viability, which warrants offsetting considerations, may cause agencies to avoid areas with low returns such as remote and inaccessible areas often inhabited by very poor or disadvantaged populations. One consideration is the expected level of demand for credit services. It is important for two reasons. First, fixed costs associated with the establishment of branches imply that when demand is lower than some minimum threshold level, credit delivery becomes very costly. Second, the marginal impact of financial services on participating households is likely to be highest in areas with strongest credit demand. For these reasons, branch and service placement decisions are likely to respond to the level of physical and market infrastructure and the general economic buoyancy of the area, all of which fuel credit demand. Related to this is earning adequate returns from activities financed by credit since this affects repayment rates. Thus, a tendency exists to avoid areas with low returns such as inadequate infrastructure or susceptibility to natural disasters and other risks. Loans for financing highly market-dependent activities are not suitable for remote areas.

Another consideration is the cost of supplying services. In this regard, at least three factors influence the location of branches. First, credit transactions raise security concerns; proximity to police stations or other law-enforcement establishments is important. Second, proximity to branches of commercial banks is also important, since the microcredit agencies frequently do not provide banking services. Third, to the extent that salaries and other compensations of the credit agencies staff do not provide rewards/incentives for more remote locations, managers are likely to prefer locations that have fairly well-developed services (education, market, and health).

A recent study of the placement of three important NGOs/MFIs in Bangladesh shows that even though the placement of branches is attentive to poverty considerations, they are nevertheless more likely to be established in locations with better access to transportation and communication infrastructure. Hence, it appears that MFIs are geared more toward the poor who reside in relatively well-developed areas than the poor in more remote and less-developed regions.

Much has been said as to whether the microcredit programs can reach the low end of the rural poor. In general, the factors responsible for the inclusion of nonpoor in the microcredit programs and the exclusion of the poorest may be said to operate from two sides: from the side of the organization (supply side) and from the side of the poor households, who are the prospective clients (the demand side).

Regarding the "demand side" restrictions on reaching the poor, it is suggested that some of the poor "self-select" themselves out of the credit programs' membership. They do not consider themselves creditworthy; they suffer from a lack of self-confidence and absence of entrepreneurial ability or feel that it is too risky for them to borrow. The poor, especially the poorest, do not always come forward to apply for credit, as they do not know or not believe that credit is really available to them. In Bangladesh, 49 percent did not join because they felt that they would be unable to repay the loan; about 25 percent did not join because to do so would violate social norms; 13 percent wanted to join but were not accepted in the groups because they constitute high risks for a variety of reasons. Under the circumstances, it may be necessary to motivate them and convince them; when they see for themselves that their poor neighbors or households have participated and benefited, they are inclined to avail themselves of the opportunity.[2]

To understand the problems that lead to a lack of demand for credit among the poorest, one must understand the constraints faced by the poorest that impede the proper use of finance. Inadequate control over any or a combination of factors such as land, labor, and capital may limit demand. Extreme poverty may be associated with lower labor force endowment and adverse dependency ratio. The availability of an adequate family labor force helps effective use of microcredit for generating a higher return; it helps to diversify the sources of income, thus making loan repayment through monthly installments easier. Frequently, families with only female workers face various obstacles in running an enterprise. Women burdened with the responsibility of small children and/or a sick husband are the most adversely affected. The fact that the poorest families in the rural areas often do not possess a house and/or land may prevent them from obtaining microfinance. The size of the homestead area may become a binding constraint for the entry into a self-employment activity or expansion of the scale of activity. A rickshaw puller requires a safe place to keep his rickshaw at night and thus the space around the house is important. The poorest often find shelter in a rich relative or neighbor's house, but they will not be allowed to keep their livestock on someone else's land. The number of livestock one can keep will depend on the availability of space. Similarly, an activity such as paddy processing requires a place for drying paddy. Frequently, the poor women are seen using a part of the highway for drying paddy.

Lack of equity capital is also a constraint; although it is not essential to obtain credit, this may influence the rate of return from enterprises through the following channels. First, when the borrowed fund is supplemented with one's own capital, the capital labor ratio will be higher. This will increase the rate of return to labor. Second, the use of one's own capital in a microenterprise may affect the choice between fixed capital and working capital. This in turn will determine the type of activity one can pursue. The borrowers who do not possess any capital are likely to use a smaller proportion of the borrowed fund as fixed capital. The limitation arises because repayment usually must be made within one year through monthly or bimonthly installments and most repayments start within a month of receipt of the credit. If the entire borrowed fund is invested in fixed capital, the regular repayments as well as the interest payments must be made from current flow of income right from the beginning. If an individual has some capital of his own, then a part of the fund may be used for loan repayment, at least during the first few weeks. Moreover, households without other sources of income are likely to invest in activities that require little fixed capital and yield high returns in the short run. This often limits their choice to a few processing activities and petty trading or peddling.

A few features such as compulsory weekly savings and the lack of "grace periods" for most loans deter the poorest. At the same time, these poorest households continue to borrow in the informal credit market, primarily during times of distress. The poorest borrow to reduce vulnerability to seasonal shocks and the "moderate poor" may borrow to invest in income-generating activities.

The supply side constraints on the inclusion of the poor may arise due to (1) the objectives and policies of the microcredit agencies, and (2) the procedures and the strategies used to achieve these objectives.

With the major stated objective of most microfinance programs as "poverty alleviation and extending credit services to the poor," they will be encouraged to include those just below the poverty line, who are termed as borderline poor, because it is easier to lift the "middle poor" or "borderline poor" above the poverty line and thus to demonstrate success in poverty alleviation.

Along with poverty alleviation, the other overriding concern has been to achieve financial sustainability and to reduce dependency on donor or external funds. A two-pronged approach may be useful to achieve financial sustainability as follows: (1) reduce cost through increasing the scale of credit activity and (2) ensure that the good repayment practice already achieved by most NGOs is sustained.

The inclusion of the poorer households may raise the cost of credit agencies and therefore inclusion of the nonpoor may be warranted by the following factors:

1. Loan size for the poorer borrowers is usually smaller than the loan size in demand by the borderline poor. The poorest do not borrow for fixed investment since they do not have other sources of income from which they may repay the initial installments at frequent intervals. Also, smaller loan

size of the poorest is also due to their lack of confidence about handling a larger amount of money. Since the cost of administration is the same for each loan, the cost of lending per unit is higher for smaller loans.[3]

2. A longer time is required than in the case of the nonpoor or the border-line poor to motivate the hardcore poor to participate in the credit program; this raises cost. To achieve economies of scale, there is a tendency to increase the number of borrowers to be covered by each field staff and this is more easily accomplished by including some nonpoor households in the program.[4]

In the early stage of the introduction of microfinance institutions, much time was spent on the mobilization and conscientization of the borrowers prior to extending credit. The need to extend membership and to expand the volume of lending has led them to shorten this period so that less time can now be spent in motivating the prospective clients as well as in the initial training and the social mobilization process. This change in practice discourages or leaves out the poor borrowers since they are the primary beneficiary of the longer process of training and motivation.

3. Due to compelling need on the part of the very poor to frequently divert a part of their loans toward purchasing food, and also due to the smaller scale and lower efficiency of economic activities pursued by them, it is feared that the extension of credit to this group may worsen the repayment performance.

4. It is alleged that in a few cases the members of the "target group" themselves wanted to include some influential nontarget group households to help them tide over loan repayment difficulties within the group. Also, households from the "nontarget group" were expected to improve the sustainability of credit agencies by depositing greater savings. The poorer the households in a group, the smaller will be their regular savings. Savings of the members of credit agencies are an important source of financial resources and these are used for relending. An emphasis on sustainability and self-reliance enhances the importance attached to members' savings.

5. Forming borrowers' groups for the purposes of group-based lending by microcredit agencies may also act as an obstacle to the entry of the poorer households into such membership groups. The collective group responsibility requires that a member who is unable to repay is motivated to find ways to do so; in some cases the group members are denied new loans if one member defaults or cannot repay within the stipulated period. The poor members are unlikely to include a poorer one who may become a burden to them.

6. Powerful factions may exist within the village who may disrupt the microfinance institutions if they do not get access to credit. Thus, the "target group" approach may need to be flexible enough to incorporate a num-

ber of socially influential households in order to maintain some link with all socioeconomic classes in the village. It is difficult to determine how important this factor is in practice.

Impact on Poverty Reduction

In assessing the impact of microcredit-financed microenterprise on poverty, it is important to remember that microcredit, provided ostensibly for productive microenterprise, is usually used for other purposes consistent with household welfare maximizing behavior; this includes financing consumption (indirect asset protection rather than asset creation because otherwise the existing asset will be sold to finance consumption); repayment of existing informal debt; and investment in housing and education. For long-term impact a borrowing household must have a source of reliable income from which to make regular weekly payments, must be free from pressing debt, have good health, and have freedom from imminent expenditures or have enough savings to cope with them.[5]

In Bangladesh, between one-fourth and one-third of loans have been used fully or partly for purposes unrelated to production in the following order of importance: subsistence expenditure on food and clothing, loan repayments, tubewells for drinking water, purchase of homestead land, and release of mortgaged land. Usually, the credit agency officials discuss the feasibility of the project with the loan applicant and once the loan is granted they often informally monitor whether the bulk of the money is invested in the proposed project. The borrowers know that if they use a large portion of the loan for a totally unrelated purpose, then they will face problems in obtaining additional credit (Zaman, 1999).

The three features of microcredit—small loan size, weekly payments, and short maturity (one year, for example)—tend to rule out long-term investments when longer-term maturity and larger loans are desirable. Each loan, therefore, has only an incremental or short-term impact on enterprises and household income and a series of loans are needed to make an income-raising impact above the poverty line.

Various estimates are available regarding the impact of microenterprise on poverty. In Bangladesh, for example, it was shown that 5 percent of the Grameen Bank borrowers rose above the poverty line each year by borrowing from the bank, whereas the corresponding percentages were 3 and 6 percent, respectively, for Bangladesh Rural Advancement Committee (BRAC) and Bangladesh Rural Development Board (BRDB). The extent of poverty gradually declines as the landless households obtain repeat loans (Khandker, 1998). Similarly, a direct comparison of household expenditure among the borrowers and the nonborrowers covered by eight NGOs showed that household expenditures of those who obtained more than one credit was 17 percent higher than that for those who did not borrow. However, the first time borrowers obtained a 5 percent increase in household expenditure over the level of the nonborrowers (Rahman, 1996).

Access to microcredit helped the entrepreneurs accumulate nonagricultural capital, including ownership of house, which contributed substantially to productivity among low-income households. Microcredit had a positive impact on savings and asset accumulation (Khandker and Chowdhury, 1996). Current savings of those who obtained microcredit were many times higher than the savings of those who did not get credit. The amount of savings increased sharply with the number of loans obtained by a household. Among the types of assets, the greatest increases occurred in the ownership of a house, transportation, and livestock.

A study of BRAC in Bangladesh revealed that microenterprise financed by borrowing more than a threshold amount (e.g., 10,000 taka) made a significant contribution to poverty reduction among the moderate poor. Borrowing more than 10,000 taka (the mean loan size for the "10,000 plus" category is 13,090 taka) raises a moderate-poor household's consumption per adult equivalent by 18.8 percent compared to an identical nonborrowing BRAC member. Also, households that had borrowed more than this threshold amount spent significantly more in terms of nonland "productive assets" (poultry and livestock in particular) during the one year prior to the survey compared to members who had borrowed less than 10,000 taka. Sharp growth occurred in productive assets for third-time borrowers, compared to first-time borrowers (Zaman, 1999).

Significant improvements in welfare once an enterprise crossed a certain loan threshold could be possibly interpreted as the result of switching from traditional, low return on-farm activities to higher return off-farm activities over time. This also could be because initial loans were often used for consumption purposes, repaying debts, and repairing homesteads while subsequent ones were used for investment purposes. Thus, providing capital to poor households improves their socioeconomic conditions, but only when loans are large enough so that investment in both on- and off-farm enterprises can earn significant returns.

The previous study indicated that the extremely poor, proxied by those with less than 0.5 acres of land, did not appear to benefit significantly from borrowing, even for those who had borrowed more than 10,000 taka. Given the fact that the extremely poor were more risk-averse than the moderate poor, they were more likely to have borrowed for traditional low productivity activities with a view to switching into riskier higher return activities over time.[6]

Through asset creation, microcredit reduces vulnerability to shocks caused by natural or man-made calamities, sickness, or death in the family. Building up a household's asset base, which can reduce vulnerability through a number of channels, is an important form of self-insurance against crisis. To start with, some assets can be readily sold to meet immediate consumption needs. Second, asset building can improve creditworthiness, thereby improving a household's borrowing capacity during a crisis. Third, a larger and more diverse asset base can reduce covariant risk. The potential reduction in vulnerability thus takes place through a number of ways, e.g., the creation or expansion of one or more income-earning assets, improvement in housing conditions, income and consumption smoothing, savings for the lean season, and the emergency assistance provided by many microfinance organizations during periods of acute natural disasters such as the recent floods in Bangladesh.[7]

Increases in income or consumption (i.e., reduction in poverty) can occur if credit is used for an income-generating activity and that activity generates returns in excess of repayment obligations. A credit-financed investment does not reduce poverty if it does not generate a significant net profit, even though it creates an asset that can reduce vulnerability. Loan repayment takes place through a reduction in consumption and not from the returns to the investment. A temporary reduction in poverty can also occur if credit is used for noninvestment purposes such as repaying existing debt, improving housing, or meeting social obligations. However, future consumption must be sacrificed to meet repayment obligations.

Empowerment of Women

Many studies have been conducted on the impact of microcredit and microenterprise on women's empowerment.[8] However, the studies in South Asia do not provide a uniform conclusion. Despite the mixed results within and across studies, some positive impact on women's lives has occurred. The main benefits derived by women seem to be

1. greater involvement in income earning activities;
2. increase in awareness about social-, economic-, and health-related issues;
3. increase in the adoption of family planning methods; and
4. increase in girls' education and school enrollment rate. (Rahman, 2000)[9]

Many of the studies indicate positive impact in respect to some other indicators. For example, a few of the studies obtain a positive impact on certain types of decision making, especially decisions regarding small purchases, expenditure on food and clothing, daughter's education, etc. Most studies show an absence of positive impact on decisions regarding large expenditures and sale of large assets.

Two types of negative findings have been extensively discussed. First, most female borrowers cannot retain control over the loan, and second, family violence does not decline and may increase with the length of membership in microfinance institutions. Some controversy surrounds the interpretation of control of microcredit obtained by women. It is not expected that in all cases women would have full control of the borrowed fund because economic activities in the rural areas are usually family activities. The use of a loan in such activities usually means sharing the control of the loan. Moreover, when the male members are not eligible for membership, it is very likely that the loan would be routed through women but would be used by male members and in such cases it is not expected that women would retain 100 percent or even a major control of the loan. Therefore, studies show that microcredit might lead to women's empowerment, even if the loans are used by male members of the family.

The studies, which show a positive association between microcredit and violence against women, are subject to differing interpretation. First, it appears that studies, for example, in Bangladesh during the mid-1990s   but not the early studies—do show such results. Second, there is no analysis of the possible cross-relationship between the occurrence of violence and other indicators of empowerment. Since women continue the membership of NGOs/MFIs despite the violence inflicted on them, the perceived net benefit of program participation is likely to be positive.

The other question relating to the impact of credit on violence is whether it is the failure to utilize credit productively and the consequent hardship or the sharing of the income generated by the credit-financed enterprise that leads to violence. In the former case, the obvious answer is to help women to utilize the loan productively to earn high income. In the latter case, microcredit must be supplemented with support from organizations committed to the advancement of women's rights.[10] In any case, more in-depth research on the incidence of family violence among the microcredit recipients is required. Violence against women has been a much wider problem in many countries with or without access to microcredit.

FINANCIAL SUSTAINABILITY

The financial sustainability of the microcredit agencies relies upon their administrative and managerial efficiency, depending upon efficient cost control, appropriate accounting, and auditing procedures. It should be noted that as outreach expands, administrative costs increase due to

1. initial cost of opening new branches,
2. costs of recruitment and training of new staff to man additional workload, and
3. time lag in the attainment of efficiency by the new staff up to the level of the existing staff.

It takes approximately three to five years for the new staff to reach the required organizational level of efficiency. It will take considerable time for the economies of scale from the expansion of operations to be realized and for the extra costs of expansion to be recovered from the interest income (Gibbons et al., 2000). Sustainability, of course, also depends on the interest earnings from microcredit. The appropriate lending rate to attain self-sufficiency depends on

1. unit costs of administration,
2. loan loss,
3. cost of funds,
4. required/desired capitalization rate, and
5. expected investment income.[11]

Should financial sustainability imply that a microcredit agency is able to meet not only (1) the cost of loans and (2) its administration, but also (3) the cost of social mobilization and awareness building among the poor? The latter encompasses the ability of the poor borrowers to understand the opportunities of and gains from self-employment, the value of savings, and their ability to manage elementary bookkeeping. As the borrowers graduate to more complicated and familiar nonfarm activities, they need to be given training in new technology and marketing. The tasks of social mobilization and training are not credit inputs. It is unlikely that returns from lending operations will, in general, meet the additional costs of noncredit inputs.

Financial sustainability conflicts with the objective of targeting the poor have been discussed in the section on targeting. The following examples illustrate the projects that are able to meet the terms and conditions that usually accompany the provision of microcredit. This is important to analyze how the microcredit provided for different varieties of nonfarm activities, with long gestation lag and moderate returns affecting their financial sustainability.

A study of microfinanced enterprises by the Centre for Agriculture and Rural Development in the Philippines (modeled after the Grameen Bank) indicated that annual returns from investment calculated as gross returns before payment of installments and interest total more than 100 percent, 117 percent on average. With the interest rate at 39 percent, it left 78 percent with borrowers as returns from their wages and profits (Hossain and Diaz, 1997). This is quite a high rate of return and is adequate to meet the payment obligations on account of loan repayments and interest payments.

The discipline of weekly repayments and high interest rates may imply that frequently payments are made in the initial weeks from the household cash flow, which sometimes requires its members to tighten their belts; this resembles a classic form of savings based on self-denial for future gain. For example: a $46 loan for hog fattening payable over six months requires weekly installments of $2.32 of which $0.21 is interest: two piglets cost $46, each at $23. After about six months, the fattened (mainly on household scraps, vegetables planted in the house garden for that purpose, and commercial feed supplement) pigs can be sold for about $92 each, giving an attractive lump sum return and net profit estimated at around 100 percent on average (Gibbons et al., 2000).

For households too poor to tighten their belts, loan activities such as petty trading or small shopkeeping that result in the quick and frequent generation of additional income are more appropriate. For example, in an Indian case, petty trading of bangles and cosmetics by poor women requires a working capital of only about 1,000 Indian rupees. If this is borrowed at an interest rate of 20 percent (flat) for a term of 20 weeks, with principal and interest repaid in equal amounts weekly, then the required weekly repayment is 60 rupees of which 50 rupees is principal and 10 rupees interest. Usually the women sell house-to-house and village-to-village, carrying their wares in a basket on their heads, six days a week. They gross about 100 rupees a day or 600 per week of which about 120 rupees are net profit. Half of this goes for repayment.

For larger loan amounts, the weekly repayments can be kept small and manageable by lengthening the loan term. For example, a popular loan activity in South Asia among the poor is the purchase of a moderately yielding, say 3 kilo per day, milk buffalo, which can be purchased pregnant for around 6,000 rupees in India. If a loan of the whole amount is made available for that purpose to a very poor women at 20 percent interest (flat), or an effective rate of around 40 percent, on a declining annual rate for a term of two years with 100 equal weekly installments of principal and interest, each payment would amount to $[6,000 + (6,000 \times 0.2) + (3,000 \times 0.2)] = 7,800/100 = 78$ rupees. The 3 kilos of milk could be sold daily for approximately 12 rupees per kilo. This means that the weekly repayment money of 78 rupees could be earned in two to three days, leaving the income from the other four to five days to reduce the poverty of the household. The risk of the buffalo dying can be covered by livestock insurance at a premium of 4 rupees per week, or 100 rupees per year. Over the two-year period, the total cost would be 160 rupees, which could be paid from the sale of the milk. However, as the buffalo produces milk for only about nine out of twelve months, the borrower must save or engage in some other income-generating effort for the remaining three months. To fill the gap, clients frequently purchase a second buffalo as soon as they can. With two milk buffaloes they can have a good, steady income throughout the year (Gibbons et al., 2000).

The above examples explain why livestock, poultry, petty trading, and small shops predominate the activities that are usually financed by microcredit—when small amounts are lent with frequent payment of installments.

To achieve financial sustainability it is frequently necessary to customize the types, terms, and conditions of microcredit to the needs and opportunities of the poor to ensure the productive use of loans in the context of (1) their overall credit needs (so that microcredit for income generation activities is not diverted to other uses) and (2) investment opportunities which yield returns high enough to meet the interest and repayment obligations. For example, in one case it was found that a one-year term for the first loan was not suitable for clients primarily engaged in tertiary activities such as petty trading with shorter business cycles. It reduced the period to six months; at the same time, to discourage the clients from resorting to traditional moneylenders in times of economic distress or income shortfalls, a multipurpose loan was introduced; thus, a loan up to $132 (at 2,000 exchange rate), compared with the initial average loan size of around $80, could be used for any purpose after six months of membership in the program. After some years, an additional type of loan was introduced (called Loan Accelerated Program) for successful members who have been with the program for many years, allowing them to draw on an overdraft account based on the need of their business, subject to all the disciplines of the regular loan program (Hossain and Diaz, 1997; Gibbons et al., 2000).

With the passage of time, scope for investment in quick yielding, high-return nonfarm activities diminishes; nonfarm enterprises seeking credit will be increasingly those that have longer gestation lags and modest returns. This would require a modification of the prevailing lending modes and procedures, including adjusting the level of interest rates to suit the unfolding circumstances. This brings one to the broader issue of general development related supplementary measures and nonfinancial inputs needed for the development of rural microenterprises. The role and modus operandi of microfinance would be seen as part of a comprehensive framework for the promotion of microenterprise.

NONFINANCIAL INPUTS

In many nonfarm activities, finances need to be combined with such noncredit issues as transfer of technology, training facilities for workers and entrepreneurs, and support for marketing. Credit tends to be relatively more important in activities that do not require any demanding skills and in which backward and forward links are not problematic. Examples are many forms of processing where working capital requirements are high, or in transportation services where the initial fixed capital outlay is large. Retailing and wholesaling are also working-capital intensive (ADB, 1997).

However, noncredit inputs such as design, product development, market information, and marketing assistance are usually much more important for a large number of manufacturing activities, including handicrafts. Generally speaking, noncredit inputs and support services are particularly important for growth-oriented microenterprises and activities with relatively numerous backward and forward linkages, such as manufacturing. Trading and services usually require fewer nonfinancial services, with the exception of training in services that are skill-based.[12] The importance of primary and secondary education in the development of the nonfarm sector needs to be stressed. The educational level of the household head correlates with the probability of starting an enterprise. Households with primary education start new enterprises at a higher rate than those without such education. Entrepreneurs with secondary and higher education are more successful in ensuring growth in their enterprises (Shilpi, 2003).

Given the variety of noncredit inputs required for various activities, including marketing and technology, only an organization with a detailed understanding of and considerable experience in a particular subsector is in a position to identify all the inputs and support services required. It takes considerable field experience to pinpoint all the bottlenecks impeding an activity because many limitations reveal themselves only when implemented. The constraints change as implementation proceeds.

The microenterprise promotional programs need to place greater emphasis on the subsector approach, which provides a framework to identify the constraints and opportunities linked to a particular final product or raw material. Training programs have not been very successful, partly because technical training has often been provided in a very generic way rather than directed to the specific needs of one or a few subsectors. The case with marketing assistance and technology transfer is similar.

There are a number of subsector-specific promotional public organizations in most countries, such as silk boards, dairy development corporations, and handicraft training-cum-production centers. Sector/subsector specific functional organizations include

1. research institutes (for example, in leather and food processing),
2. common service institutions such as testing and other quality control facilities (for example, in diamond and gem processing "parks"), and
3. training centers.

In many cases a need exists for organizations in the private sector or among the NGOs to complement government-sector institutions. An important mechanism available both for analyzing and catering to the subsector requirements are subsector-specific NGOs. Usually operating in a smaller area, the NGOs frequently have the advantage of detailed local knowledge, better motivated staff, flexible procedures, and a focused commitment to the micro sector. They can be engaged in the diagnostic and analytical task, which is itself a continuous process. Second, they can facilitate enterprises' access in a sector/subsector to the existing support services offered by the government; thus, they can function as intermediaries and perform what is usually referred to as the "linking role." In Bangladesh, for example, BRAC works with the Directorate of Livestock and Poultry to arrange for training women in the backyard poultry industry. Third, creating business links with services provided by the private sector is another aspect of this role. Promoting links of small microenterprises with larger firms through franchising and subcontracting has been a longstanding objective in microenterprise development.

NGOs providing noncredit services tend to be less numerous than credit NGOs, partly because many nonfinancial services require a high degree of specialized business acumen (such as in marketing) or knowledge of specific technologies. Subsector-specializing NGOs cannot be created overnight. However, many NGOs also follow a holistic approach. They recognize the importance of nonfinancial services and they require and need to receive technical assistance in acquiring the necessary skills and staff within the context of well-thought-out and timebound subsector plans. It is usually difficult to recover the costs of services such as training and technology transfer; it is hard to quantify their benefits though they are very real and of a long-term nature.

They can channel inputs and services already available elsewhere to microentrepreneurs, who, when organized into groups, can be reached more easily. The subsector-oriented institutions, be they governmental or NGOs, may organize producers into groups (often cooperatives) so as to achieve economies of scale in the delivery and reception of services. They can also link microenterprises with private exporters. These institutions can establish standardized descriptions for various product categories and disseminate them through national and local business associations. They gain acceptance for their quality standards in the export markets; they may help strengthen mechanisms to disseminate information on input suppliers and output buyers, setting up databases through local networks.

A distinction can be made between what is called supply-driven and demand-driven assistance (Boomgaard et al., 1992). The supply-driven assistance is provided regardless of whether the enterprise or the subsector requires it, or whether there is adequate market demand for the enterprises or the subsector to utilize the type of assistance provided. In other words, supply-driven assistance is not directly linked with the requirements of the market for the product in question. Demand-linked or demand-driven assistance, on the other hand, originates because enterprises have a contract with buyers for the supply of a product, and the fulfillment of the contract requires that they utilize specific types of assistance. Examples for this kind of assistance include the following examples. When large urban enterprises buy products for their use from small-scale enterprises, either urban or rural, they often provide financial and technical assistance in the form of product design, specialized raw materials, and production technology. The subcontracting arrangements between urban large enterprises and rural small-scale enterprises also constitute an example of demand-linked or demand-driven assistance that flows between different-sized enterprises and with different expertise or specialization. These arrangements are especially popular for activities as in the case of metal fabrication and textiles in which the key components of the production process can be partitioned, and the more labor-intensive components can be contracted out to small rural firms that have relatively cheap labor costs. Subcontracting requires good rural infrastructure and communication networks, effective laws and institutions to facilitate the enforcement of contracts, and a favorable growth environment for the industrial sector in general (Hayami, 1998; Islam, 1997).

It is also possible to link promotional measures provided by the government to microenterprises with the government procurement of products and supplies. In many countries, the government procurement of goods and services constitutes a substantial portion of market demand. This is a potential market for small-scale enterprises that can provide a great stimulus, while at the same time ensuring that efficiency in production and marketing techniques is greatly improved. The traditional way of encouraging or promoting small-scale enterprises through the government procurement is to give a preferential price or to reserve a certain percentage of the total government procurement exclusively for purchases from small-scale enterprises. This policy provides a "reserved or protected market" but does not necessarily help improve the efficiency of small-scale enterprises in terms of quality, costs, or marketing methods.

However, in the demand-linked or demand-driven system of providing government assistance to microenterprises, the government procurement or the purchasing agencies need not be under any compulsion or obligation to purchase from such enterprises. The agency in charge of providing assistance or promotion should become the supplier of goods and services on behalf of the microenterprises to the government departments or purchasing agencies and guarantee the quality, price, and timely delivery of the requisite supplies. In other words, the governmental promotional agency would act as a contractor to ensure supply to the purchasing department or agencies, which are free to cancel the contract if the agency is unable to fulfill the contract, just as for any contract between private sellers and buyers.

One major obstacle to government procurement is the difficulty in dealing with a large number of small-scale suppliers and ensuring that quality is uniform and that products are delivered on time. The transaction costs of dealing with one or two large suppliers are considerably less from the viewpoint of the government bureaucracy. The intermediary role played by a different government agency charged with the task of promoting and providing assistance to the microenterprises circumvents the problem of high transaction costs incurred by the purchasing or procurement agencies. The promotional agency, in turn, deals with a large number of suppliers and ensures the fulfillment of the contract. One way of reducing transaction costs is for the promotional agency to deal with organized groups or associations of suppliers rather than with individual suppliers (Tendler and Amorin, 1996). There is a case for public subsidies to cover the transaction costs involved in entering into contract with small firms. The government could finance a special program to be run by the promotional agency to encourage the formation of associations or organizations of small firms. Furthermore, microenterprises can also pay for the services rendered; the purchasing agency of the relevant government departments can provide the usual commissions to the promotional agency for the benefit it gains from bulk purchasing and handling.

This approach accomplishes three important things. First, it links small firms to a customer who is committed to purchasing large quantities of a product. Second, by securing a contract, it brings together support and promotional agencies of the government with microenterprises, ensures that the agency provides training at the firm site rather than in classrooms, and solves problems as they are identified in the course of production. Third, this approach helps the agency to discover specific critical bottlenecks and to learn how to overcome them. The technical experts of the assistance agency concentrate on the problems brought to them by clients. If necessary, they can take the problems from the site of the enterprise to their head office or laboratory to do research and find appropriate solutions. In various ways, the support agency can ensure the government's purchasing departments that buying from small firms is no more costly or burdensome than buying from big firms (Tendler and Amorin, 1996).

Under the supply-driven assistance program, neither the microenterprises nor the support agency of the government is subject to the tests of market demand. The support agency traditionally delivers standardized services—business advice, training, production assistance, or credit—to as large a number of firms as possible. The agency provides only generic services common to all enterprises, not client-specific services, and therefore is less effective. The success of the demand-driven approach depends on the contract or agreement with groups of firms providing identical products and on the payment being made to each producer upon the delivery and satisfactory inspection of products of the whole group. This is crucial to the reduction of transaction and monitoring costs of the government purchasing agencies. This creates the necessary peer pressure and shifts the monitoring function from the support agency to the group.

The importance of training is universally acknowledged, but its effectiveness remains little understood. The most common types of training in small- and medium-size microenterprise programs are

1. management-oriented or "business" training (in such skills as costing, accounting, bookkeeping, business plan preparation, and so on);
2. production-oriented or technical skills training; and
3. entrepreneurial development training.

Two other categories of training sometimes identified are (1) credit-oriented training, and (2) general community development or pre-entrepreneurship training. The latter targets potential entrepreneurs and focuses on more general skills such as literacy or leadership. The credit-oriented training of poor borrowers, on the other hand, in credit procedures, group discipline, and savings obligations can be considered an essential part of the "social preparation" or social intermediation component of poverty lending projects, rather than constituting a distinct training activity.

Training in business skills, unlike training in technical skills, is relevant to a large number of diverse activities so that enterprises from a variety of subsectors can be brought together conveniently to receive the training. Training in skills is important for ensuring the survival, if not growth of, existing enterprises as it is for engendering new enterprises. However, training in technical skills tends to be very activity or subsector specific, and trainers with the right specialization are usually not available. Training in technical skills is thus difficult to organize. Indeed, the only way to organize it may be by using the subsector approach, bringing together a large number of participants from the same subsector, such as handicrafts, sericulture, poultry, and livestock. The example of the government promotional agency described previously providing job/product specific training linked to marketing is one way of providing effective training that is required by microenterprises.

Entrepreneurship development training (focusing on the motivational, attitudinal, and behavioral aspects of entrepreneurship) is prevalent in South Asian countries. In India, for example, the Entrepreneurship Development Institute (EDII) in Ahmedabad has initiated a number of programs and institutions. Every state in India has an Entrepreneurship Development Institute which is expected to reach down to microenterprises, apart from catering to the needs of larger enterprises while still recovering its costs. Proponents of entrepreneurial training do not claim that entrepreneurs can be created. However, they do believe that it is possible to develop entrepreneurship in persons who have the latent potential.

Such programs are explicitly not designed for small and traditional "livelihood" types of microenterprises. For such enterprises, pre-entrepreneurship training programs envision a "preparatory" literacy course for those who may later show interest in an entrepreneurship course. The literacy course addresses community issues relevant to their lives and provides grassroots management training to producer groups. The training material is mostly pictorial and involves simulation exercises, focusing on gender, empowerment, and enterprise management issues.

Similarly, under the Rural Industries Project (RIP) in India, started in 1993, the clients were chosen for their entrepreneurial potential rather than socioeconomic background, although 40 percent of those covered were below the poverty line and about 23 percent of the entrepreneurs were women. It attempted to provide the full package of services to its clients, including market analysis, project formulation, escort services dealing with the banks and equipment suppliers, and links with technology providers, trainers, and others. Average investment was about Rs 30,000 (about $900) in the ratio of 7:2:1 as loan, equity, and capital subsidy. Manufacturing accounted for about two-thirds of the activities and nontrading services for one-third.

Most of the entrepreneurs already had some previous experience in the activities covered by their enterprises as apprentices and wage employees. If their enterprise failed, exit was relatively easy for them as they could go back to wage employment. About one-third of the enterprises, mostly started by relatively better-educated entrepreneurs, sought to create new markets for their products and services (such as tire carts, duck hatcheries, metal furniture, computer training, and fabric painting). The success rate of the enterprises seeking new markets/products tended to be lower (only about one-third had succeeded), but their profitability was higher when they did succeed.

Should the recipients of noncredit assistance or services pay for such services? For livelihood enterprises run by the poor, cost recovery is difficult until they grow into bigger and more profitable enterprises. However, in microenterprises run by the nonpoor, the case for cost recovery is strong. Cost recovery, at least partial recovery, is important not only for ensuring that clients value the services provided, but it also helps to find out which services clients do value. Services for well-to-do clients should cover a higher proportion of costs than those for the poor. However, for nonfinancial services, cost-effectiveness and net social benefit may be more important in the short run than full cost recovery.

Chapter 3

Bangladesh Case: Review and Analysis

MAIN FEATURES OF MICROENTERPRISE/
FINANCE PROJECTS

Grameen Uddog (GU)

The Bangladeshi handloom weavers had been suffering for some years from the competition of powerloom products and of imported garments. GU, a nonprofit enterprise, was created in 1992 with initial funding from Grameen Bank, and supplemented later on by loans from commercial banks, to help reorient the production pattern of a selected group of weavers, many of whom were Grameen Bank borrowers, located in the heart of a weaving community region.[1] GU linked the weavers with the export-oriented garments industry in Bangladesh through a "contract weaver" or a "putting-out" system for the supply of a type of speciality cotton fabrics, featuring attractive checkered designs called Grameen Check.

The design specifications were demanding and were enforced uniformly, triggering "breach-of-contract" fines in case of failures. The scheduling requirements of export shipments by the garments industry were to be met in time. The first point of contact of GU overseas was with the Bangladesh garments industry, which exported through the intermediary of the buying agents or offices in Bangladesh of overseas customs.

The operations of GU are currently organized along the following lines. The head office of GU is responsible for marketing, while the regional/area offices are in charge of disseminating knowledge as well as processing and distribution of yarn among the local (unit) offices. The local or unit offices identify the weaving enterprises, as well as ensure that the fabrics are woven to meet specified quality and design and are delivered on time. The level of weavers' skill, wage rates, their excess capacity, and past performance in a locality are the criteria used for the allocation of work load between different unit offices which in turn select the "weaver-manager" of a particular enterprise for a certain work order. A weaver-manager who owns a number of looms and manages them with family and/or hired loom workers contracts to produce a certain specified quantity and quality of output for delivery at a certain time. In the early days, some priority used to be accorded to weaver-managers who owned five or less looms. However, with competition and price pressures increasing over time, considerations such as skill, specialization, timely delivery, quality, and cost efficiency are predominant. While GU does not provide any general training as such, a *taant karmi* or handloom extension agent on the GU payroll helps weaver-managers to implement each new design on the fabric and thus provides valuable learning opportunities.

Microenterprise Development Unit (MEDU) of the Agrani Bank

Started in 1995, MEDU's objective is to generate rural wage employment for the landless and women through the development of and loan assistance to microentrepreneurs. Therefore, it lends out of funds received from a donor such as the International Fund for Agriculture Development (IFAD) and the government, to those who are not so poor but who employ the poor in their enterprises. It provides larger loans than are usual with microfinance institutions (MFIs).

Two channels for disbursing loans to individual borrowers exist: (1), through *direct lending* by the Agrani Bank, a commercial bank; and (2), *indirect lending* through NGOs/MFIs that function as "credit retailers."

Branches of the Agrani Bank with their proximity to rural clients in areas with good prospects of nonfarm activities and with adequate staff to handle this additional new task are selected for this activity. The clients are typically owners of existing microenterprises, with proven managerial and technical skills and five years of minimum schooling, who wish to expand, upgrade, or diversify their business that currently has an annual net income between 10,000 and 30,000 takas ($200 to $600 at the then prevailing exchange rate) with no outstanding liability or default with other credit institutions. The loans are collateral free up to Tk 50,000 and are provided at an interest rate of 12.5 percent for fixed capital and 14.5 percent for working capital. Working capital loans are normally repaid within a year while fixed capital loans have repayment periods of between two and five years with a grace period of six months.[2] Each branch is provided a target to fulfill in terms of (1) average loan size, (2) amount of total disbursement of loans, and (3) number of enterprises to be covered.

In indirect lending, a partner-NGO is first identified, which subsequently lends to its members. The target group of potential clients includes people with skills and capacity to operate microenterprises, including those who are undergoing apprenticeships or working as skilled laborers and want to start their own business. The unemployed landless, who seek self-employment in preference to wage employment, could also qualify. Indirect lending is thus intended for those groups who are less affluent than those eligible for direct lending. The period of repayment for working capital loans ranges between 12 and 18 months and the maximum period of repayment for fixed capital is three years. The interest rates for indirect lending are the same as those for direct lending. The NGOs receive a commission for credit retailing and thus borrow from MEDU at 3 percent less than the rate at which they lend. Disbursement targets in the case of indirect lending by the bank are set for each NGO/MFI location.

To ensure a target for disbursement, high recovery rates, and strict supervision of credit, a number of incentive packages, both pecuniary[3] and nonpecuniary,[4] are offered to bank staff engaged in direct lending. Under indirect lending, the only incentive to the NGOs/MFIs is the 3 percent commission for credit retailing.

No provision is made for assistance in marketing by microenterprises or for their training.

Kishoreganj Sadar Thana (KST) Project

The KST Project, financed by the United Nations Development Programme (UNDP), started in 1994 as a vehicle for social mobilization. Groups called village organizations (VOs) were formed to spearhead development activities in rural communities. In order to promote rural development activities, the VOs were to establish contacts with the government developmental agencies.[5] The project gradually developed into a savings mobilization and credit program with supplementary financing from commercial and specialized banks.[6] Currently, the number of members in each VO is a minimum of 35 and often it reaches 100 members. This figure is higher than the average of 5 to 30 members in MFIs. The social organizers (SOs), appointed by the project, as well as the managers of VOs, along with their presidents elected by the members, are the main players at the grassroots level.[7] The social organizers play a key role in initiating, mobilizing, and organizing the groups.

The KST credit program does not exclusively concentrate on the rural nonfarm sector and it does not explicitly target the poor. Instead, it focuses on the inclusion of all the members of the community, both poor and nonpoor, e.g., so-called vertical links that rely on kinship ties among VO members, irrespective of their economic or financial status. Allegedly, the KST VOs are likely to be more inclusive as conduits of credit delivery because the inclusion of the rural rich widens the market for credit and thus enhances the financial viability of the VOs. Also, the rural rich members contribute significantly to the VO's own savings and therefore to the financial resources available for on-lending to the member borrowers.[8] The VO's weekly meetings are expected to decide to whom to lend, how much, and under what conditions. De jure, the credit operations incorporate the elements of a group-lending model. However, de facto, both the application process and the sanctioning of credit become the domain of the VO manager. He or she is really the operating executive of the VO. Typically, the VO managers are young men and women with secondary education from relatively affluent families in the neighborhood, capable of enforcing compliance to norms of group conduct.

VO's own savings, UNDP funds, and loans from BKB (Bangladesh Krishi Bank) are the three sources of loans. The interest rate for loans is 15 percent per year.[9] The loan repayment schedule is flexible and the number of installments is linked to the purpose of the loan, except that the loans from BKB are repaid in 50 weekly installments.[10] The group does not require collateral, but the decision to sanction a loan is partly based on the accumulated savings of the VO member. In case of default by a VO member, the practice of group-based lending, such as disqualification for loans on the part of the rest of the members, is not strictly enforced. Since the remuneration of the VO manager is dependent on the performance of the applicants, it is in his or her own interest to ensure timely repayment. Managers receive 5 percent of the interest charged on loans from BKB and the UNDP resources, and 3.75 percent of the interest charged on loans from VO's own savings (the remaining being distributed among the president, the group fund, the risk fund, and the general members).[11] All of these factors create a system of incentives for KST-VO managers as well as a sense of ownership among all the VO members.

Besides the saving and credit services, KST provides training on a wide range of subjects from bookkeeping to agricultural production, but only a few of them are directly relevant for nonfarm activities, e.g., training in sewing and tailoring training.

Small Enterprise Development Project (SEDP)

The primary goal of the SEDP is the promotion of small-scale labor-intensive enterprises, including women's enterprises. It is jointly financed in equal proportions by NORAD (Norwegian Development Aid Agency) and by the Agrani Bank, a commercial bank that manages the lending operations. The borrowers, mostly existing entrepreneurs, are drawn from semi-urban and rural areas surrounding the market centers. Its location in semi-urban areas facilitates the selection and monitoring of borrowers since the use and repayment of loans are closely monitored. Potential borrowers (women are specially encouraged) must have at least half an acre of land and preferably but not necessarily some prior entrepreneurial competence in activities that use local resources, labor, and technology. Therefore, they are not usually so poor as to qualify for MFI credit nor so rich as to qualify for regular commercial credit. They should have no outstanding loan from another source nor should they simultaneously borrow from any NGO/MFI or bank. They are identified by the project staff and attend a three-day training period in basic entrepreneurial skills and social development issues during which they are assessed for their entrepreneurial potential. Topics covered in training include role of banks, keeping accounts, and management issues.

SEDP lends to individual borrowers. There is no group-based lending unlike in MEDU; there is no indirect lending through the intermediary of NGOs. The district project coordinator (DPC) of SEDP, with assistance from field officers (FOs), identifies, assesses, and approves the selection of borrowers. The local branches of the Agrani Bank are responsible for the sanctioning of loans on the recommendation of the project staff. The FOs monitor the use of loans and ensure timely recovery.

Loans are intended to be larger than the size offered by MFIs and smaller than the size provided by commercial banks. They range from Tk 5,000 and 500,000 ($100 to $10,000) and are collateral free up to Tk 250,000 ($5,000). Interest rates are subsidized and vary between 10 percent and 14 percent depending on the loan size.[12] The installments are usually monthly. The procedure for loan processing limits paperwork and takes a few days.[13] The period of repayment ranges from one to seven years with different grace periods. The longer period is intended for investment in fixed capital on a long-term basis and to promote new enterprises.

In addition to entrepreneurial training mentioned above, SEDP imparts skill training. Skill training is mandatory only for interested microentrepreneurs. Examples of skills taught include tailoring, dairy, fish culture, and poultry.

The distinguishing features of the four programs are summarized in Table 3.1.

TABLE 3.1. Comparative analysis of the projects.

Characteristics	Grameen Uddog (GU)	Microenterprise Development Unit (MEDU)	Kishoreganj Sadar Thana (KST) Project	Small Enterprises Development Project (SEDP)
Main objective	Weaving sector microenterprise development	Microenterprise development	Social mobilization, economic empowerment, and microenterprise development	Microenterprise development
Explicit promotion of rural nonfarm activities	Yes	Yes	No	Yes
Location	Weavers with traditional skills in checkered designs	Rural clients in areas with good nonfarm prospects	Rural area currently less exposed to MFIs but with past history of rural self-help groups	Rural and semi-urban areas near market centers
Inclusiveness/ targeting	Weavers selected for efficiency and timeliness, asset-based	Past entrepreneurial experience and asset owernship, not eligible for microcredit	"All-inclusive," (i.e., poor and nonpoor) within the villages selected	Past experience preferred but not necessary, not so poor landownership of half acre and more, and not eligible for NGO/ microcredit
Operational characteristics	Putting-out system; links up weavers supplying inputs to domestic garments industry that is export-oriented	Direct lending to individuals (constituting 62 percent all of loans and 72 percent of all enterprises); indirect lending through NGOs (constituting 38 percent of all loans and 28 percent of all enterprises)	Credit; targeted to individual borrowers	Credit; targeted to individual borrowers without past access to bank credit

Loan term[a]	Not applicable	Fixed capital loans 2-5 years for direct lending and 3 years for indirect lending. For working capital 1 year for direct lending and 12-18 months for indirect lending.	1-7 years in principle, but mostly between 2-3 years	2-3 years
Collateral	Not applicable	None, up to Tk 50,000	None, up to Tk 50,000	None, up to Tk 250,000
Interest rate(s)	No credit intervention and no rate of interest	12.5 percent for fixed capital; 14.5 percent for working capital	15 percent	10 percent for small loans (5,000-50,000); 12 percent for medium loans (50,000 to 250,000); 14 percent for large loans (>250,000)
Recovery rate (percent)		88	90	93-94
Is it a compulsory-saving program?	Not applicable	No	Yes	No
Is there any link between saving and credit supply?	Not applicable	Direct lending: No; Indirect lending: Yes	Yes	No
Is repayment weekly or monthly?	Not applicable	Monthly	Monthly	Monthly

TABLE 3.1 (continued)

Characteristics	Grameen Uddog (GU)	Microenterprise Development Unit (MEDU)		Kishoreganj Sadar Thana (KST) Project	Small Enterprises Development Project (SEDP)
Average value (Tk 000s)	Not applicable	41 for direct loans; 16 for indirect loans		5 for nonfarm loans; 10 for farm loans	Average: 37 (42.2 men and 28.2 women) Range: 5 to 500
Sectoral division of loans		Direct	Indirect		
Agroprocessing	Not applicable	9.0	41.2	17.5	74.4
Manufacturing	Yes	61.0	47.0	0.0	17.0
Services	Not applicable	30.4	11.8	82.5	8.6
Incentives for staff	No special incentive: Normal Grameen wage and incentive systems	*Direct lending* Incentives based on loan repayment performance of branch Nonpecuniary: Training and mobility, job advancement *Indirect lending* NGOs receive 3 percent commission		*Financial* Project staff: Higher salaries than alternative market remuneration. VO managers: Receive benefits linked to interest payments	None

a Terms refer to the duration over which repayment is to be made.

REVIEW OF PERFORMANCE

Grameen Uddog

Rural Nonfarm Employment and Income

Grameen Uddog (GU) was to revive a small segment of the handloom-weaving sector that suffered from idle loom and excess labor. The GU product (the checkered material) partially replaced imported fabrics, without significant substitution effect on the other locally produced textile products.[14] Although the scope of the program is not large enough to stem what turns out to be a generalized decline in handwoven fabric output, it has succeeded in restraining or reversing a decline for participants. The capacity utilization is higher and the peak period longer in the case of the participating weavers than that of nonparticipants.[15] GU procures raw materials in the domestic market, which draws its supplies from both production and imports. Although dyes (chemicals) are imported, GU relies on domestic yarn producers (spinning mills).

Even though the weaver-managers are free to work for others, they eventually end up running up their looms mostly to fill the GU work orders. This provides the possibility of high-income work for the weavers, on the one hand, and assured supplies for GU, on the other. This arrangement reduces the risk for weavers and at the same time makes them dependent on GU, which has no contractual obligation to ensure regular work orders. This increased vulnerability is not perceived as a risk by loom workers, because in view of new skills acquired through GU, they are likely to find alternative jobs in the handloom industry by switching to the production of traditional items such as lungis (men's long skirts), sarees, and bed covers. However, if a recession and a decline in the overall demand occurs, long-term contracts with a single buyer may be costly for the weavers.

Rural Poverty

The program participants are better off than the nonparticipants in respect of

1. school enrollment rate for children,
2. asset accumulation,
3. improvement in food consumption, and
4. access to health services.

GU strives to focus on marginal and poor weavers, i.e., weavers with five looms or less, in order to obtain high quality, regular, and adequate supplies. Emphasis, however, has been necessarily placed on skill, idle loom capacity, and timely delivery. At the time of the current survey, about 65 percent of the weavers have been found to own no more than 6 looms and about 35 percent own no more than 19 looms.

Economic viability of GU is dependent on two important factors: (1) low-cost weavers and (2) timely delivery of quality output. The choice of areas with low wage rates and thus with a higher incidence of poverty help in keeping costs down. The weaver-managers who can ensure timely delivery are likely to be those whose main source of income is weaving but who also have some minimum resources to fall back upon. GU needs dependable weaver-managers, i.e., a pool of weaver-managers from which supplies for a given work order are drawn. To ensure the timely delivery of quality products, GU is gradually settling down with a selected group of weaver-managers, with whom short-term contracts are repeated on a regular basis. Also, GU is diversifying its product range to include flannel-type cloth that is in demand all year. GU's partial move toward powerlooms, more efficient than handlooms but less labor intensive, has been warranted by considerations of timely delivery.[16]

Since the production is organized under a "putting-out" system, risks are mainly borne by GU. The *taant karmi,* or technical supervisor, undertakes the day-to-day supervision and ensures not only the quality of output and the correct implementation of designs but also timely delivery. The technical supervisors are provided training that is job specific, and skills that are not easily tradable.

Financial Sustainability

The initial costs of experimentation with a pilot project were borne by the Grameen Bank; some women members of the weaving community were among the borrowers of the Grameen Bank for other purposes. As GU has progressed, it has drawn upon financing from commercial banks.

GU had some comparative advantages on launching a new initiative of this kind. It found a niche market in supplying an essential input to one of the most promising export industries; thus it was able to receive the tax concessions that the export industries are entitled to. Also, in view of the Grameen Bank's prestige in the country, the bank found it comparatively easy to deal with the government's regulatory hurdles.

Since the size of the market is limited and a great deal of competition occurs in the output market (fabric as well as ready-made garments market),

GU, like any other marketing agent/subcontractor, is called upon to follow an aggressive marketing strategy either in search of new markets for its current product or to diversify into other products.

Microenterprise Development Unit

Rural Nonfarm Employment and Income

By lending to the rural microentrepreneurs who are in between the assetless, poor clients of MFIs/NGOs and those who are eligible for commercial bank borrowing, MEDU has clearly increased the degree of financial intermediation and expanded the market for credit in rural Bangladesh. It has helped increase the productivity of microenterprises by raising the level of capacity utilization and labor productivity.

The sectoral distribution of MEDU loans is as follows: around 50 percent to the service sector, 25 percent to agroprocessing, and 25 percent to manufacturing. The main activities in the service sector are trade, tailoring, and repair and maintenance and the agroprocessing sector includes rice mill and threshing, poultry, and dairy activities. The manufacturing sector includes wood furniture, baking, garment making, sawmilling, and food processing. Over time, a shift in favor of the service and agroprocessing sectors occurred, with direct lending shifting toward the service sector and indirect lending toward agroprocessing. Both agroprocessing and manufacturing activities are perceived to involve higher risks than the service sector. Since conventional practice by the formal banking sector tends to be risk averse, it is expected that their lending will be biased toward the service sector. Thus, although direct lending tends to prefer the service sector, NGOs seem to pick borrowers that venture into areas of higher long-term growth potential but are a greater risk.

Rural Poverty

The program, targeted at the relatively less poor but not well-off microentrepreneurs, especially in nonfarm activities, is also expected indirectly to help improve the income of the poorer among the rural population they employ. Compared with the nonparticipants, the MEDU clients have performed better in terms of revenue, capacity utilization, and employment.

Since the direct lending window requires that the enterprises are already established in business, one may expect that many MEDU borrowers are likely to be previous Agrani Bank clients but in fact very few turn out to be so. In the first year of the project, there were many past clients of the Agrani

Bank. Thereafter, the MEDU has been able mobilize a new crop of borrowers, who have not previously borrowed from any formal source of financing. This successful attempt in widening the credit market for rural microenterprises is probably due to the characteristics of its loan package: lower interest rates (than the upper limit on loans from the banking system), collateral-free (up to loans the size of Tk 50,000), and reduced transaction costs (faster processing, etc.). Even when many clients have collateral to pledge, collateral-free loans reduce transaction costs. The bank officials are considered to be well motivated and more accessible to the clients. The financial incentives and better job prospects based on performance have also contributed.

NGOs, the channel of indirect lending for MEDU, are found to be well suited and more enterprising or successful in identifying potential entrepreneurs because of their regular interaction with and intimate knowledge of the members of the local community. Even though many indirect lending borrowers have not graduated out of the NGOs' microcredit programs, NGOs are able to identify the eligible borrowers. In fact, among the borrowers motivated and mobilized by NGOs, there are many with higher levels of education than is the case with the borrowers under direct lending. In fact, in the case of direct lending, the requirement of a minimum level of primary education for borrowers is not always fulfilled. NGOs have also succeeded in putting potential borrowers in touch with the bank in order for the latter to be able to borrow under the direct lending system.

Sixty-seven percent of the amount disbursed is through direct lending, while 33 percent is through indirect lending. In indirect lending, only 6 percent of the target numbers of enterprises have actually received any loans, but 88 percent of the target disbursement has been realized, with an average loan size that is seven times the size of the target.

Under direct lending, the enterprises are more than double the target, loan disbursement is only 13 percent higher than the target, and the average loan size is less than half the target (40 percent of the target). It is conceivable that, under direct lending administered by the bank officials, the targets of disbursement are fixed at a more ambitious level—more than that for the target for the number of borrowers—since large disbursements keep the cost of administration low. This may account for the differential performance in meeting the targets for amounts of disbursement and the number of borrowers.

Under both direct and indirect lending, loans for working capital requirements (of existing enterprises) have greatly exceeded those for fixed capital.[17]

Financial Sustainability

The MEDU direct lending program has a recovery rate that is not only above the rates achieved by commercial banks but also within striking distance of MFIs. To foster loan recovery, the project includes pecuniary and nonpecuniary incentives for the MEDU bank officials that are contingent upon recovery. Group lending and peer pressure are not always necessary for high recovery rates for collateral-free loans; close supervision and monitoring by bank officials may ensure high recovery. Also, incentives for the bank officials and, in addition, monitoring by the MEDU project office have been important in reducing the hidden transaction costs in loan disbursements.[18] The project, though not financially viable at the beginning, started to make a net profit from 1997 to 1998.[19]

The guidelines for selecting a branch office are subjective enough to result in wrong selections. Wrong selections include the selection of branches that either cannot provide an adequate workforce or are located in areas where the prospect of nonfarm activities is bleak. The extent of erroneous branch selection is limited but it has adversely affected the overall performance of the program.

Kishoreganj Sadar Thana Project

Rural Nonfarm Employment and Income

The KST project is not specifically about promoting the nonfarm sector. Even so, anywhere between two-thirds and two-fifths of its total resources appear to have been deployed there. The ownership of some kind of asset, mainly land, is a prerequisite to investment in agriculture paying off. Significantly enough, poorer households with little land or assets tend to spend a larger share of their borrowed fund on nonfarm activities. The threshold of nonfarm production does not much depend on the ownership of physical assets, especially since it is possible to borrow very small amounts for working capital requirements in trading or services. The better-off people, when they borrow for nonfarm activities, borrow large amounts.

The nonfarm activities covered in the survey are predominantly milling or agro-processing (around 18 percent of the loan amount) while the remaining 82 percent is in the service sector, predominantly shops and trade. An important noncrop agricultural activity promoted by the project is poultry farming, in which it has established a network of VOs with links to outside markets.

KST has also stimulated significant changes in the occupational structure in the region; for example, a quarter of the past agriculturalists and more than one-fifth of the casual laborers took up "business" as their primary occupation.

Rural Poverty

Even though KST covers both the poor and the not so poor, between 50 and 68 percent of the VO members are landless and marginal landowners and therefore are poor. Community-based organizations are commonly perceived to promote only the interests of the more affluent groups. This bias has not been found to occur in the case of KST VOs as confirmed by the present survey as well as other studies. Some surveys have even found that 72 percent of the borrowers own less than one acre and 50 percent own less than 0.5 acres (Sen and Ahmed, 1999).

Access to credit is generally thought to have improved the economic well-being of the recipients. Also, compared with the members of other MFIs as well as nonparticipants, a greater percentage of the VO members have increased their savings and in many instances they have invested in farm machinery.

Financial Sustainability

In general, the KST credit program has good chances for sustainability and replicability. The key player in the program, the VO manager, has incentives to perform well and with an adequate monitoring system the model has good prospects.[20]

The dominant authority of the VO manager in selecting borrowers and in deciding on amounts and purposes of loans as well as for recovery may lead to the misuse of power. The guidelines for lending are vague. For example, no threshold level of savings needs to be achieved as a prerequisite for receiving a loan; training is not taken seriously. The monitoring of the credit program by the project officers is weak.[21]

The KST links producers with outside markets as in the case of poultry farming. Also, the extension agents in the VOs link the technology specialists in the KST project office with their members.

Small Enterprise Development Project

Rural Nonfarm Employment and Income

The SEDP is designed to lend mainly to the nonagricultural sector. However, noncrop agriculture is included under its purview. In its broad definition of the nonfarm sector, only two activities have been excluded from its domain, i.e., nursery and services of power tillers. Nonfarm activities receive about 63 percent of total loans and noncrop agriculture receives the rest. Rice milling is the single largest recipient of loans (around 36 percent of the number of loans) with tailoring as a close second in number of loans.[22] With its focus on the promotion of small industrial enterprises in the rural areas, SEDP has decided to drop trading from its list of eligible businesses since 1998. This restriction does not include female borrowers.[23] The transport sector has not received much assistance even though it is an important rural activity; it has received loans from the other sources of microcredit. In the latter case, the borrowers are mostly owner–rickshaw pullers or owner-renters of rickshaws and require smaller loans—smaller than the average SEDP loan. Mechanized transport might have been suitable for SEDP loans but requires very large loans—almost to the upper limit of SEDP financing, and would most likely involve new enterprises.

The overwhelming percentage of loans has gone to the existing enterprises with only about 7 percent of the total loans devoted to new enterprises. (It is less than 3 percent of the enterprises in the present sample.) Possible reasons for such a preponderance of existing enterprises are as follows:

- They have ongoing operations with associated cash flow to match, and are for that reason more creditworthy. This consideration is particularly compelling because loans often as big as Tk 250,000 are available without collateral;
- With a substantial equity at stake and being in business for a number of years, the existing proprietors are most likely not to change business or location and thus to default;
- Not much time and effort are needed to evaluate the creditworthiness of existing enterprises; bank staff can lend large amounts in the aggregate, and ensure timely recovery. Loans to new enterprises require more rigorous training in skill and management than are provided by the SEDP.
- In many rural market towns, space for new shops/businesses requires frequently the payment of a large lump sum up front. Hence, most

new enterprises concentrate on home-based poultry, livestock, fisheries, paddy processing, and trading that do not require ownership of floor space, a factory, or a shop.

The loans are used primarily for working capital, rather than fixed investment, and none of the borrowers used them for consumption purposes. The return per unit of capital after paying the costs of labor and other inputs (including family labor) was at least 128 percent. The highest returns accrued from cottage industries and restaurants, followed by food industries. Service sectors and large-scale trading yielded low returns.

Regarding the two training components, the mandatory entrepreneurial training does not attract the existing entrepreneurs who do not need them, while new entrepreneurs who can find them useful are very few. Holding of training at district headquarters discourages the distant borrowers. Women find the training more useful, being less experienced. In the case of skill training, there was a need for more specialized trainers.

Rural Poverty

The SEDP borrowers are mostly land-owning households and most of the borrowers are literate and have some schooling.[24] Those who obtain larger numbers and larger sizes of loans are more educated and own larger areas of land. Landownership helps raise the loan size because the borrower is more confident about his ability to service the loan from diversified income sources and because land is the most commonly used collateral.

Women constitute 26 percent of the total number of borrowers and have borrowed 17 percent of the total amount lent. Thus, women have a much smaller average loan size than men.[25] Eighty percent of the female borrowers are from households with no more than half an acre. Three-quarters of the male borrowers are from households with at least half an acre of land. Most male borrowers have some existing businesses and they need medium-sized loans. Unlike the poor and landless households, in the nonpoor households women are more reluctant to undertake activities outside the household, because it is not socially respectable to do so.[26] The lending to the women borrowers is directed toward the poor households and has, therefore, a direct impact on poverty. Since SEDP targeted the middle entrepreneurs who are to provide wage employment, it is expected that expanded employment opportunities provided by microenterprises would help the rural poor. Most of the enterprises employ both family labor as well as wage labor. The enterprises use generally less than ten workers (including both hired and family labor). About 4 percent of the enterprises are fully based on hired labor (i.e., hiring more than five regular employees).

Financial Sustainability

Targets for the total number of loans and the total amounts to be disbursed have both been more than fulfilled. Loan recovery is about 93 to 94 percent. Such recovery rates compare quite favorably with the MFIs, both in the regions and nationally. Due to close monitoring of the loan, most of the loans are utilized for the purposes for which they are ostensibly taken. The program is free of corruption and kickbacks. In practice the rate of interest is about 10 percent lower than that of MFIs and is also lower than the interest rate of commercial banks.

SEDP has succeeded in reaching its objective to a large extent by motivating its staff for making the loan-granting process simple and accessible as well as for closely monitoring the uses of credit and enforcing its recovery. The lesson seems to be that it is possible to create bankable clients among semi-urban and rural borrowers by instituting good banking practices. Although SEDP has overfulfilled the targeted number of loans, the average size of loans is smaller than the target. From the management point of view, it implies lower risks.[27] From the point of view of the borrowers, smaller loans imply lower interest rates and easier administrative procedures.

SEDP is an experimental project with temporary staff and therefore financial sustainability is not its primary objective. The number and average size of loans have not expanded much over time. This is because the availability of prospective entrepreneurs with technical and managerial skills is quickly becoming exhausted, and also partly because the market for most of the nonfarm activities is rather limited. Moreover, the expansion of SEDP may have been somewhat restrained by its narrow geographical concentration and an extension into new locations may have expanded its business.

Also, there is a need to develop new entrepreneurs with training in entrepreneurial and managerial skills. SEDP needs to identify, through market studies, prospective new enterprises and activities, and to encourage and stimulate the new entrepreneurs to undertake such activities. The provision of loans at subsidized interest rates as low as 10 percent is partly intended to encourage the entry of new entrepreneurs into the sector. Apparently this is not a sufficient incentive. Additional promotional efforts are required.

CONCLUSIONS

GU is an example of an imaginative approach toward linking the poor, rural nonfarm producers with markets, both domestic and foreign. GU successfully supplies the critically important ingredient of entrepreneurship and management. Undeniably, even for an organization of Grameen's visibility and stature, it took a significant amount of entrepreneurial "gap-filling" to make this viable.

A marketing agency model, such as the GU case, is open for replication by any private entrepreneur. Ideally, any new production in rural areas that substitutes for imports or finds new markets creates net rural employment and income. Because the production is organized by poor artisans (in this case poor weavers, in some cases employing wage labor), such programs benefit the rural poor. Although these activities can be undertaken by public agencies and the private sector, the NGO/MFIs in Bangladesh seem to have developed some amount of comparative advantage: first, they may access financial resources at a relatively low cost;[28] second, the presence of multiple programs under their auspices allows them to enjoy economies of scope.[29]

Support services for promoting such activities may be provided by specialized agencies. In Bangladesh, one example is the USAID's (United States Agency for International Development) Agriculture Technology Development Project (ATDP), which has provided credit and technology support to the agro-processing industry mostly located in urban centers, and has linked them with rural growers of agricultural produce, either directly or through the network of NGOs. It has provided consultancy and technical support to this industry to enter export markets. In addition, it has helped establish "contract" farmers' groups to grow and supply fruits to the processing industry.[30] Although the model is intended to promote urban-based agroprocessing industry, it may also be used to promote rural nonfarm employment. Similarly, an international NGO (Hortex Foundation funded by World Bank), identifies and links up buyers in the international market with local NGOs which are provided with extension and marketing services.

A community-based approach such as the KST model, which is designed to provide credit for both farm and nonfarm programs involving VO managers as credit retailers, may have much potential for mobilizing rural savings and administering credit to a wider segment of the rural population. However, in order that such an approach lends to the poor, it is necessary to disburse relatively small-size loans, targeted to the poor. Otherwise, most lending is likely to end up to the nonpoor. Second, experience indicates that an adequate system of supervision and monitoring should be set in place because otherwise a potential risk exists for financial irregularities. The VO managers belong to the better-off strata of rural population, have discretional decision-making power, and are not adequately monitored.

Two questions remain. First, why is it necessary to design a community-based model (including the poor as well as the nonpoor) instead of supporting the existing MFI initiatives (for only the poor)? There seem to be two arguments in favor. First, the VO has attracted large savings from the nonpoor, unlike the MFIs that concentrate on the poor. These savings provide additional resources to KST to enlarge its lendable funds. Second, it promotes village leadership, self-management, and encourages self-reliance among the villagers who can undertake other rural development activities, apart from the provision of credit.

Could a commercial bank hire individuals in place of a VO manager to act as its retailing agent on payment of a commission? This seems more difficult at the present moment because of the low level of integrity and efficiency in the management and supervision of the prevailing commercial banking system; with the improvement of the management of the commercial banks, which is under way, such an approach may be possible.

Given that the targeted group-based lending is currently being pursued quite extensively through numerous MFIs, a supplementary approach in the form of a community-based lending program through KST-style organizations may be worth experimenting on larger scale and in other locations. A few additional steps in this connection may be suggested. First, a consortium of VOs may combine or pool their resources and take joint responsibility of funding larger projects; second, local NGOs may be entrusted with the responsibility for organizing and supervising the formation of the VOs and their consortium body; third, an independent institution may oversee the NGOs engaged in such promotional activities. In addition, this institution may help establish a network of entrepreneurs who market rural nonfarm products and who may be provided with market intelligence service and financial assistance, if needed.

Lending in rural areas is predisposed toward the service sector. It may be because it is less risky. However, under the system of indirect lending, the partner-NGOs under MEDU have been able to pick borrowers that venture into areas of higher long-term growth potential. SEDP has been able to promote entrepreneurs in semiurban areas, providing significant assistance to the manufacturing sector and the agroprocessing industry. In both cases the emphasis has been on existing entrepreneurs.

If the objective is to promote microenterprises in the "missing-middle group," the question is: Why are the commercial banks in general unable to undertake this activity as a part of their normal operations? The justification for starting a new program in MEDU/SEDP style could be that it is believed that the regular operations of the commercial banks are biased against microenterprises. This could be either because the profitability of the sector is low and therefore banks find them risky, or because, even when projects are profitable, entrepreneurs cannot access the commercial banks for smaller collateral-free loans than the regular commercial bank loans.[31] In this regard, the major achievement of the MEDU/SEDP programs has been to prove that collateral-free loans of the required size to individual borrowers is a feasible option. An additional market for bank credit has thus been created due to lower interest rates and more client-friendly bank officials. In turn, it is possible to create a core of client-oriented bank officials if a system of attractive incentives linked to disbursement and loan recovery can be devised. However, it is not clear how much subsidy is involved in the lending operations of the SEDP and MEDU, if one considers all the elements of costs in their lending operations. In the case of MEDU, the subsidy elements are involved with respect to commission provided to NGOs for indirect lending and the overhead expenses of the bank. In the case of SEDP, the excess of the project costs plus the opportunity costs of the donor grant over the lending rates constitutes the subsidy.

A relevant question is how to ensure the sustainability and future prospects of the foregoing experiments/pilot schemes. Is it possible to ensure the ownership of these projects, in particular, those that are currently dependent on temporary donor financing? The Grameen Bank Foundation and the Grameen Bank, largely acting as a corporate body, own the activities undertaken by GU. Similarly, Agrani Bank owns the MEDU program with IFAD support. But ownership can be understood also as a group of individuals who take interest in sustaining a program or project. In that sense, GBF (Grameen Bank Family) owns the GU project and the MEDU can be considered a prospective owner because it is actively involved in running the project. On the other hand, although the SEDP project staff have the major responsibility for sanctioning and monitoring loans, they are all on temporary employment. Thus it is not clear who will own SEDP. Will it be the Agrani Bank which disburses loans under the project, or what will remain of the SEDP as a project? The status of KST ownership remains quite dubious. It is a project of UNDP, in association with the government of Bangladesh. For all practical purposes, UNDP appears to have owned it. However, once the project has gone into operation, a certain sense of ownership has emerged from interest groups such as the SOs or the VO managers. It is unclear who will own the numerous UNDP grants when the project ends.

Chapter 4

Indian Case Studies: Review and Analysis

MAHARASHTRA RURAL CREDIT PROJECT

The main objectives of the Maharashtra Rural Credit Project (MRCP) to be administered by banks with resources obtained from a donor and the government of India were first, to improve access of the rural poor to financial services and make them bankable clients; and second, to promote savings mobilization among the rural poor through self-help groups (SHGs).[1] The target group consisted of households below the poverty line.[2] From those, priority was to be given to households with incomes below Rs 8,500 which comprised mostly small and marginal farmers, the landless, artisans, and tribals.[3]

In order to make the MRCP participatory, the bank field officer played a critical role in forming the village development assembly (VDA) that comprised all households in a village. The village development council (VDC) consisted of a group of 10 to 12 members in a village selected on the basis of a consensus by the VDA. The VDC meetings, held once a month, were attended by the members of the Council, NGOs representatives, commercial banks, and other implementing agencies.

Two channels of credit were used: individual and SHGs.[4] Eligible individual beneficiaries were identified at the VDC meetings, using the list of poor households.[5] For individual borrowing there was requirement of a seed money 25 percent of the loan size that needed to be deposited at the bank. There was flexibility in repayment schedules. The interest rate was 12 percent per annum. Individual loans were earmarked for specific activities.

SHGs were formed either by a women's development organization called Mahila Arthik Vikas Mahamandal (MAVIM) or the banks. In both cases, the groups of about 15 to 20 persons could be formed either directly or indirectly through contracted NGOs. Once they demonstrated financial discipline by mobilizing savings, they were allowed to borrow from commercial banks against their savings deposits.[6] SHGs lent to the members in accordance with their own rules. The SHG group leader was responsible for maintaining records. The monitoring system ensured transparency in the financial transactions.[7] Although SHGs borrowed from banks at 12 percent per annum (the same rate as that of individual lending), they lent to its members at substantially higher rates, varying from 2 to 3 percent per month.[8] Contrary to the case of individual loans, the purpose of loans by SHGs was unrestricted and consumption loans were permitted. The loans were small and repayment periods were short, normally limited to a year usually with bimonthly or monthly installments.

In both individual and group lending, borrowers were provided guidance and information on investment opportunities, skill formation, and technical services. Several agencies were responsible for the provision of these services to the borrowers: Maharashtra Centre for Entrepreneurship Development (MCED), Maharashtra Industrial and Technical Consultancy Organization (MITCON), and NGOs/MAVIM-Sahyoginis.[9] MCED focused on training entrepreneurs and assisted them through market surveys, documentation, and management counseling. MITCON provided consultancy for small and medium enterprises at affordable fees (specifically project reports) and monitored the progress of enterprises advised by it. Finally, NGOs/Sahyonginis focused on problem-solving training through group discussions relating to various on-farm activities.

SIDBI FOUNDATION FOR MICRO CREDIT

The SIDBI Foundation for Micro Credit (SFMC) project, launched by the Small Industries Development Bank of India (SIDBI), was designed to create and develop a network of sustainable microcredit institutions in the nongovernment sector. SFMC was to act as a wholesaler of microfinance. It provided loans, equity, capacity building, and other support services to NGOs/MFIs. Even though no explicit income criterion for borrowers was established, these NGOs were expected to lend to the poor.

In order to be eligible, NGOs/MFIs were to be professionally managed, to display growth potential, and to have experience of small savings cum credit programs with SHGs/individuals. MFIs might lend directly to SHGs or to individuals or indirectly through their partner NGOs and other agencies. All nonfarm activities were supported. Loans were given to NGOs/ MFIs at 11 percent per annum for a period of four years and MFIs/NGOs charged the interest rates as determined by the market to SHGs/individuals. Collateral security was replaced by reliance on good track record and capacity assessment.[10]

The findings of a study of two illustrative cases of NGOs/MFIs acting as credit retailers are summarized in the following. The first example was that of Dharampur Utthan Vahini (DHRUVA), an NGO that had the primary function of providing credit in-kind to the farmers, i.e., agricultural items such as seeds, fertilizers, etc., but later on it was given the additional responsibility of disbursing microcredit in cash for nonfarm purposes to individual borrowers. There was no explicit income criterion for the selection of beneficiaries. Individual beneficiaries were screened initially by village-level workers, other NGO representatives, or by DHRUVA's own representatives. All beneficiaries turned out to be male. The interest rate was 12 percent per annum with equal monthly installments and a repayment period of two years.

The second case was that of the Indian Institute for Entrepreneurship Development (IIED) that motivated and organized prospective women borrowers/beneficiaries into SHGs.[11] The group members/beneficiaries received credit in-kind for two activities, agarbattis (incense sticks) and tailoring. In the case of agarbattis, after a training period of three to six months (with no payment made to the trainees), IIED provided raw material to the group leader, who distributed it to individual members. The interest rate was 12 percent per annum. There was no restriction on the sale of the product, but usually IIED bought it back. The costs of the raw material and interests were subtracted from the sales revenue and the balance was distributed among the members in accordance with the output sold. The second group was engaged in tailoring; one sewing machine was given as a loan to an individual beneficiary and the other two were kept in a common facility for the free use of other beneficiaries who bought their own raw materials and marketed their own products. The borrower-tailor was to pay the loan in 15 equal monthly installments at an annual interest rate of 12 percent.

In both MRCP and SFMC, credit was channeled to individuals and groups at the same interest rate but while in MRCP credit was dispensed by banks, SFMC relied upon MFIs/NGOs. In MRCP, all activities were eligible for the use of credit, while SFMC restricted it to nonfarm activities. MRCP emphasized community participation, i.e., village associations for identifying the beneficiaries and to a limited extent for loan monitoring. SFMC relied largely on MFIs/NGOs, leaving the mode of credit delivery to the discretion of retailers (thus credit in-kind was also an option). In MRCP, awareness building and training were provided by NGOs and importantly by other specialized institutions, while SFMC relied on MFIs/NGOs.

The two cases provided different approaches toward microfinance for the development of a microenterprise sector in India; a comparison of their characteristics and features is provided in Table 4.1.

TABLE 4.1. A Comparison of MRCP and SFMC.

Features	Maharashtra Rural Credit Projects (MRCP)	SIDBI Foundation for Micro Credit (SFMC)
Explicit promotion of rural nonfarm activities	Yes (both on-farm and nonfarm)	Yes (restricts the use of credit to nonfarm activities)
Main objective	To improve access of the rural poor to financial services, and promote savings mobilization	To create a network of microfinance institutions in the nongovernment sector and ensure that this network is viable
Location	Maharashtra (11 districts covered)	Areas covered by the 14 beneficiary MFIs
Target group	Households below poverty line (annual household income = Rs 11,000 at 1991-1992 prices)	No explicit criteria of poverty but NGOs expected to lend to the poor
Operational characteristics	• Cash credit and technical support • No marketing function • Individual lending (individual beneficiaries nominated by the VDC) and SHGs lending (groups formed by MAVIM or commercial banks or NGOs contracted by both of them) *Individual lending:* • 25 percent of the loan deposited at the bank as seed money • Interest rate: 12 percent • Flexibility in repayment schedules • Limit for individual loans Rs 25,000 (as an estimate, in the sample the average individual loan was Rs 15,600)	• Credit wholesaler for MFIs/NGOs providing loans, equity, capacity building, and other support services • No restriction on the mode of delivery *For MFIs:* • Interest rate for MFIs: 11 percent per annum • Minimum loan assistance is Rs 1 million and 10 percent plus accrued interest as security (being replaced by "good track record and capacity assessment") • Repayment period is four years, including a grace period of six months

	SHGs: • Required to mobilize savings before they access credit • Interest rate: > 12 percent, usually 2 to 3 percent per month (24 to 36 percent per annum) • Flexible number of installments and repayment schedule is flexible (normally, monthly installments and repayment with one year) • No specific limit for loan sizes (either for the ultimate borrower or for SHGs) (as an estimate, in the sample loans for SHG members ranged from Rs 500 to 4,000) Weekly or bimonthly contribution of savings	For borrowers: • MFIs free to charge additional rates over 11 percent to individual/SHG/partners • Individuals: Fixed equal monthly repayments • Maximum loan for individual/SHG member is Rs 25,000 (but there are provisions for repetitions) No marketing function
Incentives for clients	*Nonpecuniary:* Investment opportunities information, skill formation, and technical advice • NGOs/field workers train SHGs in initial stages for one to three days • MCED trains entrepreneurs for 12 to 13 days • MITCON provides consultancy for SMEs at affordable fees	*Nonpecuniary:* In theory, training is an option
Incentive for the staff	*Nonpecuniary:* Training of bank staff, VDC, and field workers	*Nonpecuniary:* Training programs for the MFIs by technical and management institutes

Source: Author.

MAIN FINDINGS

Maharashtra Rural Credit Project (MRCP)

Rural Nonfarm Employment and Income

The major proportion of MRCP credit did not go to the rural nonfarm sector; it received only 40 percent of the total MRCP sanctioned loans. Beneficiaries favored individual loans for productive purposes (including nonfarm activities) because SHG loans were too small and interest rates too high.[12] The scale of activities was larger for the nonpoor. Those who borrowed from SHGs for productive purposes used the loans mostly for meeting working capital requirements or small additions to productive capacity, as opposed to financing new ventures. All except two individual beneficiaries retained the assets acquired through an MRCP loan.[13] What prevented the poor from expanding their business was mainly their inability to obtain larger loans, while the (relatively) affluent were constrained by the unavailability of additional household labor.

The range of activities financed by microcredit for both individual and group borrowers was very wide.[14] Although the activities were perennial, they were nonetheless affected by two factors: One was the dependence of the poorest on nonfarm activities mostly during agricultural slack periods; Second, much of trading activity was interrupted during the monsoons (four to six weeks). The choice of activity depended on whether borrowers had previous experience, or whether it was a traditional family occupation or if it was easy to combine with household chores. A common consideration for all was obviously the size of the potential market demand. The returns on investment depended on the activity chosen; however, the better-off with larger loans had higher returns.[15] The additional income in the case of both individual and group loans was used mostly for consumption and education of children. In general, the additional income enabled the beneficiaries to raise their level of consumption and to repay their loans. While a few activities had both backward and forward links, most of them had strong forward links, i.e., their outputs were used as inputs to facilitate other activities; also, they were linked to the growth of market demand in the village.

Rural Poverty

Although MRCP was designed to target the poor, targeting in the case of individual loans was unsatisfactory, as the number of the nonpoor was nonnegligible.[16] The poor had meager assets and little education.[17] Further-more, the nonpoor had received larger benefits because they secured larger loans with higher returns.[18] The exclusion of the poorest among the individual borrowers was not just a result of their limited repayment capacity, but also of several additional factors, such as social exclusion and limited awareness,[19] lack of reliable data on households below the poverty line, and collusion between the implementing agencies and the affluent.[20] The guide-lines for the selection of beneficiaries was often manipulated by influential persons within the VDC.[21]

The targeting through SHGs was better.[22] Since the distribution of loans among the members was decided upon collectively in accordance with the priorities of the group, the amounts of loans better reflected their respective needs. The participation of the local NGOs and the field workers helped im-prove targeting. The self-selecting mechanism of SHGs attracted the poor because the loan amounts were small and their use was unrestricted. Suc-cessful groups were linked to the banks; they demonstrated effective leader-ship, group discipline, and transparency in decision making as well as networking ability with banks and VDCs.

SHGs were instrumental in channeling credit to the poor rural women and in improving their self-confidence.[23] They combined income-generat-ing activities with household activities and enjoyed greater flexibility. Even when women had limited autonomy in the selection and use of assets, access to credit enhanced their status both within and outside the house-holds.[24]

A concern common to both individual and group lending related to the role of training. The training components of MCED and MITCON ap-peared to be inappropriate for the majority of the rural poor, as their focus was largely on small and medium enterprises. Moreover, the duration of training was too long and the trainees had no financial compensation. This tended to discourage the poor.[25] On the other hand, the training provided by NGOs/Sahyoginis was deemed more useful, although its appropriateness for the poorest could not be assessed. However, the Sahyoginis or field workers of the NGOs were overburdened; their remuneration was inade-quate and not linked to performance. They had no incentive for inducing the poorest to participate in the MRCP scheme.

Financial Sustainability

Under MRCP, while the commercial banks lent at 12 percent per annum, the average cost of funds for them was 9 to 10 percent. As this spread did not cover all their costs, the National Bank for Agriculture and Rural Development (NABARD) provided full refinancing facilities at 6 percent per annum.

In the case of group borrowers, there was a concern that SHG's link to banks was slow; after a group was formed, one to three years elapsed before it could borrow from a bank—far in excess of the stipulated or expected period of six months. This was mainly due to cumbersome procedures, the relative unimportance of MRCP loans in the banks' portfolios, and the lack of incentives to field officers. There were a few negative incentives for the personnel of the banks that were disbursing the MRCP loans. First, managers were assessed on the sum of their total deposits plus advances, thus MRCP loans were a low priority for them; second, as field officers engaged in MRCP loans were cut off from mainstream banking, their performance was not perceived as likely to promote their career prospects.

Small amounts of investment tended to have initial high returns but once new enterprises emerged or the existing ones expanded, the returns tended to decline.[26] The central concern was the limited potential demand for their products and services in stagnant villages.

SIDBI Foundation for Micro Credit (SFMC)

Rural Nonfarm Employment and Income

In the case of loans to individuals by the NGO (DHRUVA), most of the loans financed diversification or expansion of existing enterprises and not start-ups and therefore, it had a minor role in the selection of projects.[27] Nonfarm activity was a full-time occupation for more than 80 percent of the beneficiaries (mainly because of the seasonality of agriculture and its limited income earning potential) and less than 20 percent had a stake in agriculture.[28] Of the a wide range of nonfarm activities, the majority was linked to trade in consumer goods. If new skills were needed, they were quickly acquired in most cases.[29] Because of the geographically isolated location of the survey areas, there was little competition so that the returns were high;[30] however, there were fluctuations in income over time, especially seasonally; at certain times revenues fell to 50 to 60 percent of the monthly average.

Half of the individual borrowers either saved or reinvested the surplus. Given strong forward linkages of the microenterprises with the village economy, their expansion was closely linked to the growth of agricultural incomes. In the case of group beneficiaries, the returns in both agarbatti making (i.e., making perfumed sticks) and tailoring were significant. On average, the income was a large proportion of the household income (about

38 percent). The surpluses above repayment obligations were mainly used for household expenses, school education, and accumulation of savings. The two examples resembled the traditional "putting-out" system but not based on the workers' traditional skills/occupations. The choice of activities was dictated by preferences of the relevant NGO/agency—derived possibly from their assessment of market prospects and profitability of respective products. The individual preferences for self-employment on the part of the participants were not the dominant consideration.[31]

Rural Poverty

Although the repayment rate was good, the loan disbursal was quick, and there was periodic monitoring, serious targeting failures occurred in the two examples of SFMC given previously. In the case of individual borrowers, all were (relatively) affluent and women were excluded. The exclusion of women was justified by the project on the grounds that other government programs concentrated on women (although women were rarely aware of the existence of such programs). Women were excluded mainly because they were considered bad risks; bias existed against women inside and out-side households and women also had limited awareness about the plan among them.

Limited participation of the poor was due to a combination of self-exclusion and institutional factors. Self-exclusion of the poor was mainly due to illiteracy, lack of awareness, and apprehension that they might not be able to repay the loans as scheduled. Institutionally, the actual procedure used for selecting the beneficiaries also played a role. In respect of the agriculture-related activities of DHRUVA (supply of fertilizer, seed, etc.), the field officers tended to concentrate on (relatively) affluent landholders. Thus, it was likely that in the case of SFMC's individual loans for nonfarm activities, the officers also concentrated on the (relative) affluent economic strata who owned land, especially since there was no verification of economic status and ad hoc committees had the final say.

In the case of groups, women who benefited through SHG were not the poorest. The IIED staff used training as a screening device for the selection of borrowers rather than to help the poor members overcome their diffi-dence, build awareness, and impart basic skills.[32] Arbitrary selection could not be ruled out because the selection of beneficiaries was not subject to ap-proval by any local body nor was the list displayed in a public place. Within the household, there was little evidence of women enjoying autonomy in in-come disposition or a reallocation of responsibilities between male and female members. However, there was a greater recognition of their income-earning potential, helping to broaden their scope of activities and involvement in community affairs.

Financial Sustainability

DHRUVA was financially viable because the institutional mechanisms for screening and monitoring that were designed to ensure high repayment rates had relatively low costs. The size of loans ranged from Rs 5,000 to Rs 20,000, with an average of Rs 9,800. The rate of interest was 12 percent per annum and the repayment period was two years.[33] Where the monitoring and screening costs were high, as was the case with SHGs in general, IIED was presumed to have indulged in an extraction of rents from the participants. A lack of transparency was evident in financial transactions.

CONCLUSIONS

A comparative analysis of the four sample surveys (two for each scheme, individual, and group) is presented in Table 4.2.

A few lessons from a policy perspective may be obtained from a comparative analysis. The returns on investment were subject to fluctuations though lower than that of farm incomes. Ways for reducing the variability of nonfarm incomes through a different mix of activities or through more diversified market links warrant careful analysis.

In both cases the targeting was unsatisfactory, especially in the case of individual borrowers, even though MRCP was specifically to target those below the poverty line. However, in SMFC there was no specific criterion for targeting of the poor but the NGOs were expected to lend to the poor. All the individual borrowers under SFMC were nonpoor but, in the case of MRCP, 53 percent were nonpoor. In this latter case, the community involvement in identifying the beneficiaries may have made a difference despite collusion among the locally influential persons.

In both cases, moderate income gains accrued to group borrowers but in the case of MRCP-SHGs there were additional benefits of self-managed group activities, i.e., self-confidence, assertiveness, and some widening of spheres of activities. It was questionable whether a limited group activity as under SFMC-IIED (such as making of *agarbattis*) imparted the same degree of self-confidence as that derived from a successful SHG involved in mobilizing savings and accessing loans from banks.[34]

Community participation seemed to make a difference in targeting credit to the poor, especially poor rural women. In the case of MRCP, the VDAs and VDCs might become redundant if reliance is placed on village pan-

TABLE 4.2. Comparative analysis of sample surveys.

Features	MRCP		SFMC	
	MRCP-individual (VDCs)	MRCP-SHGs	SFMC-DHRUVA	SFMC-IIED
Rate of interest	12 percent annually	24 to 36 percent annually	12 percent annually	12 percent annually
Loan sizes	*Poor:* Rs 5,000 to 25,000 (average: 10,000) *Nonpoor:* Rs 8,000 to 30,000 (average: 15,600)	Rs 500 to 4,000	Rs 5,000 to 20,000 (average: 9,800)	Maximum Rs 10,000
Repayment rate	*Aggregate: 90 percent* "High"	100 percent	92 percent (all 12 except one)	*Aggregate: >90 percent* (buy-back)
Proportion of nonfarm activities	Aggregate total <40 percent (fewer for productive purposes in SHGs)		"High"	100 percent
Nature of nonfarm activities	Wide choice: *Poor:* selling snacks, vegetables, dry fish, cutlery, buying buffaloes, and making picture frames *Nonpoor:* buying goats, sewing machines, welding equipment and grinder, and trading in textiles	Wide choice: tailoring, and training in vegetables and pulses (includes ceremonial/ consumption purposes also)	Wide choice: majority in trading consumer goods and the rest in repairing, flour milling, and other services	No choice: *agarbatti* making and tailoring

TABLE 4.2 (continued)

Features	MRCP		SFMC	
	MRCP-individual (VDCs)	MRCP-SHGs	SFMC-DHRUVA	SFMC-IIED
Purposes of loan	Assets were bought (no information available about purpose of investment)	100 percent expansion/diversification of business	67 percent expansion/diversification of business; 33 percent start-ups	100 percent new business (*agarbatti* and tailoring)
Types of capital investment	Information not available	Used mainly for working capital and small additions to production capital	In general, low investments (only two had high fixed investments)	*Agarbattis:* 100 percent working capital *Tailoring:* 79 percent fixed capital and 21 percent working capital

Source: Author.

Tailoring: 79 percent fixed capital and 21 percent working capital.

chayats (local governmental institutions) for identifying the beneficiaries. In the case of SFMC, a potential complementarity exist between the role of NGOs and village community organizations, including the panchayats. The poor might not often be informed about potentially profitable nonfarm activities; they might be informed by NGOs or microcredit agencies.

The village panchayats might be more sensitive to the needs of the poor and thus improve targeting and transparency if the panchayats are fairly and frequently elected, include wide representation of poorer and disadvantaged groups, and are accountable and transparent in their operations.[35] Collaboration between NGO and panchayats under SFMC and between panchayats and banks under MRCP might take different forms. It was possible for panchayats to be responsible for approving the selection of beneficiaries under both the projects as well as monitoring the use or repayment of loans, displaying the selected list in a prominent place.

The choice of delivery mechanism could make a difference for the poor. Three delivery mechanisms were available in the Indian context: direct lending to the poor by the banks, SHG-bank link program (as in the case of MRCP-SHG), and lending by banks/financial institutions to MFIs/NGOs which in turn lend to individuals and SHGs (as in the case of SFMC). Devices such as SHGs reduced transaction costs for both borrowers and lenders. However, whether the bank-SHG link was more cost effective than the bank-NGOs-SHGs link was to be assessed.

Were some SHGs more successful than others and why? It seemed that the quality of leadership and NGO support, previous experience of group activities, and cohesiveness did matter. Also, group cohesiveness was not necessarily synonymous with homogeneity of social background (e.g., caste affiliation) and/or economic status, as was seen in the case of MRCP sample survey; the mixed groups performed well and social heterogeneity might not be a barrier to group cohesiveness. The optimal size of SHGs needed to be examined, keeping in mind the twin consideration of cohesiveness and the need for large savings.

In the case of MRCP, while the training programs of MCED and MITCON were not very appropriate for microenterprises—often based in households, the training activities of NGOs/Sahyoginis needed strengthening by broadening the range of their expertise and by providing incentives linked to performance.

Similarly, in the case of SFMC, training should focus on skills that were more appropriate for the activities that were financed; training by MFIs/NGOs was more functional than that by specialized agencies. The opportunity costs of training might be high for the poor in the absence of any compensation.

Potential existed for promoting group or joint activities among the poor. Average returns were likely to be higher if the skills of group members were complementary and there were economies of scale in production/marketing. Moreover, the sharing of risks in view of the variability of returns would limit welfare losses.

Repayment rates were high in cases of both group and individual loans. Periodic monitoring and some flexibility in repayment schedules helped reduce the possibility of default; flexible repayment schedules would be more attractive to the poor. The flexibility in the mode of credit delivery, such as credit in-kind, as in the case of SMFC's, encouraged or allowed innovations.

Individual loans were not linked to savings under MRCP and SFMC and thus no financial swap occurred. In the case of SFMC-SHGs, there was no saving requirement for the credit in-kind; in the case of MRCP, loans for SHGs were linked to their savings. Savings requirement helped to instill financial discipline among the poor who had no experience in independent financial decision making.

As SHGs started to handle large amounts of public funds (over Rs 1 million), a case for regulation consistent with the autonomy and flexibility of SHGs was made in order to protect the interests of members. A need was recognized for efficient monitoring of MFIs/SHGs through such means as more proactive involvement of the financing agency in monitoring, access to reliable field evidence, establishment of performance indicators, and appropriate incentives for NGOs linked to performance.[36]

Chapter 5

Overview of Country Experiences

INTRODUCTION

Normally, in the case of enterprises which are based on local markets and which envision no upgrading of technology, priority is assigned to the supply of credit. MRCP and SFMC in India as well as KST in Bangladesh are the kinds of programs engaged in this type of enterprise in a locally relevant market segment. SEDP and MEDU in Bangladesh are trying to work in the segment that is relevant for the second category of enterprises with wider markets and some upgrading of technology. GU in Bangladesh cuts across both categories and integrates very closely the delivery of all the three inputs. Though marketing is the principal concern, credit and extension services are essential secondary inputs. These different characteristics are shown in Table 5.1.

TABLE 5.1. Characteristics of specific microenterprises.

Types of activities and production	Case studies in Bangladesh	Case studies in India
Low markets Low productivity, low capitalization	Kishoreganj Sadar Thana (KST)	Maharastra Rural Credit Project (MRCP) SIDBI Foundation for Micro Credit (SFMC)
Larger markets Moderate capitalization	Microenterprise Development Unit (MEDU) Small Enterprise Development Project (SEDP)	
Domestic and/or international markets Putting-out system of production	Grameen Uddog (GU)	

Source: Author.

PATTERNS OF NONFARM LENDING

Table 5.2 presents the proportion of credit disbursements on nonfarm activities in the samples drawn from different programs.

Loans to nonfarm enterprises (strictly defined to exclude such noncrop agricultural activities as livestock, poultry, fisheries, and forestry) constituted between 30 and 40 percent of total loans in the case of MRCP and KST. MRCP provided loans for expenditures on consumption as well as festivities.

SFMC, MEDU, and SEDP had a very high proportion of nonfarm loans as they were specifically designed to promote nonfarm activities, even though in several cases they defined nonfarm activities to include noncrop agriculture.

Virtually everywhere, services and trading accounted for a clear majority of microenterprises in Bangladesh, except in the case of SEDP and MEDU. Several factors might have affected the preference for trade and services:

- Services and trading might be attractive because they needed little skills and limited capital; there was a local demand to be tapped which was visible and known (or could be ascertained quickly at low search costs due to the presence of community networks). Entry into the sector was very easy.
- However, many manufactured products had to be marketed further, nearer the small towns or even the cities; as the marketing sphere widened, more complexity in product design, technology, or marketing skills was necessary as well; also the required size of capital investment, especially the proportion of equity capital needed for profitable operations, was higher.
- Rural manufacturing might also be constrained by competition of imports from smaller towns and cities. Wherever the rural-to-urban roads lowered marketing costs, this conferred a premium on those who could scale up faster and produce more cheaply. In a sense, rural manufacturing could better survive in backward villages than in infrastructurally (transports and communications) well-endowed villages. Since locally produced nonfarm goods were in general lower in quality or design and in price, they were in ready demand in less developed, poorer villages.
- A "herding mentality" existed as evidenced by a lively growth in the short run of the number of trading and service enterprises, which followed one another in quick succession.

- Price and income risks in agro-processing (such as rice processing), caused by weather-induced variation in agricultural production, and sometimes aggravated by market-destabilizing public interventions, also played some role.
- In the choice of activities, considerations such as past experience, family occupation, or experience of friends or local enterprises already operating in the area weighed heavily with the borrowers. The possibility of combining nonfarm activities with other household responsibilities was also an important concern.

High effective rates of interest and frequent installments might not stimulate demand for loans by the poor to be invested in new nonfarm enterprises. For example, with an investment of $5,000, an interest rate of 2 percent per month and a repayment period of one year, in order to cover the interest payment and the monthly installment of the principal, a return of $517 per month was necessary. Unless nonfarm enterprises were likely to generate such a high and stable level of net return to cover both debt service obligations and a reasonable return on labor, a small investor was likely to feel inhibited to participate. The SHGs or NGOs in India and Bangladesh that restricted themselves to very small loans and frequent inflexible installments in the face of variable cash inflow might not have been most effective in promoting a wide range of rural nonfarm activities. In general, the case studies indicated that promoting nonfarm enterprises was dependent mostly on relatively large-sized loans at affordable rates of interests (between 12 and 14 percent annually).

Moreover, many nonfarm activities, especially outside the category of trading activities, required both fixed and working capital, preferably with some contribution by the borrowers to equity capital, so that the fixed interest burden was not too high and the equity share of the entrepreneurs provided a cushion for the risks borne by new enterprises. An institution like KST in Bangladesh that brought the savings of the rural nonpoor and the poor together into a pool of resources available for lending for various amounts and purposes might help to expand the market for financial services.

In both Indian and Bangladeshi programs, one recurrent feature was the overwhelmingly large share of lending to existing enterprises. The proportion of lending to new enterprises was quite low: for example, in Bangladesh, this was in general as low as 7 percent in the case of SEDP (3 percent in our sample). This was true not only of microcredit agencies targeted toward the poor but also of those institutions like MEDU and SEDP in Bangladesh that lent to the not-so-poor. This phenomenon restricted the extent

TABLE 5.2. Ratio of nonfarm enterprises in the portfolio of microfinance agencies.

Nonfarm characteristics of the program	Indian programs		Bangladeshi programs				
	MRCP	SFMC	GU[a]	MEDU Direct lending	MEDU Indirect lending	KST	SEDP
Proportion of nonfarm activities (percent of disbursement)	40	High to 100	100[a]	High to 100		39	63
Sectoral division of disbursements (percent)							
Agroprocessing	–	–	0	9	41	18	74
Manufacturing	–	–	100	30	47	0	17
Services/trade	–	–	0	61	12	82	9

Source: Author

[a]Even though GU is not a credit program per se, it is featured in this table in the sense that nonfarm activities get all the program input.

of diversification and expansion into new activities through loan financing. Several hypotheses could be suggested for explaining this phenomenon:

- It was easy to identify promising entrepreneurs. They had ongoing revenue streams for that reason and represented lower levels of credit risks. Even if marginal returns from any possible investments were low, the enterprise had a cushion from existing businesses' earnings. The time and resource cost of fulfilling the mandatory loan disbursement target by the credit agency was likely to be smaller in the case of existing enterprises.
- When entrepreneurs were in business for a number of years, they were likely to have a substantial equity at stake and therefore were unlikely to default.
- Existing businesses often faced a binding working capital constraint. Reliable and quick access to small and short-term loans were provided by the microfinance institutions to meet this need. New enterprises required larger loans because they required both fixed capital (long-term loans) as well as short-term working capital.
- Also, the diversification into new activities and new enterprises required, as the experience with MEDU and SEDP in Bangladesh showed, an expanding supply of prospective entrepreneurs with managerial and technical skills.

INSTITUTIONAL MECHANISMS FOR CREDIT DELIVERY

The credit needs of the nonfarm sector could be met by a variety of institutional mechanisms. The credit institutions needed to acquire specialized skills to identify and appraise rural nonfarm projects and to devise, where necessary, appropriate credit instruments, both short and long term. As long as the volume of nonfarm lending remained small compared with the volume of overall lending, commercial banks (as in the case of SFMC in India and MEDU and SEDP in Bangladesh) could leverage its existing staff, logistic resources and overhead costs as it took on the added responsibility of lending to the nonfarm sector. This might enable the banks to achieve economies of scope/scale.

In terms of the credit delivery mechanism, alternative ways can be used. In some cases, commercial banks lent directly to borrowers without the interventions of any intermediary. These credit delivery mechanisms are shown in Table 5.3.

TABLE 5.3. Methods of credit delivery.

Credit delivery mechanism	Case studies in Bangladesh	Case studies in India
Banks lend to individuals directly.	Microenterprise Development Unit (direct lending)—MEDU-Direct	Maharastra Rural Credit Project (individual)—MRCP-VDC
	Small Enterprise Development Project (SEDP)	
Banks wholesale credit to MFIs/NGOs [MFIs/NGOs lend to individual borrowers]. Banks lend to MFIs/NGOs [MFIs/NGOs form SHGs and wholesale credit to SHGs].	Microenterprise Development Unit (indirect lending)—MEDU-Indirect	SIDBI Foundation for Micro Credit (individual)—SFMC-DHRUVA
Community-based organizations or SHGs mobilize resources from banks and donors to lend to the poor.	Kishoreganj Sadar Thana (KST)-VOs	
Banks lend to SHGs of the poor which, in turn, lend to group members.		Maharastra Rural Credit Project (SHGs)—MRCP-SHGs

Source: Author.

Table 5.4 presents, among other things, the pattern of interest rates among all these programs. The focal point here is the rate of interest that was paid by the ultimate borrowers, as distinct from the intermediaries.

The interest rates charged to the final borrowers under the various delivery mechanisms did not fully cover the

1. costs of funds and that of administration,
2. social intermediation (awareness raising, confidence building, and learning saving discipline) and
3. a certain amount of training (in numeracy, literacy, and on banking practices and procedures), incurred by the lenders (be it banks, NGO/MFIs, or SHGs).

The usual margin between borrowing and lending rates in the case of commercial banks was about 6 to 7 percent in both countries. The intensive monitoring and supervision involved in collateral-free loans to poor borrowers added to the costs of administration of the credit agencies. The additional functions (apart from the provision of credit) were undertaken either

TABLE 5.4. The interest rate spread in the sample programs in India and Bangladesh.

| | Indian programs | | | | | Bangladeshi programs | | | |
| | MRCP | | SFMC | | | MEDU | | | |
	Individuals	SHG members	Individuals	SHG members	GU	Direct lending	Indirect lending	KST	SEDP
Cost of funds to participating banks	6	6	6	6	na	7	7	7	7
Cost to participating banks of fund administration	6-8	6-8	6-8	6-8	na	6-8	6-8	11[a]	6-8
Interest rates paid by NGOs/MFIs	na	12	na	11	na	na	11.5	not known	na
Interest rates paid by SHG leaders	na	12	na	12	na	na	na	na	na
Interest rates paid by individual or group members	12	30	12	12	na	14.5[b]	na	15	11
Non-interest cost of loan application to individual borrowers	10		10		na		2.5	not known	2.166
Effective rate of interest to individual borrowers	22	32	22	na	na	16.5	na	na	13.2

Source: Author.

[a]This is arrived at by adding 5 percent on collected service charges paid to the VO managers as a de facto administrative cost to the 6 percent, which is standard administration cost of a typical specialized credit institution in Bangladesh, such as BKB.

[b]This is the weighted average of rates of interest for working capital and fixed capital loans, using the relative shares of the two kinds of loans as weights.

na = not applicable.

by the credit agencies or by other supporting institutions or were shared among them. In some cases, social intermediation and training functions were undertaken wholly by the supporting nonbank institutions (NGOs or semipublic agencies as in the cases of MRCP and SFMC) without sharing them with the credit agencies (MFIs or banks).[1]

In the case of MRCP-SHGs in India, SHGs were organized predominantly by a development agency for women, i.e., Mahila Arthik Vikas Mahamandal (MAVIM), NGOs, and the participating banks and the costs of organizing and then motivating the members were borne by the project. In the case of MRCP-VDCs, the banks met the cost of organizing the VDCs for identifying and screening individual borrowers. In KST-VOs in Bangladesh, the functions relating to social intermediation, group formation as well as the screening, monitoring, and supervision were shared between the project staff and the VO manager. In the case of SEDP in Bangladesh, screening and monitoring functions were shared between the project staff and the bank management and for MEDU they were performed entirely by the bank. The sharing of responsibility, costs, and functions among various actors or agents in the process of credit delivery partly enabled the credit agencies to keep the interest rates low.

The loans in the case of SEDP and MEDU were intended to be larger than the usual microcredit loans. However, they were not much larger and their size was about the upper limit of many microcredit agencies (Table 5.5).

Targeting the Poor

MEDU and SEDP in Bangladesh were designed to promote microenterprises for the nonpoor called "the missing middle." They were not expressly targeted toward the poor. The poor were to benefit from the wage employment offered by the growth of microenterprises in the case of both MEDU and SEDP. The missing middle entrepreneurs required both long- and short-term loans that were not restricted to one-year maturity as with the MFIs in order to finance both fixed and working capital requirements with a flexible system of installments for repayment. They desired rates of interest lower than those charged by MFIs because of the different types and nature of investment; mostly they did not have adequate collateral and physical assets as security and hence required collateral-free loans.

The KST-VOs project in Bangladesh embraced all members, the poor and the nonpoor, in village organizations (VOs). The MRCP in India was, however, targeted to those who were below the officially defined poverty line, but the VDC, dominated by the affluent villagers, also included the nonpoor. In the case of MRCP-SHGs, the SHGs were better targeted,

TABLE 5.5. Size of loans on sample programs in India and Bangladesh.

| | Indian programs | | | | | Bangladeshi programs | | | |
| | MRCP | | SFMC | | | MEDU | | | |
Loans size (Rs, 000 in India; Tk, 000 in Bangladesh)	Individuals	SHG members	Individuals	SHG members	GU	Direct lending	Indirect lending	KST	SEDP
	(Sample of 14)	(Sample of 6)	(Only one agency involved)	(Only a sample of 14 involved)					
Maximum permissible	25	25	25	10[a]	na	250	250	50	500
Range actually provided	5-5	0.5-4.0	5-20	10	na	20-50	not known	4.7-22	5-50
Average	15.6	2	9.8	10	na	40.8	15.8	10	40

Source: Author.

[a] These are in-kind, provided in the form of raw material.

na = not applicable.

thanks to the effort of MAVIM field workers, bank officials, and NGOs. Direct lending to individuals by SFMC-DHRUVA selected the better-off borrowers with whose creditworthiness it was already familiar to reduce the cost of screening and monitoring. The SHGs under the SFMC-IIED organized by the IIED did not do very well in targeting the poor.

Grameen Uddog (GU) in Bangladesh was envisioned to make a difference to the income and welfare of the small weavers (typically poor) in Bangladesh and in that sense had been targeted at the poor. The samples in Table 5.6 show that the membership of most of the programs with the exception of GU did not include a majority of the poor. In Bangladesh, KST had 50 percent of its borrowers who could be considered poor, in spite of its not being especially targeted toward the poor.

An important question was whether among these alternative methods of credit delivery, some were more effective in reaching the poor than the others. In order for the poor to participate in the credit program of a bank or an NGO, it was necessary to undertake social intermediation including raising awareness, creating confidence, and inculcating habits of bank discipline among the poor. Either NGOs or SHGs with extensive local knowledge could be effective mechanisms for screening the poor. The members of SHGs or NGOs were fully conversant with the economic status of the fellow member borrowers, i.e., ownership of land and other assets, and the nature of their occupation as well as levels of income. Even when banks were to undertake the lending operations directly, the NGOs could play a crucial role in identifying the poor borrowers. Also, elected local government institutions (such as the panchayats in India) in cases in which they attain some transparency and accountability could be helpful vehicles for identifying the poor. The Bangladesh NGOs used the mechanism of grouping in a different way than SHGs in India. The Bangladesh NGOs undertook "group-based lending" rather than "group lending."

In the case of "group-based lending," the staff of the MFI/NGO, in a group meeting in consultation with the group members, made all the lending decisions. The "groups" were used by the Bangladesh NGOs/MFIs for awareness building, social consciousness raising, screening the borrowers as well as in generating peer pressure for repayment. The leadership and management decisions in the groups were all taken by MFI/NGO staff. While group members were consulted, MFI/NGO staff were not appointed or elected by group members.

In the case of the Indian SHGs, SHG as an entity borrowed from MFI or the banks, and undertook the relevant lending decisions, such as to whom to lend, for what to lend, and how much to lend, as well as the repayment of loans. SHGs in India served several additional functions in promoting (1) grassroots leadership and (2) self-management by the poor. In that case,

TABLE 5.6. The proportion of the poor among the beneficiaries of the sample programs in India and Bangladesh.

Program perfor-mance in reaching the poor (percent of the poor among beneficiaries)	Indian programs				Bangladeshi programs			
	MRCP		SFMC		GU	MEDU	KST	SEDP
	individuals	SHG members	Individuals	SHG members				
Ratio of the poor in program beneficiaries (percent)[a]	47	35[b]	0	9	68[c]	Poor excluded[d]	50[e]	5

Source: Author.

[a] All estimates in this table of the ratio of the poor are based on samples used in the case studies. They are not necessarily representative of the "universe" at issue.

[b] Thirty-three is the percentage of the poor in the sample but the information obtained from the institution covering all the borrowers shows a much higher figure, i.e., 68 percent.

[c] Sixty-eight percent of the participant weavers sampled in the background study owned up to six looms. These are almost always poor households.

[d] The target group by screening only households with incomes of between Tk 10,000 and Tk 30,000 a year effectively excludes the poor.

[e] Fifty percent of the borrowers in KST own less than half an acre of land. In Bangladesh, rural families with no assets other than half an acre of land are almost always poor.

there was no need for NGOs to serve as an additional intermediary, even though NGOs or banks or other public institutions could undertake the task of organizing/training SHGs. Over time, demonstration effect of a large number of SHGs might stimulate efforts by poor borrowers themselves to form self-help groups with a leader of their own, without much external assistance. In the initial stage, some financial assistance might be needed to meet the expenses of organization and management as was in the case of SMFC assistance to IIED.

In Bangladesh, VO in KST and its manager were very similar to SHGs in India and its leader. The manager of a VO was virtually the "credit retailer." Several features distinguished KST-VOs from SHGs. The VOs were bigger than the SHGs; in some respects, the manager worked as an extension/development worker for borrowers. The manager served as a conduit for organizing the training of members with the help of experts hired by VO or for linking up the VO members with experts in the relevant government agencies responsible for training and technical assistance.

It was recognized that the development of SHGs was a time-consuming process. They had to gain experience in the mobilization of members' savings as well as in managing lending operations with high recovery rates.

High Recovery Rates: Group versus Individual

The periods for which loans were granted and the system of repayment by installments (number and periodicity) varied among microcredit programs. The repayment period ranged from one year to four years; the period of repayment for working capital was for a shorter period (e.g., one or two years with more frequent installments), than that for fixed capital investment, which extended up to four years.

In SEDP and MEDU in Bangladesh, repayment and installments for individual loans were flexible within limits for the "missing-middle" borrowers. Installments in MRCP and SFMC in India were monthly (and not weekly as was customary with MFIs in Bangladesh). KST had flexible installments depending on the purpose of the loans. This was a departure from the past practice of many microcredit agencies with one-year loan and weekly repayment installments—a rigid system which could not be met by many borrowers. The objective of the rigid system was to install discipline among the borrowers and to encourage savings to pay off debt but it also ruled out many nonfarm activities that did not generate exactly the same pattern of cashflow as was required by the time pattern of installments. However, the more flexible was the system of installment and repayments, the greater was the need for monitoring and supervision.

What were the factors behind the high rate of repayment in different cases under study? The SHGs in the Indian case studies (MRCP and SFMC) had attained 100 percent recovery, whereas recovery rates for individual loans were 90 percent or more in the examples cited. No collateral was required in any of the cases. Whether this was true for the overall performance of MRCP and SFMC was not known. The overall repayment rate in Bangladesh varied from 88 to 93 percent in all cases (KST, SEDP, and MEDU) and no collateral was required except beyond an upper limit.

Three factors were considered to account for the high rates of recovery of collateral-free loans as follows:

1. peer pressure exercised through groups either in "group-based" lending or group lending through SHGs;
2. prospect of repeat loans to the borrowers who had the alternative of borrowing only from private moneylenders at very high interest rates; and
3. strict screening in the selection of borrowers and close monitoring and supervision.

Seldom one member of a group was denied credit because of the default by another and the importance of peer pressure in this sense had declined over time. The screening could also be done by MFIs/NGOs with the help of village-level organizations, including local government agencies. The members of a group had a detailed or intimate knowledge of one another's economic activities and status and thus served to evaluate and screen the prospective borrowers as to their creditworthiness.

Experience in both India and Bangladesh indicated that high loan recovery rates could be obtained either in individual lending or in group-based lending provided that the MFIs/banks could ensure strict screening, supervision, and monitoring by a system of incentives for the loan officials. In the cases of both MEDU and KST-VO projects in Bangladesh, there were built-in incentives for the staff of the credit agencies. In KST-VOs, it was further strengthened by the status and influence which the manager of the VO enjoyed in the village. The VDC/SGHs in MRCP and DHRUVA and IIED in SFMC in India closely screened the potential members with particular attention to their repayment capacity and their record of creditworthiness. The prospect of repeat loans encouraged and helped to build up a long-run relationship with the credit agency and thereby discouraged default. A default tended to exclude repeat loans at affordable rates so much needed by the poor borrowers.

Savings by the poor borrowers for microcredit agencies were stipulated for at least two reasons. The first seemed to be the need to inculcate the habit of savings among the poor and thus to promote resource mobilization for the microcredit agencies. The second rationale was that members' savings would indirectly serve as a security against default. The borrowers with accumulated savings in the MFIs developed a keen interest in the MFI's viability. As the case studies show, wherever group-based/SHG lending occurred, as in MRCP and SFMC in India and KST in Bangladesh, member/borrowers were required to save. In the case of individual borrowing in MRCP and SFMC in India and MEDU and SEDP in Bangladesh, savings were not required. The individual (relatively better off) wealthier borrowers in the case of MEDU and SEDP, however, already demonstrated their ability and willingness to save and invest in the expansion and diversification of their enterprises. Most of the enterprises were started with members' own savings, supplemented in many cases by loans from friends and relatives. This helped to generate confidence of the credit agencies in the willingness and ability of the borrowers to repay.

NONCREDIT INPUTS: SKILL, TRANSFER OF TECHNOLOGY, AND MARKET ACCESS

The nonfarm sector confronted multiple limitations, both financial and nonfinancial, such as low productivity, limited market (including appropriate marketing institutions), high risks, and high transaction costs. Different measures were needed to remove the various constraints; specialized institutions needed to be developed to meet a few of these specific needs.

GU in Bangladesh provided an example of one agency combining the provision of credit and technology for the manufacture of a new product (Grameen Check), including marketing the output. GU itself employed trainers and specialists to carry out the training for the weavers but if the intention was to introduce a new product among a target group of producers who did not have previous experience and expertise, then a more sustained and an extended period of training as well as the recruitment of more skilled trainers would have been required.

In most cases, an MFI/NGO or a credit agency would not have the expertise or would not find it economical to hire a large number of technical experts among its staff in a multitude of skills and technology. When each particular skill must be imparted by a specialized set of trainers, the training is best provided by independent consultancy firms or training institutions that enjoy the benefits of economies of scale or by an NGO specialized in training activities only.

In the case of MRCP-DHRUVA in India, the training for the individual borrowers was carried out by independent government or semigovernment institutions. The MRCP-SHGs were trained by the relevant NGOs who were organizing the groups. In the case of SFMC-IIED, the NGO, IIED, not only organized the groups but also arranged for the training and the technology transfer to be provided by others.

In the case of KST-VOs in Bangladesh, with the exception of agriculture-related loans (livestock and poultry) in which training was provided by hired specialists, its main objective was to bring the KST borrowers in contact with the specialists/experts in the government to train and impart required skills to selected members of VO. The latter could in turn train the rest of the members. Thus, the KST-VOs trained village-level extension workers were to link up the members with the experts in government agencies.

The entrepreneurship development training in SEDP in Bangladesh was a prerequisite for obtaining loans and included training in the role of bank loans, accounting/bookkeeping, and management of microenterprises.

When lending was predominantly directed toward financing existing enterprises or activities requiring very little skills, capital, or the provision of working capital, there was not much need or scope for transfer of technology. However, in the cases of both MEDU and SEDP in Bangladesh, where the target group was the nonpoor middle-income borrowers with previous experience as microentrepreneurs, and borrowing was predominantly for existing enterprises, the need for technical assistance was also marginal.

In various illustrations provided previously, the cost of training and institution building was mainly paid out of resources of the commercial banks, microcredit agencies, banks, or from the grants provided by the government. However, most NGOs/MFIs providing microcredit to the poor received subsidies from the agencies financing them and do not cover all the costs from interest charges. Thus, the subsidies helped to finance the costs of social mobilization, as well as of minimum technical assistance. It was not clear whether the training costs incurred by the IIED in India were not collected from the borrowers by way of the margin between prices charged for inputs and paid for outputs. Similarly, it was not clear how far GU in Bangladesh covered the costs of organizing and training the weavers through the price differentials between inputs and outputs. However, SEDP provided specialized skill development training for new entrepreneurs.

APPENDIX: RURAL EMPLOYMENT SUPPORT FOUNDATION

Palli Karma Sahayak Foundation (PKSF) was set up in 1990 by the government of Bangladesh with the overall objective of providing the rural poor, the landless and the assetless people, with resources for the creation of self-employment. Although PKSF was established and funded by the government, it remains an autonomous, independent, not-for profit organization that was kept outside the government bureaucracy.[2] PKSF makes its own policies and develops its own management practices. PKSF receives grants and loans from local and/or international sources as well as lends and approves grants.[3]

The specific objectives of PKSF are twofold: first, it provides various types of financial help and assistance to nongovernment, semigovernment, and government organizations, voluntary agencies and groups, societies, and local government bodies so that their partner organizations (POs) can promote income and employment opportunities among the most economically disadvantaged groups in the society. Second, it assists in strengthening and improving the institutional infrastructure of the POs. It undertakes a research program related to the evaluation of the impact of its activities on poverty alleviation.

It does not directly lend to the landless and the assetless people in the rural areas; rather it reaches its target groups through POs—the delivery mechanism for reaching the poor. It implements an institutional development program for the POs that trains PKSF and PO staff, develops a management information system (MIS) and provides interest free loans to POs for buying assets and equipment such as computers or motorcycles, etc. It favors no particular model for its POs and encourages innovations and different approaches based on experience.

PKSF selects for its assistance those organizations that have experience in managing credit programs for the poor. It judges experience and performance of microfinance institutions on the basis of such general considerations as

1. number of years of experience,
2. amount of loan disbursed,
3. number of members and borrowers,
4. recovery rate of loan,
5. adequacy of skilled salaried staff, and
6. credibility of the sponsors.

In addition, they are appraised on the basis of financial sustainability, operating efficiency, and portfolio quality ratios. Among the detailed criteria specified for judging the eligibility of POs are a credit program for self-employment and income-generation activities of the landless and assetless;[4] demonstrated positive performance in microcredit programs with high recovery rates;[5] satisfactory management information system (MIS); and an adequate accounting system. The area in which the PO operates should be poverty-stricken (rural areas are given preference) but at the same time, it should have a good communication network, banking facility, and easy access to a market so that the borrowers can utilize their loan profitably.[6] If a microcredit organization applying for assistance is found deficient, it is kept under observation; suggestions are given for improving its performance. If performance is found unsatisfactory, the application is rejected.[7]

The loan to POs is collateral free and the terms and conditions of the loans vary depending on the volume of operations of POs. For organizations operating over a small area (OOSA), the amount of the first loan granted to them must be smaller than Tk 100,000, although the amount of subsequent loans is not restricted. The services charge varies from 3 percent to 4.5 percent per annum depending on the amount of loan.[8] The loan repayment period is three years, with a grace period of six months. The loan must be repaid in 10 quarterly installments, along with service charge, within the remaining 30 months.[9] In the case of a big PO operating over a large area (BIPOOL), the service charge is 5 percent per annum. The loan repayment period is ten years, with a grace period of four years. However, the service charge must be paid by the PO on a semiannual basis during the grace period. The loan is to be repaid in 12 semiannual installments, along with the service charge, within the remaining six years.

The approval of successive loans to a PO depends on the satisfactory utilization of the previous loan, a high loan recovery rate at the field level (>98 percent), regularity of loan installment repayments, and potential for expansion of the loan program.

Since PKSF provides collateral-free loans, monitoring of credit programs becomes crucial. POs monitor their programs at the field level and PKSF monitors the programs both at the field and office levels. Several complementary steps are taken to monitor the activities of POs, especially the credit program and fund management.[10]

Personnel is considered by PKSF as key to its success. It recruits graduates of above-average academic results. This policy has greatly contributed to the quality of services delivered, working culture, and the advisory role played by PKSF. It follows an elaborate screening process to recruit quality officers and support staff in order to ensure transparency. It provides theoretical as well as field experience and on-the-job training to its staff on various approaches and aspects of poverty alleviation programs, auditing techniques, and management.[11] Training in accounting and MIS is provided for accountants and credit coordinators. During field visits, regular discussions are held with the organizers and field staff, identifying problems and providing solutions. Thus, these practical training sessions during routine visits to each PO have been an effective way of training.

Starting with 23 POs in its first year of operations, PKSF enlisted 189 POs by the end of the year 2000 and disbursed Tk 7.94 billion (about US$147 million) as a revolving fund to its POs over a decade. This fund enabled POs to disburse Tk 28.13 billion (about US$522 million) as microcredit funds at the field level.[12] PKSF's recovery rate has been 98.34 percent and loan recovery rate of POs at the field level has been 98.33 percent. The most popular use for the loan among borrowers is small business or trading, since 35 percent of them use their credit in this activity.[13]

POs have been found to be effective in reaching the poor, selecting the right target groups, and delivering the desired services. The rural poor men and women have been capable of managing money and improving their income.[14]

There have been studies to evaluate the impact of the MFIs financed by PKSF on the social and economic status of the borrowers.[15] A rapid assessment performed by the World Bank in 1999 concluded as follows: income increased for 97.3 percent of the surveyed borrowers (all the sample borrowers were women), 1.78 percent reported no change, and 0.34 percent (two borrowers) reported fall in income; improvement in the quality and quantity of food intake by family members was reported by 88.59 percent of the borrowers; improvement in clothing was reported by 87.85 percent and 28.44 percent of sample borrowers reported that they could increase their land ownership (Ahmed, 2000). They acquired on an average 0.31 acres of land. Improvement in housing conditions was reported by 75.26 percent of borrowers; improvement in child education was reported by 75.41 percent of borrowers; improvement in sanitary conditions was reported by 68.74 percent of borrowers, and improvement in the quality of life was reported by 94.96 percent of the borrowers.

Recent evaluations have indicated that the social mobilization efforts of POs (social awareness, literacy leadership training, and legal awareness) have some effect on the quality of life. The use of bathing and laundry water as well as sanitary toilets has provided significant health improvement among the members of PKSF/POs compared with the nonmembers (Zohir et al., 2001). Also, closing the gender gap regarding school attendance has improved, as well as access to health care and women's roles in household decision making and fertility regulation.

The participation in the PKSF microcredit program has not only led to a reduction in poverty but has also conferred benefits on the less disadvantaged among the poor (extreme poor). In PKSF-financed participating organizations, the decline in the poverty count has been around 12.4 percent (Zohir et al., 2001). In coping with various crises (such as national calamities, income decline due to illness, and withdrawal from the labor market and unanticipated extra expenses (medical treatment or social ceremonies), the participants fared better than comparable nonmembers. They have had fewer occasions to resort to sales or mortgage of assets or dissaving compared with nonparticipants (10 percent compared with 20 percent). While 11 percent of the participants could not cope at all with crises, about 16 percent among the nonparticipants could not cope at all.

Conclusions

PKSF covers the cost of its operations by means of service charges, increasing loan disbursements, reducing operating expenses, and maintaining high recovery rates.[16] As an apex, second-tier organization, the PKSF model shows potential for replication.[17]

Recently, the future role of PKSF has been identified as follows:

1. Continuing to provide loan funds should be the main role of PKSF.
2. PKSF should help train all the staff of POs for further improvement of the capacity of POs, which will be the basis for sustainability.
3. Continuous advisory service will also be an important area of assistance.
4. PKSF should have action research not only in microcredit but also in other related areas of poverty alleviation.

The PKSF example shows that apex organizations do not need to start operating only when microcredit institutions are already relatively well developed. Apex organizations can help create new microcredit institutions and help them obtain operational and financial sustainability.[18] PKSF has played a role of both financial intermediary and market developer. Apart from its independence from political and bureaucratic interferences, it has been free to make its own policies and develop its own management practices. Its example shows that several features are important for success of such apex organizations, including competent staff and careful selection of the POs; developing their capacities to deliver the financial services to the poor; and ensured source of funds.

Chapter 6

Conclusions:
Microfinance for Microenterprise

Microenterprises in the rural nonfarm sector, based on family labor and local markets—often called livelihood or household enterprises—and micro-enterprises based at least partly on hired labor and markets in both rural and semiurban/rural towns, require policies and programs that focus on several inputs: credit, technology, training and skills, and marketing facilities/opportunities.

Microcredit for the poor can be provided to rural enterprises by rural banks or by NGOs/MFIs which could receive special funds/resources from the government, external donors, domestic capital markets, or savings mobilized from its members and nonmembers. The credit can be provided either directly on the basis of individual lending or through groups. The choice depends on the circumstances of each country, including the stage of development of the NGOs, the degree of penetration of banks into rural areas or the growth of viable SHGs.

GROUP VERSUS INDIVIDUAL LENDING

Group lending has two variants: one variant is that of a group of households constituting what is usually called self-help groups (SHGs) under an elected leader to mobilize and manage the savings of its members; it borrows as an entity from MFIs/NGOs or banks and relends its own and borrowed resources to its members, monitors and supervises use of loans, and enforces repayment on the basis of joint responsibility.

The second variant is "group-based" lending in which the MFIs/NGOs not only mobilize and create the groups, but also manage them. The NGOs/MFIs lend to the individual members of a group, monitor and supervise the use of the loan in consultation with the group, and above all enforce repayment on the basis of joint responsibility of the members of the group, i.e., nonrepayment by one borrower to be compensated by other members.

The group lending in both its forms overcomes the informational asymmetry involved in lending to small borrowers; the potential members on the basis of intimate knowledge of one another's economic status are able to judge their eligibility and creditworthiness. Initial screening of members' creditworthiness in group-based lending contributes more to regular repayment prospects than the fear of penalty involved in joint responsibility. Close supervision and monitoring of the loan-related activities of the borrowers by the groups is the second most important aspect of the group-based lending.

There is a possibility that under group lending the borrowers who are capable of increasing their loan size may be constrained by their peers who do not want to bear the risk of liability for loans much larger than their own. This may lead to a dropout of borrowers over time as some of them are ready to borrow larger amounts. Under the circumstances, the average number of members per group is expected to decline over time as some members shift from group-based borrowing to individual borrowing.

The SHGs, in addition to being mechanisms for mobilizing and managing savings, and intermediaries to channel finance from MFIs/NGOs, can potentially perform other useful functions such as

1. promoting leadership at the grassroots level;
2. self-management by the poor;
3. identification of common problems and common action for their solution;
4. serving as a link with the government agencies, local and national, to effectively utilize their extension services; and
5. organizing training and skill improvement for their members.

They can also undertake joint/collective economic activities or projects, achieve economies of scale, and use larger loans than is the case for individual loans. They can promote specialization as well as interrelated activities among members, taking advantage of special skills/abilities of individuals.

Experience indicates that the most appropriate size of groups should be governed, on the one hand, by the range of functions entrusted to them, and on the other hand, by the administrative costs that the NGOs or others organizing or managing SHGs are able to bear.[1] The smaller the group, the more cohesive and homogenous it can be; mutual trust and confidence are likely to be fostered more easily. In a more densely populated region, mutual interaction and familiarity can be promoted among a larger group because the physical proximity of the members is greater than in a sparsely populated region.

Specialized apex institutions for MFIs like PKSF in Bangladesh can be established for channeling resources from public funds or capital markets to the microcredit institutions such as MFIs/NGOs or directly to self-help groups. In addition, the apex institution can build up the capacity of the MFIs by (1) evaluating and monitoring their functions and performance; (2) providing them with training and other capacity building services; and (3) setting guidelines and standards for helping them to achieve financial sustainability.

Experience shows that it is possible to successfully divert microcredit to poor borrowers without the intermediation of NGOs or SHGs, but on the same terms and conditions (and no collateral), as with group or group-based lending. In the first place, information regarding the economic status and creditworthiness of the prospective borrowers can be supplied by local NGOs or other community organizations and local government institutions acting in a transparent and accountable way. Their recommendations regarding the borrower's creditworthiness and economic status can form the basis of appraisal and decision by the microcredit agency. The second requirement is the close supervision of the borrowers and monitoring of the loan-related activities to enforce repayment. The personnel of the credit agency may be provided sufficient incentives, both financial and otherwise, with this end in view. Third, the establishment of a long-term relationship with borrowers based on the prospects of repeat loans contributes to high recovery rate.

In some instances, it may be possible to find local agents who can be used as "credit filters" to individual borrowers. They could be influential men and women in the locality, able to recommend individuals for loans, and who are paid a fee or a share in recoveries. It is a moot question whether traders or traditional moneylenders could not be hired by microcredit agencies or banks to undertake the tasks of (1) the identification of eligible borrowers, (2) the supervision and monitoring of loan utilization, and (3) enforcement of repayment. At the same time, in this system, a need exists for the supervision and regulation of the individual intermediaries by the credit agencies or banks.

TARGETING THE POOR

In order to ensure that SHGs or individuals borrowing from the microcredit agencies are drawn from among the poor, various criteria or indicators of poverty can be used to characterize the eligible poor. Land-based targeting has been widely used for identifying the poor, i.e., those with less than 0.5 acres of land, because it is usually expensive and time consuming to undertake income and consumption surveys of the rural population to identify the poor. Apart from landownership, other indicators used include status of housing and occupation.[2] Concentrating on the poorer areas in the first instance and the poor households in the second stage help in excluding the nonpoor. However, some risk of inadequate coverage of the poor is involved since there may be people in rich areas who are worse off than the poor in poor areas. Consultation with the borrowers' groups who usually know one another's economic circumstances well facilitates selection on the basis of the previous criteria.

The local officials of the microcredit agencies usually need to play a crucial role in targeting the poor. Incentives to and training for local officials with this end in view are essential. However, the pressure on local officials to increase the number of borrowers who succeed in crossing the poverty threshold, as a measure of their performance, may create the perverse incentive of targeting those who are just below the poverty line. An increase in the latter's income—even a small increase—is likely to put them above the poverty line. A substantial increase in income would be required to lift those who are far below the poverty line out of poverty.

To promote microcredit among the poor, i.e., to make them willing and able to borrow, certain preconditions need to be met. This is especially true for the poorest who have low motivation and a lack of self-confidence. Limited awareness and social exclusion add to their apprehension to borrow and to be able to repay. Increasing their awareness about credit availability and about the possibilities of productive use of capital is essential. They need to learn the practices of the microcredit agencies and implications of credit discipline. The NGOs and the SHGs have proved to be an effective mechanism for "social intermediation."

TO REACH THE POOREST: AN ILLUSTRATION

In microcredit programs in general, efforts are made to rule out potentially risky clients. Preference is given to borrowers from households with stable incomes and multiple earning sources on the expectation that even if program-funded activities do not generate sufficient funds and profits, the borrowers are still able to make repayments. Therefore, there is a concern that destitute households will either consume the loan amount or income generated by the loan, or are too poor to make regular repayments from activities that do not generate immediate income. For the poorest group, it is frequently suggested that grants should be the most relevant and legitimate form of assistance. After graduating to a certain level of income, it is useful for the poorest to take recourse to microcredit.[3]

In Bangladesh, BRAC (one of the biggest NGOs) has a program for the poorest called IGVGD (Income Generation for Vulnerable Group Development Program). The poorest are identified as those who meet the following criteria:

1. widowed, divorced, abandoned female heads of households,
2. households owning less than 0.5 acres of land; and
3. earning less than Tk 300 per month ($6 at exchange rates of late 1990s or early 2000s).

A discussion of the principal features of this innovative program to reach the poorest may provide some useful lessons as mentioned in the following text.

The program combined food transfer (provided by the UN/World Food Program) with training program in skills so that a basis was created for the eventual participation of the members in the microcredit program. Moreover, the participating members made a compulsory saving of Tk 25 per month so that the accumulated sum at the end of the program (two years) provides a lump sum for investment. After a period of only two years, the participants were encouraged to join the microcredit program of BRAC; they also became eligible for other health and awareness-raising services provided by BRAC.

Training is a very important component of the IGVGD Program. This involves a significant element of subsidy in the program. It has been estimated that the average cost of training in the IGVGD Program for each woman is Tk 500. A 15 percent flat rate of interest on the loans to the IGVGD women is the same as the rate charged to regular BRAC microfinance members. The cost of servicing each client is approximately Tk 600. In the case of BRAC's regular microfinance activity, this cost is recovered from the interest earnings on even the first loans of Tk 4,000. The IGVGD program, however, provides a first loan of only Tk 2,500, and the interest collected is not sufficient to cover the costs of servicing these client. Therefore, Tk 225 are paid in subsidy to each IGVGD client for each loan. Therefore, subsidy for both credit and training services amounts to Tk 725 per client. The 18-month foodgrain for the IGVGD members is obtained free from the World Food Program. If one were to add in the cost of the foodgrain (Tk 6,000 at current prices), the total subsidy provided would amount to about Tk 6,725 for each client (approximately $135). Both the government and donors feel strongly that this represents a small subsidy, given the overwhelming majority of the IGVGD women who graduate out of a need for continuous handouts.

Three activities under the IGVGD program (training, food grant, and credit) are administered separately. As a part of an already existing program, the government is responsible for the provision and distribution of free foodgrain. Training is conducted jointly with the government. The cost of providing loans is cross-subsidized by BRAC's regular microcredit program.

The overwhelming majority (about 85 percent) of the IGVGD members receive training and support in agriculture (poultry and livestock raising, vegetable gardening, fishery production) or the restaurant and grocery businesses. The women often already have some skills in these activities, and there is a ready market for such products. Poultry rearing requires very little investment in equipment or infrastructure, and it is primarily for this reason that poultry rearing constitutes the bulk of the IGVGD training program.

In order to promote both income generation for women and the overall growth of the poultry sector, the program sets up link with the livestock department of the government to provide training, vaccinations, and veterinary support.

Different types of training are provided to ensure that every step in the poultry-raising process (from hatching eggs to selling chickens) is integrated with the next step to reduce imbalances in the system. Some women receive training on raising newly hatched day-old chickens to two-month-old chickens. They then sell the chickens and buy another round of day-old chickens. Some women receive training on raising chickens for eggs. Many women collect eggs and sell them to the hatchery or in the market. Still others sell feed. A few are trained in setting up a hatchery. The training is provided collectively by BRAC and the government's poultry and livestock department.

The program took up sericulture, a traditional craft art in some parts of Bangladesh that suffered competition from factory-produced cloth and synthetics, and the lack of new designs and marketing. It sought to engage in different steps of production in an integrated fashion. In cooperation with the relevant government agencies, mulberry trees were planted along roads. IGVGD women receiving foodgrain planted mulberry trees and guarded them from theft and destruction. They picked the leaves and sold them to silkworm rearers. The silkworm rearers, who were also usually IGVGD women, raised the silkworm from egg to cocoon. Cocoons were sold to spinners who then sold the thread to weavers. The program provided designs, controlled for quality, and sold the silk through its large, urban retail outlets as well as exporting it. BRAC provided a one-day training for tree planters cum caretakers and a five-day training for silkworm rearers.

About 80 percent of the members of the IGVGD program eventually joined the microcredit program. Subsequently, after initial experimentation, loans were provided at the end of the first rather than the second year, simultaneously with the continued food grant for another year. Since the total quantity of food rations available at the disposal of BRAC was not adequate to cover all those who were eligible, a selection needed to be made and priority was given to

1. widowed or abandoned women with young children;
2. households that recently experienced a shock (i.e., illness, death, or accidents involving a male head of household); and
3. chronically poor who sometimes went without food and those who sometimes needed to beg for food.[4]

Experience with this program over the years demonstrated that the poorest did not necessarily qualify after two years of grants and training for the microcredit program; several of them could not sustain whatever improvement in income and consumption they achieved during the two-year period. After three years or so of participation in the program, the increase in income or assets could not be fully sustained, even though some of them had by then entered the microcredit program. After the completion of the two-year program, some went back to seek free food rations. Second, even some among those who graduated to the microcredit program went back to obtain free food rations. This indicated that different individuals/households among the poorest could improve their standards at different speeds. For several, reversals would occur so that instead of the two-step approach originally proposed, it would be necessary to repeat steps. Moreover, emergency grants or loans would be necessary in cases of shocks. The poorest, therefore, would need several and not always two steps to graduate into viable microcredit clients and to be on the way to achieving a reduction in poverty. Food rations and training would need to continue for longer than two years; for others, financial transfers might be required. During and after the two-year grant and training program, it was found that the beneficiaries would need emergency loans or grants (even when they are recipients of microcredit for production purposes) and other services, especially basic health services, need to be continued during and after.[5]

In the long run, there would be households of elderly couples, single widows, destitutes without homes, disabled, mentally retarded, household heads too sick to work, as well as other socially disadvantaged persons or outcasts who would not be eligible for microcredit. They would require continued grants and traditional welfare programs for food, shelter, and health services (Matin et al., 2003).

FINANCIAL SUSTAINABILITY

Financial sustainability for microcredit agencies concentrating on the poor will take a longer time than is expected in the case of lending to the non-poor, because of the various additional costs for lending to the poor mentioned earlier.[6] As the number of borrowers increases, raising the scale of operations, per unit costs decline; also as the size of the loan increases, unit costs of loan administration decline; costs of social intermediation for new members who learn from the demonstration effects of current members may also decrease. All these factors facilitate progress toward sustainability.

The pressure of achieving sustainability in the short run has frequently induced the credit agencies to include the nonpoor among its borrowers. This enlarges the pool of savings as the nonpoor make larger deposits of savings, and this facilitates the growth of self-sufficiency in resources; it is not only the larger loans to the nonpoor but also the less burdensome appraisal of their creditworthiness, since it is based on assets/existing enterprises, that reduces the cost of loan per unit. All these factors tend to help achieve financial sustainability.

Ways to reduce transaction costs and enhance the returns on investment need to be explored without excluding the poor borrowers. Examples include granting freedom and incentive to frontline (local) officials to assess market potential and constraints as well as to identify, design, and price appropriate services accordingly and granting incentives to locate branches in remote areas and experimentation with mobile banking in some local areas.[7]

While on the subject of a strict and inflexible system of installments for repayment of microcredit, other options in terms of repayment schedule may deserve consideration, i.e., daily or monthly instead of fixed weekly installments that currently is the practice in most MFIs. Moreover, strict enforcement of repayment obligations may need to be relaxed in case of bad years, droughts, floods, and other natural disasters or temporary income shortfalls due to sickness or death of the main earner. In such circumstances, rescheduling of loans may be needed. Under these circumstances, the safety net of mutual help in the group may help for a short period. The fact that the moneylenders remain an important source of rural credit, even in the presence of nonformal NGOs and public-sector credit agencies, is at least in part a testimony to their flexibility in rescheduling or shifting one or two installments. At the same time, too much laxity may lead to loan delinquency. No easy solution is available.

At a more general level, risks of failure in repayment are likely to be reduced if a microcredit agency finances diverse activities by sector/region. In the same vein, credit may not be tied too rigidly to a particular sector, i.e., nonfarm sector. This is especially the case with the poorer borrowers engaged in a livelihood or household enterprises. This rigidity may discourage the poor. Flexibility in the use of credit is attractive to them. The use of microcredit for meeting consumption requirements for the poor is legitimate and indeed essential for consumption smoothing during the course of the year and over the bad and good years. The production credit is not viable without consumption credit or a safety net provided by essential social services. In this context, the provision of saving opportunities by the microcredit agencies is important. The saving instruments need be flexible in terms of the time periods for which they are available as well as the withdrawal possibilities for meeting unforeseen multifarious needs. Easy access to savings is essential for the poor. Most microcredit agencies have mandatory savings requirements, which are subject to rigid inflexible rules regarding withdrawal; this needs to be supplemented by various types of voluntary savings accounts, which can be more flexibly utilized by members in cases of need.

It has been suggested that even for those who are above the level of the poorest of the poor and are eligible to participate in microcredit program, a strong case exists for combining grants and loans in stages in order to facilitate the transition of the poor in time for the graduation above the poverty line. The idea is to let the poor start the initial enterprises with grants rather than loans and gradually to undertake to borrow and thus improve their viability.

The new/start-up enterprises in the initial period meet unexpected adjustment problems in terms of organization, production, and marketing; they face what is known as "teething troubles"—they require a slow process of growing up and gaining experience—during which they are unable to generate adequate cash flow so that they find it difficult to meet the inflexible repayment obligations in respect of amounts and periodicity. The startup or initial grant functions as equity capital and provides a cushion to absorb risks or inadequate cash flow. The equity finance through grant plays the same role as the "venture capital" fund plays in developed economies. This would enhance the chances of survival of microenterprises and therefore expand the flow of successful microenterprises (Pretes, 2002).

UPSCALING AND FINANCING THE MISSING MIDDLE

The provision of credit for growth-oriented microenterprises, i.e., those which are not eligible to borrow from microcredit agencies and are not affluent enough to qualify for collateral-based large loans from the commercial banks—called "missing middle entrepreneurs"—requires special arrangements. They can be served by either specialized windows of commercial banks or separate institutions established for the purpose, if the scale of operations justifies it. Since most of the prospective borrowers in this category are likely to be already engaged in microenterprises and have a credit record of some sort, it is relatively easy to appraise and examine their creditworthiness and determine their eligibility. This is all the more so if loans are merely granted for working capital needs of existing enterprises or long-term capital for their expansion to be repaid on the basis of cash flows in the short or medium term. What is crucial in this context for the productive use and the timely recovery of loans is strict and close supervision and monitoring of the borrowers.

The poor borrowers of the microcredit agencies—those with confidence and some education—may over time graduate into middle entrepreneurs as they scale up and expand their operations. The more enterprising among the employees after having acquired some technical and managerial skills may strike out on their own. Also, "middle" enterprises may be drawn from those who have often worked as employees in other microenterprises. To facilitate such scaling up, it is necessary to provide business development services, i.e., training in skills, management, and marketing, etc. Many NGOs/MFIs facilitate such scaling up by giving the borrowers larger loans until the time the latter opt out of the system of "group lending" and resort to financing from commercial banks or specialized institutions; even a few NGOs and MFIs themselves resort to individual lending. To induce "the missing middle" entrepreneurs to venture into new and unfamiliar nonfarm activities, it may be necessary to charge interest rates lower than usually charged by commercial banks. Similarly, their installment payments may have to be more flexible, depending on (a) the types of loans (short or long term, mixed or working capital) and (b) kinds of activities so that the flow of repayment obligations matches the pattern of cash flow generated by the specific activities.[8]

NOT BY CREDIT ALONE

As nonfarm enterprises move beyond the stage of simple, low skill, and low capital-intensive activities (livelihood/subsistence enterprises), noncredit inputs such as technical skill, production technology, and marketing arrangements assume critical importance. Credit alone cannot serve the purpose of promoting a wide range of growth-oriented nonfarm activities in a dynamic process of development. There is a wide variety of necessary nonfinancial inputs. They range from technology upgrading, product development, and technical training in higher skills to business and management training, assistance in market information, and development and input procurement.

Noncredit services and inputs are best provided on a sector or subsector basis because constraints, possibilities, and problems, as well as solutions, differ widely among sectors and subsectors. The governments often found it useful to establish autonomous consultancy firms or organizations to undertake the training and transfer of technology. Also, NGOs could be assisted by the government to combine their training activities run by specialist consultants for a number of activities. The provision of generic training or marketing services is not very efficient. Technical assistance should be specific to the product and the choice of the product itself should be demand-driven and, if possible, technical assistance should be provided in combination with marketing functions. The nature, context, and duration of training to be imparted need to be appraised carefully so as to meet borrowers' specific needs. Marketing intermediaries are in a good position to identify the required skills as well as training needs in view of their detailed knowledge of markets for the specific products. They could also serve to provide training. Cooperation between MFIs/NGOs/banks and SHGs or community-based organizations should also aid in identifying the training priorities of borrowers. A mismatch between skills required and those preferred by the trainers should be avoided. Several activities require minimal training as the skills required are easily acquired. A few days of training for a particular trade or activity, combined with training in basic accounting and bookkeeping may be all that is needed. Long spells of training (a month or more) away from the places of work or household increase opportunity costs of the poor in terms of time and earnings and may not be attractive for them.

Alternative ways of providing technical assistance and services can be provided by NGOs specialized in training and in marketing services, relating to particular sectors or subsectors; or they can be provided by government agencies, autonomous public institutions, or otherwise, by private consulting firms.

It is preferable for a microfinance institution to provide financial services only, and to rely on partner relationships with the nonfinancial service providers. Any subsidies should be provided by other partners, not by the microfinance institution, so that the focus on developing a sustainable institution is not diluted. Any cross-subsidization of interest rates should be carefully structured so as to not have a negative impact on the financial viability of the entire program or institution.

A pertinent question is whether the costs of social intermediation and of essential minimum training for bank borrowing and discipline, apart from job or skill-specific training, should be borne by the borrowers. If the costs of technical assistance by the MFIs are included in the interest charges, the borrowers would pay for the assistance. It might be argued, however, that charging poor borrowers venturing on a new activity for the cost of technical assistance would act as a disincentive for the borrowers. In the short run, when the poor are being inducted into the process of microcredit, such costs constitute social costs with externalities and are appropriately financed by public resources as subsidies to credit agencies. The benefits of training services for the concerned borrowers and enterprises in a sector/subsector usually spill over to all enterprises in the sector/subsector. Also, there are arguments against charging for technical assistance before its results are fully reflected in increased income. First, the borrowers are not sure how much, if at all, increase in productivity and income would occur as a result of the technical input. The poor trainees already have to bear a part of the opportunity cost of training. Therefore, payment for technical training against uncertain increase in returns that might accrue later is likely to discourage them. Second, the question of the time lag between the completion of training and the marketing of the product might lead to an increase in income. "Deferred payment," that is, coverage of training costs in interest charges or otherwise after a grace period during which the enterprises reach their full efficiency and operational scale is a likely option for small or medium enterprises if the provision of technical components is costly. Similarly, subsidization of the costs of administration (high costs in view of small-sized loans) and loan recovery is necessary in the crucial early years.[9]

The "minimalist" approach (i.e., to exclude functions of social intermediation and elementary training) to achieve wide outreach and financial sustainability as quickly as possible may not only miss the poorest but also may fail to include the "new or fresh" starters among the poor in the nonfarm activities. In this approach, the emphasis is necessarily on the borrowers who already have some experience in nonfarm activities.

The trading/industrial enterprises located in semiurban or urban areas can develop links with rural producers for the supply of inputs or marketing of output, and may provide the necessary training and support in this connection. Similarly, links can be established with international markets through the intermediation of urban trading or marketing or industrial enterprises. The urban enterprises may invest in labor-intensive supply chains in rural areas. The NGOs or rural artisans' cooperatives may link up with urban enterprises with access to international markets. The relevant public-sector promotional agencies can bring them together, and provide advice and guidance as to the potentials and benefits of such urban/rural links.

The personnel of banks and other microcredit agencies also require training, especially regarding loans to "middle-level borrowers" (large loans for complex enterprises) and in particular with reference to the start-up of new enterprises. Ideally, credit agencies or other public sector specialized institutions working in tandem with them should be able to identify and prepare a shelf of projects ready to be used in their lending operation with the mid-scale borrowers. The lending institutions need to go beyond existing enterprises and financing working capital requirements and to play a proactive role in promoting "missing middle" enterprises; they should appraise investment possibilities and explore jointly with prospective borrowers the opportunities for and design of project formulation.

Designing, experimenting with, and building financial institutions for the poor require economic resources and adequate consideration of longer-term social returns. As policymakers seek to make rational policy choices, they must weigh the social costs of designing and building financial institutions for the poor against their social benefits. Of course, some experiments in institutional innovations will succeed, while others will fail. Public policy will need to support and evaluate this experimental process and nurture those designs or institutions that hold promise for future success.

As the microcredit agencies (NGOs/MFIs or SHGs) mobilize different types of savings for both members and nonmembers, and diversify both farm and nonfarm activities, they need to observe prudential financial standards to ensure the security of deposits and to undertake lending operations based on risk assessment. There is, therefore, a need for regulation and supervision of such agencies, be they MFIs/NGOs or SHGs so that they implement appropriate standards regarding reserve requirements, accounting norms, and reporting systems. In addition, there is a scope for organizing a federation of SHGs or NGOs to supplement the bodies of the government with a certain degree of self-regulation with respect to various standards and norms.

Microcredit services are one approach toward the solution of the complex problems of poverty. Studies in various Asian countries demonstrated that microenterprises had improved the economic and social status of the poor by increasing their income, food security, employment, and assets as well as promoting gender equality through enhancement of women's positions in decisionmaking in household expenditures and family planning. Mobility of women advanced as well, with women visiting health centers and NGO offices and other public institutions. The impact on the poverty level also depends on the extent to which the very poor have benefited from self-employment microenterprises. As discussed previously, their participation has been limited due to various factors. Special measures have been under way to increase the participation of the poorest, and to the extent that they participated, they have benefited from microenterprises (Zaman, 1999).

Microcredit would be a more effective remedy against poverty and vulnerability if complemented by other interventions. Nonfarm sector producers do face fluctuations or risks and uncertainty in activities/income flows. This arises from the variability of agricultural incomes, the state of local demand, seasonal interruption of business (monsoon seasons), uncertainty of supplies from within or outside the local area, and the shifts in consumer preferences. Considering such risks and uncertainty, there is a case for insurance plans.

The microfinance enterprises directly promote the productive use of labor and capital and hence contribute to poverty reduction. But healthy and productive life depends upon improved access to health and education services. Education favorably impacts the promotion of entrepreneurship and promotes the development of skills needed for job training. Also, education has a favorable effect on health and child nutrition. Access to health services increases the productivity of workers. Ability to read the small print and to count or estimate the likely costs and benefits of new investments contribute to more productive use of resources.

These services can be supplied by noncredit NGOs or public agencies. This prevents the leakage of such services to the nonpoor, if they are provided to the poor borrowers of the microcredit agencies. They may provide efficient or innovative ways of delivery of such services. In Bangladesh, for example, BRAC has been very successful in providing nonformal education services as well as moderately successful in providing health services.

Infrastructure constraints and farm/nonfarm links are related to the prospects of sustainability of microenterprises. To a large extent, the expansion of the nonfarm sector is hampered by weak infrastructure, such as limited power supply, lack of roads, and absence of transportation or telecommunications. Improvement of infrastructure reduces rural poverty by stimulating development in both agricultural and nonagricultural sectors. Also, by linking rural with semiurban and urban areas over time it might bring a better awareness of and an easier access to employment opportunities outside the rural areas. Conversely, opening rural economies to outside market forces may threaten the survival of many local nonfarm enterprises except those providing customized products/services, traditional handicrafts, or localized repair facilities. On the other hand, if the benefits of rural transformation in the agricultural and nonagricultural sectors are widely shared, there may be more than compensating increase in demand for new and widely diversified nonfarm products. In stagnant villages with limited economic space for competing enterprises, overall rural economic growth that is stimulated by expansion of infrastructure will provide the demand stimulus for microenterprises, which in turn help rural growth and poverty alleviation.

In countries where agriculture continues to be the main source of rural income, the expansion of the agricultural sector is crucial for the growth of nonfarm microenterprises. Increased agricultural productivity reduces demand for labor in agriculture. Surplus labor is thus available to extend employment and output in nonfarm enterprises, which causes a rise in wages and incomes fueling in turn an expanded demand for nonfarm goods and services. Agricultural surpluses also help finance the expansion of the nonfarm sector, and the latter may in turn stimulate agricultural production via a greater availability of inputs at lower cost.

Chapter 7

Public Employment Schemes: Case Studies in India, Bangladesh, and the Philippines

INTRODUCTION

An important source of employment in the rural nonfarm sector has been the provision of wage employment by the public sector. The three most important avenues of the rural public-sector wage employment are

1. public works programs;
2. local employment in administrative agencies run either by national governments or, more important, by local self-governments; and
3. provision of public services in the social sector such as health, education, and social welfare services.

The first category of employment-generating public expenditures expands employment and income for the poor. The second and third categories are important sources of employment primarily for those with primary and secondary educations, who are frequently above the poverty line.

This chapter deals with public employment schemes undertaken under a variety of circumstances; first, they are intended to combat the adverse effects of natural or manmade disasters, such as civil wars, on income and employment of the poor. Second, public employment plans are implemented to offset unemployment in agriculture caused by variations such as lean seasons when agricultural operations are stalled. Third, public works projects are undertaken to provide supplementary employment to those who suffer from long-run endemic unemployment due to inadequate farm and nonfarm employment opportunities. Also, some of the rural unemployed/poor lack the necessary motivation, entrepreneurial spirit, or necessary skills or training to engage in self-employment activities and therefore seek wage employment.

Public employment projects consist primarily of building or creating rural public assets such as rural infrastructure that is not suitable for private investment. These include roads, flood protection/control plans, drainage works, irrigation, water storage devices, afforestation projects, and community fishery projects, etc. The public works program as a means of poverty alleviation rests on two arguments. One is the screening argument, i.e., work requirements tend to exclude the poor. Second is that work requirements unlike unconditional transfers do not deter poverty reducing investment, say, in human capital by private individuals. Unconditional transfers may reduce the incentive to work and thus induce individuals to choose a lower level of effort.

To the extent that the poor have alternative employment opportunities, employment in work programs provides a net increase in income that is less than gross earnings. However, the opportunity cost of earnings elsewhere is reduced by the fact that transfers are avoided in case of the nonpoor who are discouraged to participate in work programs.

The direct and indirect benefits that accrue to the poor as a result of public employment projects may be said to consist of the following:

1. gross earnings minus opportunity costs of alternative employment,
2. the share of the poor in additional income from assets created by public employment programs, and
3. other indirect benefits such as upward pressure on agricultural wages due to a rise in reservation wage of labor and increased bargaining power.

Moreover, the provision of employment and income in lean seasons or years has a smoothing effect on income or consumption variations; if the latter declines sufficiently, it may cause sale of assets or a drop in consumption. Public employment programs preempt such an eventuality.

The development of rural infrastructure has powerful growth effects on the rural nonfarm economy. In the first place, it stimulates agricultural growth by lowering transport costs and increasing access to markets; increased agricultural income in turn leads to additional demands for goods and services in the nonfarm sector. It also increases the access of rural people to nearby towns and rural market centers, enabling them to diversify and expand their consumption of urban nonagricultural goods and services.

Infrastructure development also affects the supply side of the rural nonfarm economy. Electrification, for example, is especially beneficial to small manufacturing and processing enterprises, shops, and service establishments, giving them a more reliable and cheaper source of power. Rural roads facilitate the movements of raw materials to rural towns and between villages as well as of final products to their main markets, and at lower cost. Improved access to larger geographic areas expands market size, provides economies of scale, and increases rural labor mobility to take advantage of employment opportunities in towns. Telecommunications are increasingly important in linking rural firms to their customers and to the wider economy, enabling them to provide better and more timely service (Ravallion, 1991a,b, 1989; Clay, 1986; Bird, 1994).

The analysis of the public employment projects in the following text draws upon examples from South Asia, as well as two case studies, one in India and one in the Philippines.

THE FOOD-FOR-WORK PROGRAM IN BANGLADESH

Using food resources donated to the country, the initial purpose of the food-for-work (FFW) program in Bangladesh was to provide relief for the poor facing severe food insecurity. Over the years, the focus of both the government of Bangladesh and food aid donors shifted from relief to development. Currently, the main objectives of the program are intended for the dual purpose of construction of infrastructure and generating seasonal employment for the rural poor, including mitigation and prevention of losses due to floods and other natural disasters through appropriate protective structures (Hossain and Akash, 1993; Islam, 1999).

Table 7.1 presents an overview of the FFW program financed by the World Food Program (WFP) and Cooperative for American Relief Everywhere (CARE). This provides a good idea of the process of project implementation by aid agencies and the government from its own resources. A multiplicity of government ministries and departments are involved in implementing projects financed by WFP. Currently, a large portfolio of FFW projects are financed and implemented by the government. The CARE projects are implemented by local governments.

The program includes both large and small projects such as coastal embankments, flood protection along major rivers, re-excavation of canals, construction and repairs of interior earth roads, and digging and re-excavation of small irrigation channels. The local project implementation committees (PICs) consist of the members of local governments as well as local social workers and are responsible for supervising actual construction, including payment to the workers.

A subprogram of FFW called the Post-Monsoon Rehabilitation is designed for women's participation focusing mainly on social forestry (forestry for community uses rather than commercial logging) and fisheries

TABLE 7.1. Overview of the food-for-work program under CARE and WFP, Bangladesh.

Program features	CARE	WFP
Sources of funds		
Donors	USAID	WFP/donors
Local contribution	GOB	None
Commodities distributed	Wheat	Wheat
Target group	Unskilled laborers	Unskilled laborers
Regional focus	National, 44 districts	National, 60 or more districts
Activities undertaken	Roads (80%); canals (15%); tanks, bridges, and culverts (5%)	Embankments and canals (40%), roads (30%), forestry (8%), fisheries (7%), and other (15%)
Implementors		
Who handles the grains?		
Port to LSD	DG Food	DG Food
LSD to work site	PIC	PIC
Who identifies beneficiaries?	Self-selecting	Self-selecting
Who pays beneficiaries?	PIC work supervisor	PIC work supervisor
What commodity?	Wheat	Wheat
Who proposes schemes?	Thana and district	WDB, LGEB, MF, MFL
Who reviews proposals?	CARE (100%)	Ministries/WFP (17%)
Who implements construction?	PIC: roads Contractors: bridges	WDB, LGEB, MFL
Who monitors construction?	Thanas	WDB, LGEB, MFL, and WFP/donors
Who monitors impact on beneficiaries?	CARE	WFP/GOB (selective)
Link institution		
Local level	Thanas	PIC
Central ministry	Local government	MRR, MF, MFL

Source: WGTFI (1993).

Abbreviations: CARE = Cooperative for American Relief Everywhere; DG Food = Directorate General of Food (Ministry of Food); GOB = government of Bangladesh; LGEB = Local Government Engineering Bureau; LSD = Local Supply Depot; MF = Ministry of Forestry; MFL = Ministry of Fisheries and Livestock; MRR = Ministry of Relief and Rehabilitation; PIC = project implementation committee; Thanas = local governments; USAID = United States Agency for International Development; WDB = Water Development Board; WFP = World Food Program.

development schemes. In some regions, women work together with male laborers as well in dry-season regular FFW activities. A project known as the Road Maintenance Program (RMP) is designed to provide cost-effective maintenance of essential rural roads. It is meant exclusively for the employment of women and their training in necessary skills; the predominant beneficiaries are destitute women, who are frequently heads of households and are divorced, widowed, separated, or abandoned and have no means of sustaining themselves. The wages are paid partly by CARE (90 percent) and partly by local government (10 percent), and are deposited directly into a bank account, a percentage being put into a savings account to enable them access to financial institutions and a small capital that could be used to start a small business when they leave the project.[1]

Roads have been traditionally the most predominant type of the public works program. They are the easiest type of work to undertake, especially given the limited number of PICs, the limited time available for planning, and the nature of the supervisory arrangements. Road alignments and roads' embankment formation are ideally suited to labor-intensive techniques with minimum technical supervision; roads can be used without any need for social or commercial organizations. Diversification of projects requires greater planning and organization; projects in fisheries, forestry, irrigation, and drainage, for example, call for technical inputs including thorough survey work, workers' training and close supervision, and the organization of beneficiaries.

Influential local people participate in the planning process through the mechanism of PIC in nearly one-half of the relatively smaller projects. In contrast, such local-level participation is limited in the large projects (only one-fourth), since they were technically too complicated to be amenable to extensive consultations with local people. The local-level government officials in view of the burden of their multifarious administrative responsibilities frequently are unable to devote enough time to monitoring and supervising projects—a factor which adversely affects the quantity and quality of work.

The cost of project implementation is higher in the case of CARE than in WFP. Its costs of supervision and reviews of proposals are higher. First, CARE projects that originate from the local governments require more thorough scrutiny and examination in view of the latter's limited capacity to prepare projects whereas WFP projects are prepared by the various national ministries. Second, all its projects are reviewed by CARE, whereas WFP reviews only a sample of its projects and the government ministries selectively monitor the impact of the projects.

WFP projects have often been associated with large-scale developments, designed and approved by the central ministries, whereas the CARE projects have been mostly local projects, initiated by the local people, often representing the rural elite. While the CARE projects have consisted of roads (80 percent), canals (15 percent), and water tanks, small bridges, and culverts (5 percent), the WFP projects have covered a wide variety of projects including large canals and embankments (40 percent), roads (30 percent), forestry (8 percent), fisheries (7 percent), and others (15 percent) (Islam, 1999).

Although the government ministries implement the WFP projects and monitor their construction with some input from WFP, the local implementation committee undertakes the construction of the CARE projects and the monitoring is done by the local governments. This probably is the cause of underpayment of wages and undercompletion of work, since the supervision system at the local government level is weak. This leaves room for corruption and misappropriation of resources; there is no watchdog in the form of either workers' organizations or independent local NGOs acting as pressure groups on the local government.

Compared to FFW, the cash-based RMP is more cost-effective. It avoids the cost of commodity handling (i.e., cost of physical transportation and distribution) by providing cash income to beneficiaries through bank transfer, and prevents any leakage to noneligible groups.

The FFW program in Bangladesh represents a combination of three types of targeting. First, it is self-targeting in the sense that, because of the nature of work and rate of wages, only the poorest households tend to participate. Second, in most cases, it distributes a self-targeted commodity, i.e., wages in wheat, which is considered to be an inferior good in rural areas and hence only the poorest would work for it. Third, the program is seasonally targeted. Over 85 percent of the FFW resources have been traditionally utilized during the agricultural slack seasons. However, with the rapid expansion of the irrigated boro rice and wheat crops in recent years, agricultural employment opportunities have increased during what was previously considered lean years.

In spite of the self-targeting characteristics of the FFW program, leakage of resources from the program do occur. Leaks in the FFW program occur in various ways: over-reporting work done; inaccurate measurement of the actual volume of earthwork, and underpayment to workers below what was provided in the project. The rate of underpayment in the FFW program ranges from 17 to 27 percent.

A survey of the project villages indicated that 69 percent of the participants were functionally landless: 48 percent were landless and 21 percent owned less than 0.50 acres of land. The functionally landless constituted 50 percent of the population in the project villages. Also, 80 percent of the workers were drawn from the group of casual laborers. Given the rural income distribution, the participants were drawn mostly from the poorest 20 percent of the population (Osmani and Chowdhury, 1993). About 84 percent of the participants are drawn from the lowest 22 percent of the rural population.[2]

Insofar as the income and employment impact of the work program were concerned, net employment for the FFW participants was 12 to 13 percent higher than it would have been otherwise. Seventy percent of the employment in FFW was a substitution for self-employment. The net increase in wage earnings, allowing for the opportunity costs of alternative employment, amounts to an increase of 55 percent—10 to 11 percent of annual wage income prior to the FFW program. The introduction of the FFW employment program acted to raise the maximum wage rates in the project villages. Any possible reduction in the use of wage labor outside FFW program was very small.[3]

Although under the FFW program, wages are to be paid wholly in kind, (mainly wheat) in practice the workers frequently receive payment in cash. Due to delays in the arrival and disbursement of wheat, workers are paid in the interim period in cash either in part or fully, either by members of the PICs or through the intermediation of the labor leaders.

Due to changes in seasonal patterns of crop production activities in recent years, the implementation of FFW projects has tended to conflict with the crop production activities. By raising the opportunity cost of alternative employment, incremental gains in wage income due to participation in FFW activities are likely to have diminished. On the other hand, the same phenomenon might induce an increase in participation of female labor in FFW activities. Also, in order to enhance the employment impact of FFW, it has become necessary to design and adopt projects that can be carried out in other seasons of the year, without involving a substantial amount of earthwork. For example, there is some scope to operate FFW during wet post-monsoon season through a switch from construction to maintenance activities, such as maintenance of roads (since road maintenance is often required immediately after the rain) and embankment-side trees as well as such new activities as social forestry, latrine construction, and construction and maintenance of primary schools.[4]

The long-term impact of development activities under FFW has in general had favorable effects on employment. Infrastructure development under FFW has a positive effect on output, a negative effect on input prices, and a positive effect on the adoption of improved technologies that leads to new employment generation. Physical infrastructure also has an important impact on nonagricultural incomes. Villages with an FFW program have better short-term nutritional status, which is an indication of a more secure income stream and better ability to buffer seasonality in income, food consumption, and health stresses.

Although considerable savings are accumulated in their bank accounts under RMP, few women have left the program or withdrawn their savings to become self-supporting, partly because they are not linked up with microcredit programs for nonfarm employment and training. Moreover, since most of them are destitute women selected evenly among different localities, it is difficult to organize them into groups and link them with other NGO programs.

JAWAHAR ROZGAR YOJANA IN INDIA

Jawahar Rozgar Yojana (JRY) was the largest national employment program in India that consolidated in 1988/1989 a number of employment programs initiated in the early years. It had the general objective of providing employment of 90 to 100 days per person especially in backward districts with a concentration of unemployed/underemployed persons.[5] The target group was to be those below the poverty line. It covered every village in the country; the village panchayat received technical advice and support from the District Rural Development Authority (DRDA). The expenditures were financed in 80:20 ratio by the central and state governments; the central allocation to states was based on the proportion of rural poor in the state to the total rural poor in the country.[6] The allocation of funds from the state to the districts was based on (1) an index of backwardness calculated on the basis of proportion of scheduled castes/scheduled tribes (SC/ST) in a district to the total population of SC/ST in a state and (2) inverse of a district's agricultural productivity.[7] The weights allocated to the two criteria were 50:50. Twenty percent of the funds were kept by the DRDA for interblock/village projects and 80 percent were allocated among the village panchayats that had the responsibility for the selection of projects and for their execution.

The funds were allocated directly by the central government to the district and panchayat governments, bypassing the state government that had, however, the responsibility of (1) acting as an intermediary to pass on the funds from the central government to the districts and villages, and (2) monitoring the performance of the program. The latter function did not greatly evoke the interest of the state governments that disliked the idea of responsibility without power.

JRY was intended to achieve the following objectives:

1. creation of sustained employment through strengthening rural infrastructure;
2. creation of community and social assets;
3. creation of assets in favor of the rural poor, in particular for SC/ST; and
4. improvement of overall quality of life in rural areas.[8]

The sectoral distribution of the total expenditures under the JRY during 1989 to 1996 were as follows: rural roads, community buildings, and other work (47 percent); house, house sites, and wells (42.5 percent); minor irrigation, flood protection, soil conservation, village tanks, and land development projects (6.8 percent); and social forestry (3.6 percent) (Thamarajakshi, 1997).

The conditions for project implementation were

1. ratio of nonwage components to wage should be 40:60;
2. contractors were prohibited in execution; this was to be accomplished through the elected panchayats and their members;
3. preference in employment was to be given to SC/ST, bonded labor, and women; and
4. minimum wages (as prescribed under the Minimum Wages Act) were to be paid to casual and unskilled laborers, and could be paid in kind.

Starting in 1993, payment in kind (food) was made optional depending on the price in the open market. Wages were to be paid on a fixed day in a week.

How much of the JRY program was targeted toward the rural poor? About 59 percent of the employment generated accrued to the landless laborers and 17 percent went to women in 1993-1994. The SC/ST, who were to receive preference under the scheme, accounted for 54 percent of the total JRY employment. However, by 1999-2000, participation of both landless and SC/ST declined to 37 and 55 percent respectively. The participation of women, however, increased from 17 to 37 percent (World Bank, 2001).

An analysis of the percentage distribution of the JRY workers by income classes indicates that, on an average for the whole of India, about 57 percent of the workers under the schemes are nonpoor as defined officially by the government of India, i.e., households with annual income of Rs 6,401 and above. Twenty-five percent of the workers were just below the poverty line (as officially defined), whereas the rest of the poor (ranging from the destitute to the very poor) constituted 18 percent of the workforce.

The mistargeting was partly due to such factors as excess of JRY wages over local wages, delays in payment of wages, and partial payment of wages in kind. The panchayats were not above procedural violations, i.e., use of private contractors. Under the program, projects were to be executed by the government ministries and agencies, without the employment of contractors so that full benefit of wages went to the workers. The payments to contractors constituted at least 10 percent of the cost of projects.[9] Thirty percent of the beneficiaries or workers worked under labor contracts. These malpractices could have been prevented by a greater transparency in the working of the panchayats and by the "watchdog" function of NGOs or workers or peoples' organizations which could organize to assert the poor's rights as well as to spread an awareness of the projects and their intended benefits. That these arrangements could have improved the performance of the scheme was demonstrated by the workers' organizations or NGOs in the case of JRY in one region, i.e., West Bengal (Neelakantan, 1994).

Clear-cut guidelines were absent regarding the criteria to be used by the panchayats in selecting the rural poor. It was not enough only to indicate that the JRY was targeted at the poor; in practice, the executing agencies did not follow any list of workers belonging to poor families needing employment.[10]

Under JRY, the national average wage/nonwage ratio was 54, ranging from 83 in Assam to 21 in the Punjab, depending on the composition of projects and technology used. The unskilled workers were paid Rs 22.80 per day (all-India average for male) compared with Rs 25.65 per day that was the local wage rate in off/lean season for unskilled workers. The actual wage rate among states for male in the JRY varied between Rs 33 per day to Rs 18 per day, whereas the market wage rate among states varied between Rs 44 per day and Rs 18 per day (Neelakantan, 1994).

A relevant question related to the quality of assets created under the scheme and their maintenance so that the process of income generation by the assets was sustained. The field agencies, in various sectors, found 74 percent of the projects to be good/satisfactory under JRY, whereas 16 percent were incomplete, and 8 percent were of poor quality (2 percent were unapproved per norms and were not considered useful at all). When the workers were asked about the usefulness of the projects to the poor, 97 percent considered them useful for the poor people. However, in reply to a specific question whether the projects reflected the felt needs of the local community, only 82 percent replied in the affirmative.

With the expansion of program expenditure, shortage of technical manpower at district and block levels was felt, including the need for very specialized skills such as those required for watershed-based schemes for drought areas.

After completion, projects were to be handed over to the relevant government ministries/departments, which were to be responsible for their maintenance. Only 10 percent of the JRY funds could be used by the panchayats/the DRDA for maintenance. The maintenance of the completed projects by the relevant ministries or departments was less than satisfactory; they played a limited role for the maintenance of the JRY assets. For 58 percent of the assets, the panchayats looked after the maintenance with very limited/inadequate funds and the field agency looked after 14 percent of the assets. There was no maintenance at all in the case of 18 percent of the projects.

Local participation was intended to be the defining feature of JRY. The guidelines provided that the plans for development should be discussed thoroughly in the meeting of the village panchayats; the village panchayat should appoint a committee for each village to oversee, supervise, and monitor the implementations of the projects; in order to ensure social control, a meeting of the village panchayat should be held every month, at a fixed date, time, and place to consider the issues regarding the planning, execution, monitoring, and supervision of JRY. The meetings were to be open to any member of the village community, who should be free to raise any issue regarding the program implementation. The village assembly (meeting of all the inhabitants of the village), to be held at least twice a year, was to be kept informed about the implementation progress and all related issues. The officers at the district and local level were to closely monitor all aspects of the program through regular visits to the work sites and were to provide an illustrative list of projects to be carried out under the program (Shanker, 1994).

In practice, the process of popular participation fell far short of expectations. Only 75 percent of the panchayats were elected; only 39 percent of the panchayats were trained on the guidelines and on the operational instructions of the program. The annual action plan embodying the projects was discussed in the panchayat meetings in only 61 percent of the cases (Neelakantan, 1994). Moreover, the guidelines provided that the panchayat should choose one prominent wall in each of the villages, which would be used to display the essential information, such as allocations received and projects chosen for execution. In many cases, such displays were missing.

No systematic guidance was provided to identify the infrastructural gap and how to plan for closing the gaps. The program was isolated from other rural development programs. A tension existed between what was called the "felt needs" of the community and the priorities of different departments engaged in rural development. The "felt needs," in the context of a limited participation by the poor, often turned out to be articulated needs of the rural elite. It was likely that the direct benefits from link roads to markets, irrigation, soil conservation, and land reclamation had accrued mainly to those with land and marketable surplus, even though indirect benefits from rural diversification, agricultural intensification, or nonfarm rural-sector development were shared by the poor through increased employment and from lower food prices that might have resulted from increased food production.

According to a study, employment generation amounted to 62 person-days of employment per family per year with a minimum wage of Rs 20 per day. This implies merely an income of Rs 1,240 per family per year from this source.[11] On the basis of Rs 6,400 of family income as the poverty threshold, this amounted to about 12 percent of the poverty threshold income. In poorer states (such as Orissa), the average annual income worked out at 24 percent of the poverty threshold income (Thamarajakshi, 1997). The total impact of the program was limited because resources were very thinly spread.[12]

A factor that tended to inhibit the poor's participation (especially the very poor) was the delay in the payment of wages. If wages were paid at intervals of one or two weeks and not daily, the very poor could not take advantage of this opportunity since they had no means of survival (either stocks of food or cash) in the intervening days. Part payment of wages in food was meant to reduce the worker's purchase from the market whenever the public distribution system was weak and inadequate. In the latter instance, cash wages had to be spent for buying food in the open market at a price higher than that of publicly distributed food (Parameswaran, 1994).

Genuine voluntary organizations of the poor could have helped to bring the delivery of such programs closer to the beneficiary groups. However, any organization of such beneficiaries would, at the initial stage, require finance and organizational help from outside as indeed was the case with the recent growth of NGOs.

The experience in the state of West Bengal in India illustrated both the strength and the weakness of the village-level local government engaged in implementing employment programs. The orientation of the political party (the Communist Party) in power in West Bengal was helpful in establishing village local governments with a large representation of the rural poor and access to development resources. The organization of the rural poor, i.e., workers' and peasants' organizations, was conducive to an effective people's participation in local government. The party workers elected to panchayats actively promoted the involvement of different interest groups in the implementation of the employment program. On the other hand, the local government lacked the technical and organizational strength for implementation.

While popular participation was quite effective, the professional or organizing abilities of the panchayats were not up to the task. The assistants assigned by the state government to panchayats to provide technical expertise were not adequately qualified. There was even a lack of expertise in keeping proper accounts in an appropriate format, resulting in delays in the release of funds and in one of the lowest rates of utilization of funds among the states. At the same time, cost of employment per person-day in West Bengal was lower than the average for all states—Rs 9-20 compared to Rs 12-95 average for all states (Hirway and Terhal, 1994).

MAHARASHTRA EMPLOYMENT GUARANTEE SCHEME IN INDIA

Maharashtra had the second highest average per capita income of any Indian state, with wide rural-urban disparities. It had the highest percentage of urban population in India. Sixty percent of its total labor force depended on agriculture, but agricultural productivity was very low, resulting from low irrigation potential/facilities with frequent droughts. The share of landless households was one of the highest in India.

It was in this context that the Employment Guarantee Scheme (EGS) sought to guarantee employment for a minimum of 30 days to any person seeking employment, on a continuous basis provided that the beneficiary was willing to do unskilled manual work on a piece-rate basis. Failure to provide employment within 15 days entitled a person to an unemployment allowance of Rs 2 per day. Self-selection of the poor was built into the EGS primarily by the offer of low wage. Until 1988, the wage rate was usually below the agricultural wage rate; no choice of work was offered. Also, as the guarantee held at the district level, a person might be required to travel a long distance for a few days of temporary work.

A complicated procedure that involved completing several documents and contacting different persons was to be followed before an applicant was eligible to apply. The EGS was built upon a coordination between technical and revenue departments in the state administration. Officials in the general administration at districts and in the panchayat were responsible for providing employment on demand, while the actual implementation of projects was in the hands of various technical departments such as irrigation, forestry, and agriculture. The project proposals emanating from the village panchayats to the various agencies were likely to be given a short shrift.

The scheme was entirely financed by the state, which levied special taxes for this purpose as follows:

1. tax on professions, traders, and employment;
2. additional tax on motor vehicles;
3. supplementary sales tax;
4. special assessment on irrigated agricultural lands;
5. surcharge on land revenues; and
6. tax on nonresidential urban lands and buildings.

The burden of taxes was mainly borne by a prosperous urban sector. The state government made a matching contribution to the net collection of taxes and levies. Over several years before EGS was introduced, this scheme was a subject of intense political debate and administrative preparation. There was a prolonged period of preparation or motivation/persuasion of the urban affluent to pay for the scheme, which cost up to 15 percent of the state revenues.

Over the years, the EGS was expanded to include additional components. A subscheme—Shram Shaktidware Gram Vikas (SSGV)—consisting of all development activities in a village in an integrated framework taking backward and forward links into account, was launched in 1989. The programs that predominantly benefited the poor (such as wells, horticulture, and farm forestry on land owned by marginal and small farmers) were financed by the government. For all other projects, beneficiaries had to bear 50 percent of expenditures. Beneficiaries were to share in the "equity" of the project mainly in terms of labor days invested. Horticulture linked to the EGS was introduced in 1990 at 100 percent subsidies for SC/ST/small farmers while others partly paid.

Another program called the Jawahar Wells Scheme was introduced in 1990, extending to all marginal and small farmers and focusing on soil, moisture, and water conservation with a view to minimizing recurrence and impact of droughts. Employment was to be generated in the course of construction and subsequently through increased cropping intensity. The cost was shared between the government and the individual beneficiaries. Given the large subsidy, the number of relatively affluent landholders among the beneficiaries was significant.

Over the period 1991 to 1996, the share of traditional EGS and the SSGV projects fell. The share of Jawahar Wells, on the other hand, almost doubled, thus substituting individual assets (i.e., wells) for community assets that were likely to divert the benefits of the EGS away from the poorest landless households to the moderately poor or relatively affluent owning land.

In a typical year, the scheme provided on average 100 million person-days of employment in a state with a rural labor force (workers and farmers) of 20 million persons. Were it not for EGS, the open unemployment rate in rural Maharashtra would probably have been about 2.5 percent higher. EGS was found to have exercised upward pressure on agricultural wages. There was a reduction in variability of income. Ninety percent of the users of the assets were cultivators and 6 percent were agricultural laborers. Much of the area that benefited belonged to large landholders resulting in a change in cropping pattern and increased productivity.

The proportion of labor content in the total costs of a productive project in EGS was mandated to be 51 or higher; labor content in the beginning was 60 percent.[13] As time wore on and the availability of projects meeting the criterion of specified labor content was exhausted, the required labor intensity was itself rolled back. Minimum labor-intensity requirements could be relaxed if the benefits of the assets accrued to the workers. Over time, the need arises for more skill-intensive projects, requiring in effect different categories for skills of workers. Even though the importance of EGS as a supplementary source of employment diminished in recent years, its role as an income-stabilizing factor remained an important one in the backward regions.

The overall participation fell sharply over the period 1980 to 1997. In fact, many withdrew when farm and nonfarm employment opportunities expanded. Among those who withdrew, there was a sizeable section that was better off. On the other hand, among those who continued to depend on EGS, the share of chronically poor was significant. The disincentive effects of this scheme, if any, were thus unimportant.

In its early years, EGS was more self-selecting in relation to the poor due to the wage discount relative to the market wage. With the wage hike of 1988 and reduction in budgeting expenditures, its coverage declined. Also, the poor's share declined more because of ways in which the reduction in the allocation of work was implemented, i.e., longer delays in assigning work, longer distances from the village and delays in the payment of wages—all of which affected the poor adversely and led to a proportionately greater decline in their participation.

The piece-rate system attracted lightly skilled persons who could complete many tasks within a given time, more than those with poor physical strength who preferred a slower pace of work. Assessment of quality and quantity of work in the piece-rate system caused delay which discouraged the poor. Contractors seeking private profits preferred physically strong and dexterous workers, excluding the poorest.

The current survey (2001) found that the importance of EGS as a supplementary source of income was high for both poor and the nonpoor. However, its importance as a source of income of the poor and the poorest was higher. The survey was incidentally carried out in a very depressed area with pervasive poverty without significant economic disparities. This ruled out much mistargeting. Also, with a widespread awareness of the scheme, blatant violation of wage schedules was not possible. Wages under EGS were substantially higher than agricultural wages, especially during the slack season. There was mixed response to the piece-rate system. Some, especially women, favored it because of a flexible work schedule; others favored time rate because income was assured at the end of the day with no need for elaborate calculations. Considering the opportunity cost of alternative income and employment, direct net transfer of income was 60 percent of the EGS wages; a small positive impact occurred on agricultural wages.

The location of the EGS assets was crucial in determining the distribution of benefits; it was felt that the poor would have benefited more if the panchayats played a role in the selection and location of assets; also, maintenance would have been better. As no separate financial provision for the maintenance of the EGS assets was made, their potential benefits were not fully realized. Early in EGS, the availability of a large portfolio of well-prepared microprojects, ready for implementation, ensured good performance. Later on, quality declined as the shelf of projects thinned and new projects were launched without due diligence.

The survey confirmed the findings of earlier studies that there was a stabilizing effect on income since the workers did not have to sell assets, cut consumption expenditure, or borrow at high interest rates during bad seasons/drought years. A marked reduction occurred in the scale and duration of the EGS projects, causing many to incur time and effort in search of employment in the neighboring villages.

From a broader perspective, EGS facilitated the building of coalitions among the poor through close interaction on work sites. Thus, it contributed to the growth of political activism among the poor which was likely to make the political system more responsive to their interests.

PUBLIC EMPLOYMENT PROGRAM IN THE PHILIPPINES

The Philippines pursued an approach toward public employment projects that emphasized infrastructure projects which generated productive assets, e.g., viable and durable physical assets rather than employment. However, wherever feasible, labor-intensive methods with an orientation toward the poor were chosen. Moreover, they were not designed to relieve seasonal unemployment and/or to meet short-term shocks to income and employment levels, even though in many cases there was an attempt to schedule work to avoid competition with peak labor demand for farming.[14]

Although in general, public work projects in the Philippines were meant to have favorable impact on rural poverty and unemployment, they were not especially designed to target the poor or to achieve the maximum degree of labor intensity consistent with the desired quality of projects.

However, an attempt was made in several cases to introduce the participation of the beneficiaries in the projects. These are illustrated in three case studies relating respectively to irrigation, road building, and water supply projects. The national government, the local government, and the association of beneficiaries all had, in varying degrees and different ways, participated in the financing, design, and implementation of projects. The national government participated financially in all projects, whereas in another, a water supply project, the local government was a major contributor. The beneficiaries also supplied a certain proportion of labor either as an equity contribution toward financing a project or as paid labor. In respect to maintenance and repair, in two cases, it was the association of the beneficiaries and in another, the local government was responsible.

A communal irrigation project was based on "participatory approach plus cost sharing" in that the Irrigation Association (IA) of beneficiaries was involved in planning, designing, and implementing the project in cooperation with the National Irrigation Authority, which employed social mobilization workers to help organize IA. The IA was to be established prior to the construction of the irrigation facilities and was required to cover at least 80 percent of the farmers in the area.

The IA recruited the workers, in addition to the labor that was provided as equity contribution by farmer-beneficiaries themselves. The cost sharing was mandatory. The equity contribution could be made in terms of both labor and raw materials, in any ratio the IA was willing to provide, thus allowing considerable flexibility to farmers. The materials were sand, gravel, boulders, lumber, etc. The local government's contribution consisted first in securing "right of way" for the irrigation canals/works, and second, in providing a few trucks and other relevant equipment that helped to bear a part of the capital cost of the project.

The equity contribution by the IA water users and their ensuing material stake ensured a high quality project; moreover, close monitoring of expenditures minimized the possibility of corruption.[15]

The potable water supply project was a pilot project which was implemented by a local government with a 70 percent financial contribution by the national government, the rest being the share of the local government in materials and labor.[16] The technical and management functions were performed by the local government; the design and layout of pipelines were carried out in consultation with residents.

In the water supply project, labor was recruited by the local officials from within the area by a "rotation system" so that all the users or residents of the locality served by the water supply could participate. But labor was not a free contribution; it was paid for by the local government with two exceptions. First, a few days of free labor were contributed by beneficiaries (labor supplied by everyone by rotation) during the construction phase but free meals and snacks were provided by the local government. Second, one day per week of free labor was provided during the pipe-laying phase and construction of faucets, etc.

The community was not involved in the engineering design of the project, i.e., determination of its labor intensity. In supplying the labor requirements, free or paid, all individuals following the traditional community system worked on a rotation basis. The social mobilization for the construction of the project was jointly undertaken by (1) farmers' cooperatives and (2) the local officials—two entities whose leaders often largely overlapped. The maintenance of the system was taken over eventually by the water users' association, organized by the barangay (local government) officials. The farmers' cooperatives were involved only in the construction phase.

The past traditions of community participation in local projects illustrated by irrigation and water supply projects have been breaking down in recent years with the growth of individualism and erosion of social cohesiveness.

Implemented by the national government agency, the local irrigation association (association of the water users from the irrigation project in the area) participated in the design and layout of the road project as well as facilitated the grant of "right of way" by the villagers to enable road construction; it attempted to introduce equity contribution by the villagers but did not succeed.

The recruitment and employment of labor in the road project took place under the "pakyaw" system. Under this system, the workers were organized by the labor leader, who himself worked as a laborer. He entered a contract with the local government for the labor component of the project, the equipment being supplied by an equipment or a construction company. This arrangement could be interpreted as piece-rate system for the contract between a labor leader, who managed and monitored labor and was paid on the basis of output-related performance, and the government agency, while the actual employment of labor was based on the time-rate system. Given that the leader of a labor group himself was a worker and was under the close supervision of the implementing agency, abuses such as the underpayment of wages were unlikely to be frequent.

In general, public works projects were intended to be carried out by labor-based methods or more accurately, labor-based and light-equipment supported methods. The requirement of mandatory labor cost ratio in total project costs was considered too inflexible, given the site-specific features of rural development projects. Usually both the labor groups (pakyaw system) and the small contractors were expected to use light equipment or labor-intensive technology. The large private contractors, whenever they were used, preferred capital-intensive technology, because of the high cost of supervision of labor and their easy access to cheap capital.

However, in one of the projects (i.e., in the irrigation project), the option of equity participation by the beneficiaries in the form of labor provided an incentive to supply a large percentage of their contribution in labor in the context of prevailing high un- or underemployment. Moreover, wage rates that were used to estimate the equity contribution were higher than market wages or wage rates actually paid so that the excess of imputed wages over market wages was considered as their equity contribution.

Not all projects required the same degree or intensity of beneficiary participation. The projects with public goods characteristics, i.e., irrigation, drainage, erosion control, etc., that were better managed or monitored collectively were more suitable for the approach.

Community participation was not necessarily smooth or effective in all cases. Frequently, the consultation process largely involved village leaders, either in local governments or in community organizations. Moreover, the community participation might be plagued by conflicts or the projects might drift owing to the lukewarm interest of the community. Consultation or participation might also involve delays, but in the long run benefits exceed costs.

The workers were not chosen on the basis of any poverty criterion related to income or asset, nor were the wages actually paid less than market wages in order to attract only the poor; wages frequently were higher or equal to market wages, but because projects were often located in deprived poorer districts or localities, the large percentage of the labor force was poor. The percentage of the poor seeking employment was 50 percent in the irrigation project and between 40 and 45 percent in the road and water supply projects. In the first two projects, the proportion of the poor among those seeking employment was higher than in the comparable control group; in the third project, the reverse was true—the proportion of the poor was higher in the control group. In other words, the latter was proportionately more oriented toward the nonpoor. Also, the impact on poverty alleviation was the highest in the first project but did not make much of a dent on poverty in the latter two projects.

In a poverty-oriented program, the proportion of the poor among the workers was expected to be higher. However, in the absence of any poverty criterion and in the presence of wages that were often equal to or slightly higher than market wages, a higher proportion of the poor in the first two projects than in the comparable control group was partly due to institutional factors. Most of the labor force were cultivators and were poor; also, whenever other workers outside the family were recruited, they were likely to be poor relatives. The irrigation project was located in a very depressed area with highly fragmented landholdings; furthermore, the poorest workers had the option of supplying their equity contribution in labor. The income of the bottom 10 percent and the bottom 15 percent increased by two and a half times and two times respectively after employment in the project, whereas the income for the whole group increased by 50 percent. The average employment among the bottom 10 to 15 percent was much higher than the average employment for the whole group.

This was not the case with the other two projects. Employment in the road project was not biased toward the poor: only 9 percent in the workers group compared to 7 percent in the control groups was poor. Average employment time was short and the poor did not get employment of much longer duration than that for the average worker. Due to the multiplicity of labor groups in the road project, rapid turnover occurred, so that for any one worker, the length of employment was limited as the applicants had to take turns to accommodate as many as possible. Wages were not low enough to discourage the nonpoor, thus employment leaked to the nonpoor. Because of the rotation system, which was designed to involve all of the beneficiary community in the water supply project, both the poor and the nonpoor participated. Although coverage of the poor expanded, leakage to the nonpoor was significant.

Chapter 8

Review of Country Experiences of Public Employment and Policy Conclusions

EXPERIENCE REVIEW

In South Asia, the focus on public employment schemes (including choice of nature, size, and location) was on providing jobs for the unemployed poor, while attempting to ensure that the employment projects created productive and growth promoting assets. In the Philippines, on the other hand, emphasis was primarily on productive infrastructural projects, while trying to implement them so that they generated employment for the poor.

The country experiences show that the poor potentially could gain from public employment schemes on two counts: (1) participation in the work of building and maintaining (repairing) projects and (2) enjoyment of benefits accruing from output or services generated by the projects. The extent of participation depended on the labor intensity of employment projects and inclusion of the nonpoor among the employees.[1] There was an attempt in India to enforce a labor intensity ratio, a mandatory proportion of labor input vis-à-vis materials that would maximize employment, whereas in Bangladesh and the Philippines no attempt was made to enforce such a predetermined labor-content ratio; however, in the Philippines it was stipulated that projects were to use labor-based methods and light equipment only. In fact, it was frequently found in the Philippines that the labor intensity achieved was likely to be less than that achieved without compromising the quality of work.

Instead of requiring mandatory labor content ratio for (one single) project, it might have been more efficient to require such a ratio for a broad set of projects or programs on the whole.[2]

In the public employment schemes in the Philippines that were located in a region where land reforms had been implemented and where there was not much maldistribution of income, the poor seemed to have benefited from additional employment in their construction as well as from the flow of output and services generated by the projects. This was also true of projects in

India that were located in the depressed areas with a higher than average level of poverty.

Until 1988, by paying wages below the minimum wage or market wages, the EGS in India succeeded in attracting the poor. By following a similar low wage policy, the FFW program in Bangladesh was successful in reaching the poor but provided no guarantee of employment. The wages under JRY in India were at market or higher than market wages and thus attracted the nonpoor. Similarly, the requirement in the later years under the EGS to pay minimum wages, combined with a reduction in total expenditure, affected the poor. Projects that did not confer benefits on all but catered only to a group of beneficiaries tended to favor the rich in the distribution of benefits.

The time rate of wage payment was prevalent in the Philippine and Bangladesh cases, whereas the piece rate was used in EGS. There was no clear-cut division between the rich and the poor in terms of preference of one or the other. In view of (1) delays in the verification of the quantity of work under the piece-rate system and (2) ensured income under the time rate at the end of day, the latter often was, as in India, preferred by the poor.

Similarly, a variety of considerations affected the choice between cash-for-work versus food-for-work. In the Philippines, the women workers preferred payment in kind because payment in food enabled, facilitated, or encouraged a higher food intake for the children. On the other hand, payment in cash was likely to be subject to a greater leakage through underpayment than payment in food. Above all, be it cash or kind, a payment system on a fixed date at infrequent intervals was found to be, in all countries, not conducive to the participation of the poor, who needed payments at frequent intervals.

The role of the national and local governments and the role of the associations of users, workers, or beneficiaries differed among various programs in the countries in the design, implementation, and maintenance of employment projects. The execution and implementation were in some cases undertaken by the local governments (JRY in India) and the maintenance was left to the state/national governments. In other cases, execution was the responsibility of the state government and maintenance was left to the local government (EGS in India). In some cases, in Bangladesh both the execution and maintenance were the domain of the national government and in others both tasks were left to the local governments with the help of NGOs. The systems had differential implications for cost and efficiency, especially when executive responsibility was not matched by technical competence, as was often the case with local governments. In any case, the split responsibility between various national/state departments as well as between local and national/state agencies created difficult coordination problems.

In most cases, provisions for repair and maintenance tended to be inadequate. Faced with limited financial resources relative to the large demand for projects all over the country the tendency was to undertake new projects such as constructing roads in new areas rather than maintaining roads in old areas. This helped spread given resources over wide geographical areas—a step that was politically convenient or popular—but left few resources for quality improvement or durability of the project.

The nature of participation of beneficiaries or workers in the employment schemes varied widely between countries and projects. Under the EGS, the formation of the annual project portfolio was done by a local committee which included the local and state politicians and the officials, but the workers or beneficiaries were not represented. Under the JRY system in India, a provision was included for people's participation in decision making by local government. Through their representatives in local governmental or workers' organizations the beneficiaries were intended to keep watch on the implementation of local projects. It was not much of a success in practice. There was no cost sharing in the EGS by the beneficiaries except in SSGV projects, which provided that the beneficiary villages would make a contribution of 50 laborers per day during the construction of the project.

In the Philippines, the beneficiaries and workers played various roles in the public works projects. In one irrigation project, the beneficiaries made equity contributions, participated in the project's identification and design, oversaw construction, and were responsible for maintenance. In another case, the users shared cost to a very limited extent in a local government project, and were responsible for repair and maintenance. The cost sharing/equity participation helped to identify projects that responded to the requirements of the beneficiaries, who were also interested in their successful implementation, high quality, durability, and adequate maintenance.

The pakyaw or labor contract system was unique in the Philippines but variations of the system existed in other countries as well. By assuming responsibility for supervision and monitoring, the labor contract reduced burden on the government bureaucracy. If the labor contractor was also the leader of a group of workers and was a worker himself, he was likely to employ the poor. In the Philippines, the locally elected village officials, in addition to the implementing agency, assisted him in the selection of workers.

In India, EGS was considered to have contributed to political activism and organization among the poor. First, the concentration of a large number of workers in one place increased their interaction and could potentially help them overcome social and other differences. Second, EGS as a fallback option provided a measure of security and generated self-confidence that was needed for the poor to organize. Third, politicians eager to secure the support of EGS workers attempted to build independent organizations comprising such workers. Consequently, multiple channels emerged to represent workers' interests, and the political system became more responsive to their needs.

CONCLUSIONS

Public employment schemes provide income and employment opportunities at times of seasonal unemployment and act as safety net measures to mitigate the effects of short-term shocks caused by natural disasters or economic fluctuations. They help alleviate the long-term endemic unemployment of the poor who are unable to participate in the labor market. This was demonstrated by the past experience of South Asia and in recent years in the aftermath of the East Asian economic crisis.

In order to maximize the impact of employment programs on the poor, they should be provided at a time when employment opportunities are limited and in locations close to home. The provision of some kind of child care when women do extra work may prevent nutrition loss among children.

The wage rate must be low enough to attract the poor with very little alternative work opportunity and thus reduce its attraction to the nonpoor. Also, self-targeting may be further encouraged by using location or household size to focus on the poor. It is generally accepted that public employment programs should be geared toward higher coverage rather than providing above-market wage rates. To deny coverage to some to be able to offer higher wages to others may be perceived as unfair. Provisions should be made to accommodate poor workers' frequent physical difficulties. When a large proportion of the poor cannot participate in public works due to ill health, it would be desirable to organize public works projects contributing to health improvement such as primary health care centers, control of insects, etc.

Also, to reduce the leakage of employment benefits to the nonpoor it is necessary to use other criteria such as landlessness, absence of homestead, destitution, women-headed households, etc., for eligibility for employment in public works. Extensive publicity about the criteria for employment as well as the role of NGOs or workers' organizations as watchdogs is likely to help in targeting the poor. Whenever economic disparities in an area are low (for example, in an area with successful land reforms) and where a high degree of awareness exists about available public employment schemes, the targeting of the poor is likely to be more successful. The location of employment projects in extremely depressed areas with high unemployment is likely to confer large benefits on the poor.

It is preferable to have payment methods that are more suitable for the poor. Piece rates are sometimes favored by women. Individuals who prefer longer hours to more intense work do better under the piece-rate system; also several members of a large and perhaps weak household can share the work under the piece-rate system. Workers who are frail and unskilled tend to favor time rates because income is guaranteed at the end of the day. Prompt payment and a flexible payment system, rather than fixed payment by date, week, or month, suits the poor better as they lack cash reserves to fall back upon. The piece-rate system requires a greater degree of supervision and administration—difficult to ensure whenever administrative capacity is limited. The construction projects, unlike the maintenance of roads or irrigation, can be split into parts or components, standardized in terms of quantity and quality and are therefore more appropriate for piece rate. In addition, it is necessary to minimize the "transaction costs" to be incurred by the poor such as additional payments to contractors or officials, or manipulation of wages or material costs. The workers' organizations or unions, if they are effective, can help alleviate such problems.

An appropriate balance must be kept between the wage and nonwage components: project quality suffers if skill or raw material components are reduced below the optimum. Moreover, labor intensity may need to be considered for the employment program as a whole, not for each project. Some capital-intensive projects may be linked with or lead to other labor-intensive projects. Over time, as many simple labor-intensive projects are completed, the public employment schemes must rely on projects that need increasing skill levels. The public employment schemes should provide, whenever possible, opportunities for training and skill development for the workers. This will confer long-term benefits on the poor.

In most cases, employment projects benefit both the poor and the not so poor. However, a road, irrigation, or drainage project that increases the output of the large farmers may benefit the poor if the increased farm production leads to additional employment. It is possible, however, to improve the long-run impact on the poor of the public employment schemes by choosing, where possible, projects such as schools or hospitals in the disadvantaged rural areas or irrigation canals or roads in regions where the poor are concentrated to benefit the poor on a long-term basis.

If all projects are designed to generate additional income and employment exclusively for the poor, the political support of the rich for the projects may erode and thus may be worse for the poor in the long run. The choice between projects that benefit the poor and those that benefit the nonpoor depends on the particular circumstances of a country as to the balance of sociopolitical interest groups and on the nature of the projects.[3]

It is important to consider systems of implementation that reduce the costs of search, screening, and supervision of labor engaged in public works. The choice between alternative systems such as public works executed by government agencies or by private contractors under the latter's supervision cannot be made in the abstract. Both systems may be inefficient or leave scope for corruption. The choice depends upon circumstances. In all situations, there are problems of interdepartmental/interagency coordination. It is generally held that public employment schemes are better left to the local government institutions, which can assess and respond better to the needs and circumstances of local population on a priority basis. This requires that the local governments are representative, democratically elected, and are accountable to the local population and act honestly. For large projects with engineering and technical characteristics that exceed the capacity of the local governments, the task of financing and implementation must be shared jointly with the central governments.

The financial contribution by the local governments will increase the interest and raise the stake of local governments in the quality, durability, and viability of projects. Local governments may need to be strengthened in respect of technical expertise in project design and preparation. They need to command adequate managerial or administrative capacity in cooperation with national or central agencies.

The employment schemes should be well prepared ahead of time if they are to act quickly to prevent distress at times of drought, floods, or other seasonal hazards. As experience with food-for-work projects in many countries confirms, a shelf of preplanned projects, ready to be implemented when needed, is the most important condition for success. Advance preparations are needed because potential scheme participants must be informed and sometimes must be taught specific skills.

The organization of beneficiaries/workers at the grassroots level or of labor unions organized under village leaders or committees to participate in the design and implementation of projects is likely to improve the pro-poor orientation of projects. They can act as watchdogs to monitor and supervise the executing agencies at the national or local level, and thus help to check misappropriation or misdirection of resources. Frequently such organizations need to be promoted, if necessary, by providing financial assistance to them in the early stages of project implementation. The participation by the beneficiaries in the financing of public works projects whenever feasible in terms of cash or labor or materials strengthens their stake and incentive in the discharge of the previously mentioned functions. Moreover, the NGOs may be helpful in this context. They can

1. mobilize labor for pressing the government regarding the demand for employment projects,
2. spread the awareness of the public employment schemes among the poor illiterate workers, and
3. help check malpractice, leakage, and occasionally help local governments in solving implementation problems.

The long-run development impact of public employment schemes crucially depends on the adequate financial and administrative provisions for repair and maintenance. The local governments are probably the most appropriate agency to bear this responsibility. It is at the local level that the consequences of inadequately maintained projects are felt most directly. Also, the equity participation by the beneficiaries, in cash or labor, may promote their active interest in the quality of projects and hence in an adequate provision for long-run maintenance.

Complementary projects/measures increase the economic and social benefits of public works projects. The benefits of a road greatly increase if farmers are able to produce a larger marketable surplus through the provision of inputs and credit. An irrigation project is likely to yield higher returns when improved seeds are provided at the same time. High female participation and income in their hands may favorably affect children's health status but at the same time may reduce the time for child care and thus more than offset the favorable impact on the children's health. Similarly, complementary health/nutrition/child care schemes for the poorest communities may facilitate women's employment outside the household. The access to primary education or nutrition in general increases the productivity of labor engaged in public works.

The economic and social benefits of public employment projects are maximized if they can be integrated within the framework of overall rural development programs and projects. After all, infrastructural projects are essential/critical for the development of a wide range of directly productive private or public projects in the nonfarm rural sector such as rural agriculture, rural industry, and services. Attempts should be made as much as possible to integrate public employment schemes with development projects.

PART II:
COUNTRY CASE STUDIES

Chapter 9

SIDBI Foundation for Micro Credit

Raghav Gaiha

INTRODUCTION

The Small Industries Development Bank of India (SIDBI) launched a new initiative, SIDBI Foundation for Micro Credit (SFMC) in November 1998.[1] The objectives were: (1) to create a network of formal institutions from the nongovernmental sector, and (2) to ensure that the network of institutions so formalized is strong, viable, and sustainable. In pursuance of these objectives, SFMC not only provides loans but also equity, capacity building, and other support services.

To be considered eligible for assistance, MFIs must be professionally managed and display growth potential.[2] Another is that they must possess small savings experience cum credit programs with self-help groups (SHGs)/individuals. An innovative feature is that there is no restriction on the *mode* of delivery of financial services. The MFIs may on-lend directly to SHGs or to individuals or indirectly through their partner NGOs and other agencies. All nonfarm activities are supported.

Minimum loan assistance for an MFI is Rs 1 million. Maximum loan amount for an individual/SHG member is Rs 25,000. There is provision for annual need-based repeat assistance. Loans are given to MFIs at 11 percent per annum. MFIs are permitted to charge market-determined interest rates on loans to their partners/SHGs/individuals. The repayment period is four years (including a moratorium of six months). Term deposit receipts equivalent to 10 percent of the loan amount, together with interest accrued are required to be pledged by MFIs as security. Collateral security is being replaced with greater reliance on good track record and capacity assessment.[3]

With a view to enabling MFIs to transform into more formal entities, SFMC plans to provide equity support where permissible. In addition to helping them consolidate and expand their operations, this may make it easier for them to leverage funds from other national and international financial institutions. Eventually the MFIs are expected to build up their equity from the savings of their members, thus imparting ownership to poor borrowers/savers and reducing dependence on external support.

Assistance is given in two forms, financial and training. The financial assistance is *general* or *specific*. The first includes grants to MFIs to ensure adequate and intensive capacity building in operational, organizational, and managerial areas. The second comprises need-based assistance to client MFIs to cover the program expansion costs in order to make them sustainable in the long term. In addition to financial assistance, SFMC provides financial support to technical and management training institutes for conducting orientation and training programs for the staff of MFIs.

The claim that SFMC as a wholesaler of microfinance is a somewhat unique experiment rests on the following features:

1. financial resources are available to it on tap from SIDBI;
2. enjoys operational autonomy and flexibility;
3. shift of emphasis from collateral requirements to track record and potential for expansion;
4. capacity building of MFIs; and
5. dissemination of innovative microfinance practices.

A review of the credit delivery process and its impact on the poor may yield insights into the performance of this experiment.

The objective of the present study is to attempt an assessment of the credit delivery process to individuals and groups. An attempt is therefore made to assess the targeting of credit, repayment rates, rates of returns on nonfarm investments and their variability, and deficiencies, if any, in the process of credit delivery. Of particular importance is the participation of socially excluded groups, e.g., tribals and women and the extent to which they are able to mitigate the severity of their poverty. The analysis is based on primary data collected through detailed interviews of households, both participating and nonparticipating, in two districts. Officials were involved in the implementation of this and other related antipoverty interventions, panchayats, and participating NGO/MFI representatives. One of the two districts chosen was Navsari—a predominantly tribal district. The second district, Sabarkantha, was also backward but socially more differentiated. While Navsari had only individual beneficiaries, Sabarkantha had only group beneficiaries in the SFMC scheme. The selected individual beneficiaries (19) were scattered in seven villages in Navsari, and the group beneficiaries (14) were distributed in four villages in Sabarkantha.

The Navsari villages were mostly tribal, with weak infrastructure and little social differentiation. The communities were thus homogeneous and cohesive. The Sabarkantha villages, on the other hand, were easily accessible and electrified. The social differentiation on caste lines was marked, as those belonging to upper castes seldom visited low caste areas.

As far as the control group for individual beneficiaries was concerned, the procedure was straightforward—four nonbeneficiary individuals were randomly selected from the lower strata in the sample villages. For the second control group, since there were hardly any groups that did not function satisfactorily, we picked six individuals who were not involved in SHGs but whose socioeconomic backgrounds were similar to those of SHG members.

MAJOR FINDINGS

Individual Beneficiaries

All 12 individual beneficiaries were males. Their average age was 36 years. Except for two in the age group 45 to 60 years, all others were younger. All (except one) belonged to tribal households. Six were illiterate, three had primary education, and the remaining three had middle-level education. All beneficiaries were relatively affluent as their incomes were well above the poverty threshold.[4] The majority were engaged in selling consumer goods, and the remaining in repairing, flour milling, and other services.

Seven microenterprises were located in the premises belonging to the beneficiaries, three in the homes, one in a hired place, and one did not furnish any information. Two used electricity. For ten beneficiaries, it was a full time activity. Eight had been engaged in microenterprises for 1 to 3 years, three for 3 to 6 years, and one for 17 years. The average investment was Rs 48,266. The average (gross) revenue was Rs 58,542.[5]

Except for two beneficiaries who had a stake in agriculture, all others had shifted to nonfarm activities completely.[6] Seasonality of agriculture and limited income earning potential were the main reasons why they switched to the more profitable nonfarm activities As the nearest town was 10 to 15 kilometers away and there were a few scattered habitations in an inaccessible terrain, some beneficiaries were quick to realize that (virtually) monopoly profits could be earned in some activities. Although availability of SFMC loans was cited as an additional reason, few of the beneficiaries started a new enterprise through such loans. SFMC loans were used mostly for diversification/expansion of enterprises. In all cases (except one), the decision to switch to a nonfarm activity had more to do with self-initiative than family background. Expansion of a business or its diversification in a few cases was helped by counseling from the commercial officer of Dharampur Utthan Vahini (DHRUVA)—an NGO. If any new skills were needed, these were quickly acquired in most cases.

A wide range of nonfarm activities was observed, from general provisions to oil engine/TV repair facilities. As noted already, the majority was in trading consumer goods. Although there was some variation in monthly demand, most were perennial activities. Trading in consumer items, for example, was usually brisk but peaked during the festival season of October to February. Oil engine/hand pump repairing facilities, however, could be classified as seasonal, since much of the demand was concentrated in the summer months when wells were used for irrigation.

Three enterprises were entirely family-run businesses without any hired workers. But even those with hired workers relied on family support when those managing the business were away to procure supplies or perform some other urgent task.

Most of the enterprises (eight) were financed either from personal savings or loans from relatives or banks. However, diversification/expansion was financed by SFMC loans. In the remaining, start-ups were financed by SFMC loans. So SFMC had a more important role in diversification/expansion than in start-ups.

Most beneficiaries became aware of SFMC through ayojan samitis organized by DHRUVA.[7] Upon contacting the gram sevak or village-level worker (VLW) of local government or the NGO representative or the ayojan samiti, their names were forwarded to DHRUVA. The loans ranged from Rs 5,000 to Rs 20,000, with an average of Rs 9,800. The rate of interest charged was 12 percent per annum. The total amount (including the interest) was repayable in two years.[8] Only one complained of any difficulty in repayment of loans. The loan disbursal was quick. Nine beneficiaries obtained loans within 15 to 30 days of submitting their applications while the remaining three did so between 30 and 60 days. Since loan disbursal was quick in most cases, neither start-up nor diversification/ expansion plans were delayed.

With two exceptions, in all other cases the investments were low. In retailing of consumer goods, fixed investment ranged from Rs 6,000 to Rs 8,000 while working capital ranged from Rs 4,000 to Rs 11,000; in engine oil and spare parts, the fixed investment was Rs 7,500 and working capital was Rs 21,000; in oil engine/hand pump repairing, both fixed investment and working capital were Rs 10,000 each; and, in the TV repair facility, the fixed investment was Rs 15,000 (and the working capital was not reported). The returns were high—presumably partly because of an element of monopoly. The gross monthly revenue in consumer goods, for example, ranged from Rs 8,000 to Rs 15,000, and in tempo service (power driven rickshaws or small passenger vehicles), it was Rs 10,000. However, there were fluctuations, with revenues falling to 50 to 60 percent of the monthly average. The surplus over and above all expenses (including loan repayment obligations and "compensation" to family members) ranged from 10 to 15 percent.[9] Five saved the surplus; one reinvested it in the business and another used it for renovating his house. Indeed, some were confident that the returns would grow over time, as there was no prospect of competition in their business activities, enabling them to finance future expansion from their own funds. This might have been somewhat overoptimistic, however.

The fact that monthly returns could be as low as 50 to 60 percent of the monthly averages suggests that the returns were highly variable. The underlying factors were:

1. highly variable agricultural incomes,
2. interruption of business during the monsoon,
3. uncertain supplies, and
4. shifting brand preferences (in consumer goods).

There was, however, no consensus on their relative importance. Just three respondents agreed that variability of nonfarm incomes was largely a manifestation of variability of agricultural incomes. Indeed, business risks were underestimated, if not largely ignored. Risk-bearing capacity was not considered important for survival in business, nor was much importance given to the protective role of insurance. (In fact, a few respondents were dismissive of its usefulness.) As perceptions of risk were conditioned by the circumstances of the respondents, the following observations might be helpful:

> As most of them were relatively affluent, they had the option to fall back on their savings to compensate for a drop in business income;
> Since the business portfolio was mixed, it was possible to substitute one source of income for another;
> Limited understanding of the potential benefits of insurance among the illiterate or those with just a few years of schooling (recall that 9 out of the 12 beneficiaries were either illiterate or had primary education) was not unlikely;
> Finally, a somewhat naive presumption that their (near) monopolies were immune to market forces imparted an exaggerated sense of economic security.

Although available estimates confirm that variability of nonfarm income is lower than that of crop income, it is nevertheless a matter of concern. For those dependent on the former—especially the poorest—it involves substantial welfare loss in the absence of income-smoothing options. Besides, unanticipated income losses can propel households into permanent poverty. So, to the extent that variability of nonfarm incomes can be reduced through a different mix of activities, or through more diversified market links, it warrants careful analysis.

Except for a few (e.g., flour mill), the backward links for most other enterprises were weak. Retailers of consumer goods obtained their supplies from neighboring towns (e.g., Vansda, 20 kilometers away from Raibor). However, forward consumption links were strong, as the enterprises catered largely to the demand from their own villages. Both input and output markets functioned well, without any glaring imperfections (no evidence of interlinked transactions).

Some constraints were more pervasive than others—erratic power supply affected flour milling and oil engine/pump/TV repairing, while limited transport facilities and absence of *pucca* (hard surface) roads affected virtually all enterprises as buying and selling were restricted. Although general optimism prevailed about self-financing of expansion/diversification of existing enterprises, a few respondents pointed to inadequate working capital. Mixed response was given regarding the related issue of the likely effects of withdrawal of SFMC on enterprise development. A few feared adverse effects on new enterprises, while others were optimistic that SFMC would be replaced with an alternative source of financial assistance.

It is arguable that the exclusion of the poor—especially the poorest—and women from individual beneficiaries was to some extent deliberate. Women were excluded because they were presumably considered bad risks. To some extent, such biases *outside* the household were reinforced by those *within* the household. A few respondents were opposed to their wives taking over their enterprises as it would involve interaction with outsiders. Some others were emphatic that their wives could not handle technical operations. A third reason for their exclusion was the small size of the enterprises. Yet, despite these arguments, *most* of them relied on their wives to manage their businesses when they were away performing urgent tasks such as buying materials. It is therefore not far-fetched to suspect a deep-seated resistance to female autonomy and empowerment.

The poor were excluded due to combined factors such as illiteracy and, associated with that, lack of awareness of promotional schemes for nonfarm activities. Another factor was the fear that they might not be able to repay the loans as scheduled; however, it is arguable that the procedure used for selecting the beneficiaries had far more serious implications. Since an explicit income criterion was not employed, all that was required was a nomination by the ayojan samiti (the planning committee of the village), comprising elected representatives of beneficiaries of Abhyutthan Yojana

(Glorious Awakening).[10] Launched by DHRUVA, the latter disbursed credit in kind—usually in the form of seeds, fertilizer, and pesticides—up to a limit of Rs 10,000. As it would suit DHRUVA to concentrate on relatively affluent landholders to ensure repayment of loans on time, and since it offered them an easy line of credit, the proportion of affluent landholders among the beneficiaries of this Yojana was likely to be high. The ayojan samiti—composed of elected representatives of the beneficiaries—was thus likely to be dominated by them. In that case, nomination of affluent sections for SFMC loans could not be ruled out. In short, even in tribal communities with very little social stratification, discretionary disbursal of loans in a context of limited credit facilities was likely to favor the affluent.

The fact that all beneficiaries were affluent reflects a serious targeting failure. Arguably the design and implementation of the scheme in question were guided more by timely repayment of loans than by a genuine desire to help the poor or vulnerable. The loan disbursal was quick, and the monitoring of repayment was periodic. Since most of the loans financed diversification/expansion of existing enterprises, DHRUVA had a minor role in the selection of projects. The returns were high, confirming the profitability of nonfarm enterprises. All except one could repay the loans and use the surplus for different purposes. Given the strong forward links of the microenterprises with the village economy, their expansion was closely linked to growth of agricultural incomes. Since these enterprises belonged to the affluent, the benefits of agricultural growth would accrue to them and bypass the poor.

Group Beneficiaries

As noted earlier, SFMC did not impose any particular method of credit on MFIs' delivery. Of particular interest was an unusual experiment launched recently by the Indian Institute for Entrepreneurship Development (IIED)[11] in the Sabarkantha and Panchmahal districts of Gujarat. It involved a novel experiment of group credit in kind making *agarbattis* (incense sticks) and tailoring.

Forming SHGs was the first step in the agarbattis project. During the initial six months, they were trained in making the product.[12] After the training was completed using an SFMC loan, IIED provided raw material to the group leader, who was responsible for its distribution among the members. Each member was entitled to a maximum loan (in the form of raw material) of Rs 10,000 and the group to a maximum amount of Rs 100,000 (for a group of ten women). The interest rate was 12 percent per annum. Although sales were not restricted, usually IIED bought back the product. The cost of the raw material and interest were subtracted from the sales revenue. The balance was distributed among the members in accordance with the product sold.

A total of 14 beneficiaries affiliated to a group in each of the four villages were interviewed. Ten beneficiaries (belonging to three SHGs) were engaged in making agarbattis while the remaining four (belonging to the fourth SHG) were engaged in tailoring.

The channeling of credit to the tailoring group took a different form. For this group, three sewing machines were bought. One machine that cost Rs 7,000 was given as a loan to a beneficiary, and the remaining two were kept in a common facility for use by the remaining three.[13] They could bring their own material, threads, etc., and share the use of the machines for free. They were expected to market their own products. All beneficiaries were women.

Thirteen belonged to scheduled castes (SCs) and one to other backward classes (OBC). All except one were nonpoor. A few in fact were somewhat affluent. Only three beneficiaries were illiterate; the rest had either primary or middle education. Eight beneficiaries operated from home and six from IIED premises (or common production facility). Eleven had been engaged in microenterprise activities for one to two years and the remaining three for less than a quarter of a year. None of the beneficiaries engaged in making of agarbattis had made any investments. The tailor who was given the sewing machine had a fixed investment of Rs 7,000 and a working capital of Rs 1,900. For all participants, except one, this was *almost* a full-time activity. Moreover, although demand peaked during certain months (in the case of agarbattis, the peak demand period was August to January), it did not drop sharply during the rest of the year. In any case, the buyback option ensured steady returns. None of the beneficiaries hired labor.

The average share of earnings from microenterprises in total household income was about 38 percent, implying that this was a major source of income. Large variation occurred in this share, however. Three beneficiaries making agarbattis earned less than 3 percent of their household incomes from this activity, as they were new entrants. Four other beneficiaries earned well over 50 percent of their household incomes from this project. The range of variation among the tailors, however, was smaller, from about 15 to 46 percent.

IIED's field staff was instrumental in motivating the beneficiaries to participate in SFMC. The fact that these enterprises could be combined with household chores added to their attraction for women. Whatever doubts they had about the feasibility of these ventures were to some extent dispelled during the training. Success of similar ventures in neighboring villages also had a strong demonstration effect. Since for all participants these were new ventures insofar as there was no family connection, it is significant that the beneficiaries did not lack the motivation to acquire new skills and use them to generate income. Indeed, even business acumen was not lacking, as a few could identify profitable activities on their own. Although none of the beneficiaries interviewed shifted from one activity to another, dropouts were reported when *actual* earnings turned out to be much lower than expected earnings. The experience therefore was not altogether favorable.

The raw material for agarbatti making was supplied in bulk to the SHG leader. Members of the SHG could ask the leader for any amount of raw material, subject to the maximum permissible limit of Rs 10,000 per member. The group leader supplied the material within a few hours. The amount released was recorded in a register. All ten beneficiaries were satisfied with both the quantity and quality of the raw material supplied.

Although members were not obliged to sell the agarbattis to IIED, it was convenient to do so. IIED had a tie-up with a wholesale dealer who bought the agarbattis at Rs 9 for 1,000 pieces. Each member received Rs 6 per 1,000 pieces net of the cost of raw material and interest (at 12 percent per annum). The average annual returns of seven beneficiaries were more than Rs 7,000. If the returns varied, it was due to variation in output. No delays in payments were reported.

Although the material was obtained from Ahmedabad, the wholesaler sold the product both locally as well as outside. Thus, forward links were stronger than backward links.

As pointed out earlier, only one of the four engaged in tailoring received an SFMC loan (e.g., a sewing machine). Their monthly earnings ranged from Rs 200 to Rs 600. Although some variation in demand occurred (it peaked during the marriage and festival season) the variation in earnings was relatively low. As much of the material was bought in neighboring towns, the backward link to the village economy was weak, but since the tailors catered mostly to local demand the forward link was strong.

There was some evidence of autonomy in the disposition of additional income through agarbatti making. Five respondents were emphatic that they made independent decisions while four admitted that it was a joint family decision. Four out of the ten in agarbatti making reported that they used additional income for household expenses and saved the rest. They could do so because their families had two to three other earning members. One used her earnings to meet the school expenses of her children (e.g., tuition fees, cost of books, and stationery). Three others could barely cover their personal expenses with their meager earnings. Three out of the four in tailoring used the additional income to cover household and personal expenses. As the fourth was required to repay the loan in monthly installments, which exceeded her earnings, she was unable to contribute to household or any other expenses.

Channeling credit to women was not resisted by men. Although only some women enjoyed autonomy in income disposition, greater recognition of their income-earning potential was realized both *within* and *outside* the household. As a result of this recognition, the sphere of activities for women widened somewhat, and they became more actively involved in community affairs (e.g., village sanitation). However, while the extended family allowed them to undertake additional responsibilities, there was not much evidence of reallocation of household duties between male and female members. The reallocation, if any, was confined mostly to female members.

Were the poorest individuals excluded? Two sets of factors were involved: one connected with the exclusion of the poorest from SHGs and another associated with their exclusion from financial assistance under SFMC. It will be argued that to some extent the poorest themselves were responsible for their exclusion, but the design and implementation of SFMC also played a role.

Five out of the six members of the control group were interested in supplementing their household incomes through nonfarm activities, but they were unable to do so for a variety of reasons. Although there was some awareness of successful group activities in neighboring villages, sharp caste divisions impeded the formation of SHGs. Other negative factors included weak motivation, diffidence, and absence of entrepreneurial skills. Failure to participate in the SFMC scheme, however, was the result of a set of related yet somewhat different factors. Even though three members of the control group underwent training under this scheme, they were completely ignorant of how to gain benefits. Although it could not be confirmed, it is plausible that the field staff of IIED concentrated on a subset of the trainees who were more motivated and displayed greater potential for the activities promoted by them. Exclusion of those not considered suitable through denial of information about the benefits and procedural aspects of SFMC was thus not unlikely. What perhaps adds to the plausibility of this explanation is that several respondents expressed apprehension about obtaining financial assistance/ loans that could easily have been dispelled by the field staff. The presumption was that loan application procedure was complicated and time-consuming, seed money requirements were unaffordable, and repayment obligations were much too stringent.

Some major deficiencies in the design and implementation of SFMC are discussed as follows.

Although in principle flexibility in the mode of delivery of credit is desirable, some inefficiencies/distortions are likely when a virtual monopoly exists in the provision of credit. Since IIED had a virtual monopoly under SFMC, and there was no transparency in financial transactions, rent extraction could not be ruled out.

Those interested in SFMC assistance could either participate in making agarbattis or tailoring as members of SHGs. There was no provision for those interested in individual loans to finance a nonfarm activity. In the manner in which the scheme in question was implemented, the beneficiaries were essentially piece-rate workers. Although they were protected from market uncertainties such as unavailability of materials, demand fluctuations, and price variability, virtually no scope was available for them to augment their skills and to acquire the experience for setting up their own microenterprises. In fact, there were several other options in the nonfarm sector such as making *papad* (lentil-based snacks), pickles, and grinding spices that the beneficiaries would have preferred but were unable to engage in for lack of financial assistance.[14] To the extent that individual preferences matter in nonfarm sector self-employment as a poverty alleviation strategy, the IIED product design is much too rigid, if not flawed.

Even though a small survey is not appropriate for assessing the targeting accuracy of IIED's credit delivery mode, it must not be overlooked that *all*, except one beneficiary, were nonpoor. A few in fact were relatively affluent. It is arguable that exclusion of the poor—especially the poorest—was not unrelated to the discretion given to the field staff. Instead of using training to help the poorer segments overcome their diffidence, build their awareness, and impart basic skills, it is plausible that they used it as a screening device. Although some basic skills were imparted, denial of information on how to obtain benefits of the project to the less motivated and energetic women kept them out of it. Presumably, this helped ensure a steady flow of products. As the selection of beneficiaries was not subject to approval by any local body such as the village panchayat, and the list was not displayed in a public place, arbitrary selection could not be questioned.

IIED was not forthcoming about the financial transactions (e.g., price of materials supplied), nor were the names of suppliers or buyers revealed. The beneficiaries only knew that IIED had a sales tie-up with a wholesaler who bought agarbattis at the rate of Rs 9 per 1,000 pieces. After deducting the cost of the raw material and interest (Rs 3 per 1,000 pieces), they were paid at the rate of Rs 6 per 1,000 pieces. However, enquiries revealed that, after perfuming the agarbattis, they were sold by IIED at the rate of Rs 8 per packet of 25 pieces. As such a large discrepancy was not explainable in terms of perfuming and packing costs, diversion of a large chunk of sales revenue could not be ruled out. Indeed, on the basis of some other irregularities that came to light, it was surmised that the rents extracted were much larger. Briefly, compared with 80 participants belonging to 14 SHGs in agarbatti making in official records, the actual number was barely 14. There were also complaints that payment for sales was sometimes delayed up to five or six months. Finally, there was no field staff, except an inexperienced trainee when our investigators visited the sample villages.

CONCLUSIONS

In some ways, the SFMC schemes were innovative. In particular, the flexibility given to MFIs in using any mode of delivering credit was indeed a significant departure from current practices. Coordination and monitoring problems occurred in the case of lending through NGOs. Although various committees existed to coordinate and monitor the activities of NGOs, these were essentially bureaucratic responses to the implementation of SFMC.[15] Their meetings were rituals and the recommendations were not mandatory. If, however, the irregularities that came to light in our survey were valid (e.g., exaggeration of number of beneficiaries, discrepancy in prices received by IIED, delays in payment of dues to the beneficiaries, absence of field staff in sample villages), it would be difficult to maintain that the coordination and monitoring were satisfactory. Some performance indicators must be devised to assess NGO performance.[16] Appropriate incentives may be offered, such as linking loans to performance, to induce them to improve their performance—specifically in cost effectively targeting the poor.

Since IIED had a virtual monopoly in providing credit under SFMC, it was able to extract rents. If the discrepancies revealed were accepted at face value, the rents extracted at the expense of the beneficiaries would be large. In a competitive setup, i.e., if there was more than one provider of credit, the rents would disappear.

The exclusion of the poor—especially the poorest—was due to a combination of factors. Of particular importance was the selection procedure employed by the implementing NGO. As the economic status could not be verified and an ad hoc committee had the final say, the selection was guided more by repayment ability than any other criterion. The exclusion of the poorest was thus not surprising. Presumably women were considered bad risks by the implementing NGO. What was worse, these deep-seated prejudices were reinforced by male resistance to wives engaging in market transactions with outsiders, except during their absence.

Among the group beneficiaries, all but one were nonpoor. As the selection was done by the NGO's field staff, some evidence suggested that they concentrated on a subset of trainees who appeared to be better motivated and more productive. The rest had virtually no information about the credit scheme and were thus easily excluded.

To the extent that rural organizations sensitive to interests of the poor existed and functioned satisfactorily, their involvement in credit interventions was likely to be helpful. A case in point was village panchayats. One argument for their involvement was their informational advantage—they were fully aware of the economic status of a household. Another was that they were more representative than ad hoc community organizations, as there are quotas for backward sections in village panchayats. A third reason was their accountability to the village communities; decisions made must be ratified in a village assembly. It was therefore surprising that village panchayats had not been assigned any role in SFMC, despite their satisfactory function in the sample villages. As a result, the two NGOs (DHRUVA in Navsari villages and IIED in Sabarkantha villages) enjoyed considerable authority—either independently or with the help of ad hoc bodies created by them (ayojan samitis)—in the selection of beneficiaries and provision of credit. However, with panchayat involvement, the targeting might improve as a direct consequence of greater transparency in the selection of beneficiaries. If any aberrations came to light (such as collusion between NGOs and village-level workers), these could be dealt with effectively in a village assembly.[17] Collaboration between NGOs and panchayats could of course take different forms, but broadly speaking, the list of beneficiaries of SFMC should be approved by village panchayats and they should have a role in monitoring the loans.

Little attention has been given to designing appropriate financial products for the poor. Most of the enterprises were financed from personal resources, including loans from friends or relatives. However, diversification/expansion was financed by SFMC loans. In the remaining enterprises, start-ups were financed by SFMC loans, so SFMC had a more important role in financing diversification/expansion than in start-ups.

Individual loans were not linked to savings. When SHGs were involved, however, since the amount of loan in kind was not completely unrelated to savings mobilized, there was a partial or implied swap.[18] Savings inculcated some degree of financial discipline among the poor.

Another unattractive feature of SFMC products was that, while one NGO (DHRUVA) offered individual loans to men only, the other (IIED) gave loans in kind only to women belonging to SHGs. In both cases, the loans were given specifically for nonfarm activities. Such rigidity in product design was likely to discourage the poor, but often even when they had access to credit they were unable to use it as desired because of contingencies such as illness or accidents. Combining microcredit with insurance was an option that warranted careful consideration.

Considering that even large MFIs might lack the resources to provide insurance services, an alternative is that MFIs act as insurance agents and insurance companies, on their part, design customized insurance schemes relating to agriculture, trade, industrial activity, and health. Access to various forms of insurance for clients would reduce risk of default and delinquency (Rutherford, 2000).

Three mechanisms of credit delivery were prevalent in India. They were

1. conventional weaker section lending by banks;
2. SHG—bank linkage program; and
3. lending by banks and financial institutions to MFIs for on-lending to ground-level groups or individuals.

SFMC uses the third mechanism.

Group formation was one of many ways of organizing access by clients and to overcome information asymmetry. Other options included regular daily visits to individuals, fixed collection points that were attended at fixed times, and various kinds of agency arrangements.

Of particular interest was a network of autonomous largely self-reliant local financial intermediaries owned and managed by their members. They mobilized their own resources, secured access to sources of refinance if internal resources were insufficient, covered their costs, offered voluntary savings deposit and credit services, and made profit to finance their expansion. On a small scale, such an experiment involving federations of SHGs was already underway (FWWB, 1998). Whether this was more cost effective than NGOs lending to individuals or SHGs warranted careful assessment. Some observations, however, could be made on the basis of a recent survey of six such federations (FWWB, 1998). They had a three-tier structure, comprising the primary group, a cluster of local groups, and then the federation itself. However, four out of the six were still dependent on subsidies. Although a federation might help mobilize a larger volume of savings and consequently provide a larger flow of credit, it could become so unwieldy that the members could lose their affinity.

The success of MFIs depended upon awareness on the part of the potential borrowers; also important was the nature and extent of training in skills provided for the poor. The ayojan samitis in DHRUVA's jurisdiction and to a limited extent the commercial officer/field worker (when available) of IIED had an important role in awareness building and in motivating a high proportion of beneficiaries. If village panchayats got involved in these schemes, promotion of SFMC schemes through gram sabha meetings would be feasible.

Illiteracy and lack of skills were considered to be the main barriers to the participation of the poor—especially the poorest—in nonfarm activities.

That illiteracy and lack of skills were constraints for the poor was of course indisputable. However, there was a risk of overstating the obvious. A few observations might clarify:

1. Many of the traditional skills (making of papad and pickles, wood-work, etc.) do not depend on literacy. These were traditional skills that were easily acquired with some initial guidance.
2. Basic knowledge of bookkeeping and accounting were vital for micro-enterprises. Given a talent for numeracy, these could be mastered easily. In case such talent is lacking, these functions could be delegated to a family member or an outsider with necessary skills.
3. To the extent that illiteracy restricted the use of conventional methods of instruction, alternative methods such as role-play or experience sharing could be used to impart skills.

Besides, if there was a strong contextual flavor in them, the receptivity would be greater.[19]

Some deficiencies in the provision of training under SFMC came to light. A few of those involved in the implementation of SFMC schemes drew attention to the unsatisfactory quality of trainers. Out of the two trainers of DHRUVA, for example, one (the commercial officer) underwent an orientation course of seven days and the other (the community organizer) did not receive any training at all. Moreover, organizations engaged in training NGO trainers were not well equipped to train them in dealing with the poorest, nor did they attach much importance to training women differently (e.g., helping them overcome their timidity in market transactions with outsiders).

Sometimes the skills imparted through training were different from those the participants preferred. Moreover, only the expenses of the participants were covered without any compensation for the time spent. Since in a few cases the training lasted a few months, the exclusion of the poorest was inevitable.[20]

While individual beneficiaries did not undergo training provided under SFMC, group beneficiaries did. Seven group beneficiaries engaged in agarbatti making were trained for three months (including two months of on-the-job training) by an expert invited from Karnataka, and three others were trained by a practitioner from a neighboring village for a few days. Those in tailoring were trained by the IIED field worker for six months. Some of the officials (as well as those involved in entrepreneurial development) advocated longer training periods. Many of these activities did not require any training at all (trading) or limited training for a few days (pickle and papad making). Just a few other activities (embroidery, tailoring) required more specialized skills that might take longer to acquire. In addition, familiarity with basics of bookkeeping and accounting was useful. But even with this additional requirement, an ideal duration for most cases may be no more than four to six weeks.

MFIs—especially NGOs—were restricted in their operations by some legal impediments. Their legal form was not appropriate for carrying out microfinance operations on a long-term sustainable basis. As charitable activities, substantial revenues earned by NGOs from lending activities would endanger this loss of charitable status. Another related restriction was that, since the NGOs were registered as societies or trusts, they were not allowed any equity capital. This interfered with their capital adequacy.

What should be the nature of a regulatory framework for MFIs, if any? Three key issues were identified: which activities of MFIs should be regulated, by whom, and in what form. The Task Force on Microfinance constituted by the National Bank for Agriculture and Rural Development (NABARD)[21] made a case for regulating MFIs, drawing attention to the interests of small savers, ensuring credit and financial discipline and institution of a proper reporting system. Regarding NGO-MFIs, the recommendations included

1. compulsory registration of all such entities;
2. those mobilizing savings between Rs 2 lakh (one hundred thousand) and Rs 25 lakh to be subject to a reserve requirement of 10 percent of savings;
3. if the savings exceeded Rs 25 lakhs, besides reserve requirements of 15 percent and registration with the Reserve Bank of India (RBI), compliance with prudential accounting norms regarding income recognition, asset classification, and provisioning may be prescribed; and
4. until self-regulatory organizations (SROs) emerged, regional offices of RBI to be responsible for regulation of MFIs.

In addition, the task force recommended setting up a microfinance development fund in NABARD for capacity building of MFIs, SROs, banks, SHGs, and others.

Chapter 10

Maharashtra Rural Credit Project

Raghav Gaiha

INTRODUCTION

The Maharashtra Rural Credit Project (MRCP), started in 1994, was designed to

1. improve access of the rural poor to financial services,
2. make them bankable clients, and
3. promote savings mobilization among them through self-help groups (SHGs).[1]

To achieve these objectives, a comprehensive strategy was formulated with the following components:

1. financial and technical support for on-farm activities;
2. technical, financial, and infrastructure support for microenterprises;
3. provision of a line of credit to enable commercial banks to support all viable income-generating activities;
4. institutional support to strengthen credit delivery and reception;
5. promotion of savings and credit operations by building the capacity of SHGs; and
6. management support for monitoring and evaluation.[2]

The target group consists of households below the poverty line (i.e., with annual household incomes up to Rs 11,000 at 1991-1992 prices). Priority is given to those with incomes up to Rs 8,500. This subset comprises mostly small/marginal farmers, the landless, artisans, and tribals.

In order to make the MRCP participatory, a village development assembly (VDA) comprising *all* households in a village is formed. This serves as a forum for a preliminary dialogue on the problems, prospects, and process of development. Out of the VDA, a village development council (VDC)—

composed of 10 to 12 members—is constituted.[3] The VDA prepares a People's Action Plan (PAP), geared to social development of the village. Special attention is given to credit requirements and support systems for fulfilling them. The VDC is responsible for the implementation of the PAP.

Two channels of credit are used, individuals and SHGs. Using the list of poor households, eligible (individual) beneficiaries are identified in a meeting of the VDC, attended by the members of the council, and representatives of the nongovernmental organizations (NGOs), commercial banks (CBs), and other implementing agencies. SHGs, on the other hand, are formed either by Mahila Arthik Vikas Mahamandal (MAVIM) directly through its sahyoginis (SYGs)/field workers or through NGOs contracted by it or by the banks directly or through NGOs contracted by them.[4] Both individuals and SHGs are charged an interest rate of 12 percent per annum. Individual borrowers are given loans for specific activities. SHGs, on the other hand, are required to mobilize savings first. After demonstrating some financial discipline, SHGs are allowed to borrow against their savings deposits from a CB.[5] The loans are distributed among the members in accordance with their own priorities/rules about loan use, amount, interest rate, repayment, and penalty. The rate of interest is 2 to 3 percent per month. Consumption loans are permitted. Eventually, when the credit worthiness of SHGs is established, it is expected that they will be able to borrow independently from CBs.[6]

Considerable importance is attached to information about investment opportunities, skill formation, and technical advice. Several agencies (MAVIM, Maharashtra Industrial and Technical Consultancy Organization [MITCON], and Maharashtra Centre for Entrepreneruship Development [MCED]) provide these services to the borrowers. Moreover, members of VDCs are trained to perform their functions efficiently, as are sahyoginis. On the other hand, bank staff are trained to deal more sympathetically with poor borrowers with limited financial skills and awareness. Under the MRCP, although CBs lend at 12 percent per annum, the average cost of funds for them is 9 to 10 percent. As this spread does not cover all their costs, NABARD provides full refinance at 6 percent per annum.[7] Lately, the consensus regarding MRCP indicates some concern that the targeting has been unsatisfactory. Toward the end of the 1990s, out of the 10,000 members of SHGs, about 32 percent were above the poverty line. One difficulty is, of course, the unreliability of the list of below poverty line (BPL) households. It is not uncommon for it to include 80 to 90 percent of households in a village.[8] But more seriously excessive coverage of above poverty line households in some areas is largely a consequence of inactive VDCs or their manipulation by locally influential persons.[9] An additional concern was that the link of SHGs to banks was slow. The reasons include cumbersome procedures (e.g., elaborate documentation), relative unimportance of MRCP loans in the banks' portfolios, and lack of incentives to field officers (FOs). Finally, despite the emphasis on market intelligence, skill formation, and technical assistance, the promotion of nonfarm activities has been unimpressive. Reluctance of banks to lend to individuals for investment in such activities—connected to some extent with the high variability of returns—is a factor. Among, SHGs, on the other hand, small loans at high rates of interest are a deterrent.

In the following section, a selective review of the MRCP is undertaken, focusing largely on the promotion of nonfarm activities as a poverty alleviating strategy. From this perspective, the following issues among others are addressed:

1. Often a presumption in the microcredit literature indicates that the exclusion of the poorest is unavoidable. The present study provides evidence to test the hypothesis.
2. Considering the fact that the interest rates charged by SHGs are substantially higher than those payable by individual borrowers, a specific concern is whether SHG members are less likely to invest in nonfarm activities, as the returns may not cover all costs (including high interest payments). Another related concern is whether the stringent repayment requirements of SHGs (e.g., bimonthly/monthly repayment schedules) further discourage investment in nonfarm activities with variable returns.
3. A few recent studies have drawn attention to the difficulties of targeting credit on poor rural women in a male dominated community. Women may be used as conduits for obtaining loans without any control over the assets and incomes accruing from them. Are there assets/activities that allow greater flexibility to women in combining them with household activities?

This study was based on detailed interviews of representatives of implementing agencies and a few participating and nonparticipating households (constituting the control group). The household samples were small. However, in-depth interviews based on structured questionnaires compensated for the small sample of the households.

Pune district was chosen largely because of its diversified rural economy. In order to complement the earlier survey, two new villages were chosen from this district, Karhati and Sitewadi. They were chosen as they contain (relatively) large proportions of groups that tend to suffer acute deprivation (e.g., scheduled caste/tribe [SC/ST]). The shares of agricultural laborers in these two villages are 26.50 percent and 7.08 percent, respectively. Karhati is better endowed in terms of infrastructural support such as roads, educational institutions, and communication links.[10]

TARGETING OF MRCP

Two aspects of targeting were considered (1) participation of the poor in the MRCP and (2) benefits accruing to them.

The poor accounted for barely 47 percent of the individual beneficiaries (i.e., 7 out of 15), implying that the majority were nonpoor.[11] All poor beneficiaries, except one, belonged to backward classes (including SC/ST and OBC). Among the nonpoor, exactly half belonged to upper castes. The share of female beneficiaries was slightly higher among the poor beneficiaries.

Among the poor, 43 percent owned land while 62 percent of the nonpoor were landowners. Moreover, the average land owned was greater among the latter (5.83 acres compared with 2.66 among the poor). None of the poor owned any other asset (e.g., bullocks) while the majority of the nonpoor (75 percent) did. Slightly more than half (about 57 percent) of the poor were either illiterate or had no more than a primary education as compared with a much lower fraction among the nonpoor (about 37 percent). The remaining among the latter had higher educational attainments. Two of the beneficiaries—one belonging to an ST another to an SC—were the poorest.[12] Both were males and possessed a few years of education. Their acute deprivation was linked to their social exclusion. By contrast, two affluent beneficiaries fully exploited their social networks in securing loans, pointing to collusion among locally influential persons and implementing VDCs and banks.

Out of the six representatives of different SHGs interviewed, two were poor. One member of the poor group one belonged to an ST. All others—relatively affluent—belonged to upper castes.[13] All respondents—including the poor—had more than primary education. Four owned land—ranging from one acre to seven acres—and two were landless. Two owned other assets (e.g., sewing machines, bicycles, grinding machine). There was some occupational diversity: three agricultural laborers, one farmer, one teacher, and one tailor.[14]

Among poor individual beneficiaries, the range of loan amounts was large, from Rs 5,000 to Rs 25,000. The average amount borrowed was a little over Rs 10,000. Among the nonpoor, the range was larger—from Rs 8,000 to Rs 30,000. Also, the average was much larger (Rs 15,600). That the difference in the average amounts borrowed is partly a reflection of the difference in the absorptive capacities of the poor and nonpoor is plausible, but more significantly it is attributable to the influence of the latter with the implementing agencies (VDCs and banks).

Among the SHG respondents, however, the loan amounts were considerably smaller, ranging from Rs 500 to Rs 4,000.[15]

The poor (individual) beneficiaries used the loans in a wide range of activities (such as selling snacks, vegetables, dry fish, and cutlery, buying buffaloes, and making picture frames). The nonpoor engaged in activities such as buying goats, sewing machines, welding equipment and grinders, and trading in textiles). Among SHG representatives, the loans were used for ceremonial purposes (e.g., Diwali), tailoring, and trading in vegetables and pulses.

All but two beneficiaries retained the assets acquired through MRCP loans. One poor tribal beneficiary sold the buffaloes within three months of buying them to repay a loan from the moneylender. Another—a relatively affluent beneficiary—abandoned his textile trade with a view to concentrating better on his other business interests.

The returns in nonfarm activities varied over a large range among both the poor and nonpoor. Some notional estimates of gross returns in activities pursued by the poor were: cutlery (about 43 percent), picture frames (120 percent), snacks/tea stall (109 percent), and vegetables and dry fish (about 128 percent). A larger range was observed among the nonpoor: textiles (30 percent), oil (90 percent), and tailoring (200 percent). Apart from measurement problems, such high returns must be viewed with some skepticism for two reasons.[16] First, the number of cases is much too small in either group. Second, initially the returns are high on small investments. With the expansion of such enterprises or emergence of new enterprises, the returns might fall somewhat sharply in a relatively stagnant village over time.[17] However, even with a downward adjustment the returns in most cases (with textiles as an exception) would more than cover interest charges and monthly/annual installments of the principal.[18] Thus, net income accrual would be more than moderate.

SELF-HELP GROUPS

Since the MRCP is distinguished by its emphasis on self-help groups (SHGs), their functioning and qualitative aspects of impact are reviewed in detail, drawing upon the experience of successful SHGs (denoted as SHGs functioning satisfactorily [FS]). SHGs are designed to inculcate financial discipline among the rural poor. Joint liability of the members ensures prompt repayment of loans.[19] Pooling of credit requirements helps lower transaction costs of the poor borrowers as well as of the lenders (mostly banks).[20]

A group leader is elected to conduct meetings and to maintain records of financial and other matters. Group size varies from 15 to 20 persons. The meetings are held once a month on fixed dates (on the second, eighth, or eleventh day). The agenda includes outstanding loans, sanctioning of new loans, repayment obligations, and special matters (e.g., contingencies). In the unusual event of someone leaving SHGs, the individual's contribution is returned after deducting a small amount for stationery and other expenses. The monthly contribution (varying from Rs 30 to 50 per person) is deposited in a bank, and a group passbook is issued. A *minimum* period of six months is required for the bank to sanction a loan. During this period (in some cases, stretching to one to two years) the performance of the SHG is closely monitored.

After a loan is secured—ranging from Rs 20,000 to Rs 25,000—it is disbursed among the members in accordance with the priorities laid down by the group. The disbursal is quick (one respondent, for example, was emphatic that she obtained a loan within two to three days of applying for it). The loan use is unrestricted. The interest rate varies from 2 to 3 percent per month.[21] For productive purposes, the rate is lower. The loan is repayable in monthly installments within a year.[22] Records of savings and loans, maintained by the group leader, are periodically verified by the group members and NGOs, ensuring transparency of all financial transactions. As all SHGs repaid their dues regularly, the recovery rate was an impressive 100 percent.

Few of the respondents were engaged in nonfarm activities. None of them was inducted by MRCP group loans to engage in such activities. In fact, the activities had started earlier. As the loan amounts were small, they helped to cover part of working capital requirements or expand the scale of an ongoing activity. No one complained of high interest rates, as the returns were adequate. However, a few of the concerns expressed related to the long waiting period for linking to a bank and rigidity in repayment schedules. The fact that none of the respondents experienced repayment problems reflected not so much adequacy of returns as initially high returns on small investments in a few specific activities (e.g., tailoring). Even with a small expansion in their scale over time, the returns might fall sharply. Their viability might thus be short-lived. Nor did prospects of sustaining group ventures through SHGs seem bright unless closer links were forged with other sectors.

Apart from the initiatives of NGOs and a few enthusiastic bank field officers, the presence of a successful SHG (e.g., the Savitribai Phule Group in Karhati) was an important factor in promoting new groups.

Although there were signs of male dominance—in choice of an activity, purchase of an asset, and disposition of income—several female respon-

dents (as well as some representatives of implementing agencies (field officers of banks, NGOs) were emphatic that their social status had improved
and that a few were in fact more actively involved in local institutions (e.g.,
panchayats). The additional income was used mostly for consumption and
education of children.

EXCLUSION OF THE POOREST

Various studies confirm exclusion of the poorest from microcredit projects. In fact, there is usually a presumption that they lack the capacity to repay the loans. The exclusion of the poorest is a result of not just their limited
financial capacity but also several other factors.

Lack of awareness among the poor strata—especially among the nonparticipants—was pervasive. Specifically, three of the poorest nonparticipants knew nothing about banks, how to obtain loans, interest rates charged,
and repayment requirements. Another three knew a little more (bank loans
carry a low rate of interest) but were apprehensive of documentation required and other formalities. Besides, ignorance of SHGs as financial intermediaries was equally pervasive. The few poorest who did participate in the
MRCP either had previous experience of participating in an antipoverty
program or were inducted by the gram sevak or village-level workers (VLW)/
field officers. The failure of this scheme in awareness—building among the
rural poor, especially among the poorest—was thus a major shortcoming.

The poorest were often distinguishable from the rest in terms of their social exclusion due to caste or other social barriers. The tribals, for example,
were usually cut off from the mainstream of village activities. If a few did
succeed in enhancing their awareness, they failed to overcome social resistance to these inclusions in SHGs.[23]

Although the poorest faced several hurdles in securing loans, the relatively affluent—usually belonging to the upper castes—succeeded easily
through their network of relationships with the implementing agencies
combined with a better awareness of procedural requirements. Three striking illustrations are described in our sample as follows.

Sudha Sudhakar Natkar—a resident of Karhati—was a teacher. Her husband was a headmaster. Their combined annual income was about Rs
85,000—highest in the sample. Recommended by the VDC, she obtained a
loan of Rs 30,000 for a cutlery business within two months. She did not
foresee any difficulty in expanding the business through another loan except that no one in her family could assist her.

Tukaram G. Bhalerao, a blacksmith in Sitewadi with an annual income of Rs 45,000, exploited his networking skills fully in securing several loans, two individual loans under the MRCP (Rs 15,000 in 1996 and Rs 20,000 in 1999), and one as a member of an SHG (Rs 10,000 in 1994). Except for the last, all other loans were used for buying equipment and material for the business. The loans were sanctioned without any delay (within a week).

Leelabhai D. Bhosale, a moderately affluent person with an annual household income of Rs 36,000, was a locally influential person. As a member of the village panchayat in Karhati and chairperson of the VDC, she had no difficulty in securing a loan of Rs 15,000 for her tailoring business under the MRCP in 1998. Subsequently, as the leader of an SHG (Savitribai Phule Group), she obtained two loans of Rs 3,000 and Rs 1,000. Her loans were sanctioned within two to five days.

Given the local power structure, the domination of the VDCs by influential persons was hardly surprising. What was surprising was blatant violation of the MRCP guidelines. The fact that more than a few of the participants were quite affluent suggests that such violations in collusion with VDC and bank officials were common. If a few of the poorest were inducted into the project, this was done merely for target fulfillment. Given a fixed outlay, collusion between the affluent and implementing agencies unavoidably involved exclusion of the poorest.

The documentation required as part of the application was costly—especially for the poor. Handicapped by their ignorance and lack of influence with local officials, they were driven from pillar to post, and ended up bribing various persons responsible for maintaining relevant records and issuing certificates.[24] Several poor participants complained of the difficulties of the documentation required. A few of the poorest in fact opted out due to apprehension regarding problems in completing the documentation.

Another requirement that posed problems for the poor included one that required 25 percent of the loan as seed money to be deposited in the bank. Besides, a practice of giving loans in the form of demand drafts in favor of suppliers precluded buying secondhand equipment—an option that some of the poor respondents preferred. Although these procedures/practices could be defended as financially prudent, the disincentives for some—especially the poorest sections—were important. Equally, although the ease of obtaining a small loan from an SHG was an attractive feature, stringency of repayment schedules (strict repayment requirements, penalties for delays) in addition to contribution of savings acted as a deterrent for some (e.g., agricultural laborers) primarily because of the irregularity of their incomes.

Except for a few cases, the VDC's recommendation was accepted without verification of the economic status of applicants. As the below poverty line (BPL) lists were highly unreliable—in most cases, the majority of the village population was included—the guidelines for the selection of beneficiaries were easily manipulated by locally influential persons, as reflected in the present sample. Admittedly, the banks were not in a position to verify the economic status of recommended cases.

Few (individual) beneficiaries received loan related advice. Given the limited contact between bank field officers and the beneficiaries, conflicting views were expressed about repayment obligations. Project management advice/guidance was conspicuous by its absence.[25]

As VDCs had an important role in identifying beneficiaries, their formation, representativeness, and functioning were closely linked to the targeting of the MRCP. The VDC in Sitewadi was constituted in 1995 and in Karhati in 1998. Both were constituted somewhat hastily at the initiative of bank officers.[26] The members were nominated in a village meeting. The VDC was headed by a chairperson, followed by a secretary and six to ten members. Executive positions were filled by nomination.

Whether the VDA was a distinct entity—distinguishable from the gram sabha (or the village assembly)—could not be ascertained. The relationship between the VDA and VDC was also unclear. Specifically, respondents were unable to throw any light on whether the VDA had any role whatsoever in the formation of the VDC and in screening applications under the MRCP for the approval of the VDC.

Although SCs/STs and OBCs (other backward classes) were included in the VDCs, their presence alone is of limited significance for various reasons. One was the fact that they were nominated and not elected, implying that some might be proxies for the more influential persons in the village community. Another was that, with a few exceptions, their general awareness of the objectives and functioning of the VDC and of their own role in it was woefully inadequate.

The VDC met once a month, usually on a fixed date.[27] The agenda was seldom announced in advance. The venue was a gram panchayat office or a school or the space under a tree. As there was no quorum, decisions were made even if some members were absent. Selection of beneficiaries was usually a ritual with the gram sevak (VLW) playing a decisive role. Since the BPL list was in his custody, and a household must be included in this list to be eligible for a loan under the MRCP, it was the gram sevak's endorsement of an applicant that counted. In general, there is often blatant collusion between the gram sevak and the chairperson, which the representatives of NGOs, banks, and official agencies are unable to check.

NONFARM ACTIVITIES

A wide range of activities was observed in the sample. These included cutlery; making picture frames and baskets; selling tea, snacks, and confectionery; tailoring; blacksmithy; sweet oil extraction; selling vegetables and dry fish, bangles, etc. None of these—except to some extent selling dry fish—was a seasonal activity in the sense that supplies are available only during certain months. However, much of trading activity was interrupted during the rainy season (four to six weeks), resulting in a sharp drop in income.

The activities undertaken by the poor and relatively affluent beneficiaries were largely similar, except that the latter operated on a much larger scale and consequently obtained larger returns. Most of these activities had strong forward links (e.g., cutlery, bangles, tailoring, confectionery, etc.) while a few others had strong backward and forward links (e.g., snacks, making of baskets and frames, sweet oil).

Although the reasons for choosing nonfarm activities varied a great deal within the sample, they hardly differed between the poor and the relatively affluent. The reasons included prior experience (selling snacks), suitability for the local area (bangles, tailoring), traditional family occupation (hair cutting), and ease of supplementing regular income (cutlery business). Whatever the stated reasons, a common consideration was the demand potential.

Many of these activities, such as bangles and cutlery, were particularly suitable for poor rural women as it was easy to combine them with household chores and/or other activities (e.g., agricultural employment).

Lack of prior experience was not a constraint, as those without it did not fare badly. Several activities such as selling snacks, cutlery, dry fish, spices, etc., could be handled without specialized experience or skills. Nor was training of the type imparted by MCED—two-week residential programs focusing on entrepreneurial skills—particularly useful in its entirety. A few respondents found marketing and bookkeeping skills useful. Both poor and nonpoor beneficiaries favored individual loans under the MRCP for productive purposes—including nonfarm activities—on the grounds that SHG loans were much too small and interest rates much too high.

Some additional insights into why nonfarm activities have not expanded much under the MRCP emerged from the responses of the members of the two control groups (one for individual beneficiaries and another for SHGs). While they were well informed about profitable nonfarm activities, their failure to participate or expand essentially reflected their failure to obtain loans individually or as members of SHGs. The failure to obtain loans was attributable to skepticism of the bank staff, ignorance of bank procedures and repayment options, and inability to raise seed money. Thus, lack of capital was a major impediment. In addition, attention was drawn to limited possibilities for group enterprises. The difficulty of getting finance for second-hand equipment was an additional deterrent. Finally, a few respondents were apprehensive of exploitation by unscrupulous middlemen.

Significantly, several poor beneficiaries had concrete expansion plans designed to exploit the potential for growth either through upgrades of the present enterprise or through diversification into related activities. A major hindrance was their inability to get another loan under the MRCP before repaying the earlier loan. In sharp contrast, although the relatively affluent had equally clear-cut expansion plans, some were unable to do so because other family members/relatives were not willing to assist them.

AUTONOMY AND EMPOWERMENT OF WOMEN

In male-dominated households, targeting credit on women could have undesirable effects. In particular, women may be used as conduits for obtaining credit with no role whatsoever in the choice of assets and disposition of income accruing from them. In extreme cases, failure to repay loans aggravates domestic violence against women. Our survey throws some light on these issues.

Among the four poor female beneficiaries, one (a widow) had complete autonomy in the selection and purchase of assets. The remaining beneficiaries, however, confirmed that the decision was a joint one, and in two cases involved the husbands and in one, the son. Thus, at best these women had partial autonomy in such decisions. Also, for the four nonpoor female beneficiaries, except for one who sold bangles, there was limited autonomy as the decisions were made jointly with the husbands. In the exceptional case (i.e., the sale of bangles), the autonomy was understandable, given the nature of the product.

Even if the decision to buy an asset was a joint one, female autonomy could assert itself through an exclusive ownership right. If the asset was registered in the name of the woman, her ownership was established. Among the four poor female beneficiaries, only the widow had the shop registered in her name. In another case, the snack shop was located on land registered in the husband's name. For the remaining two cases, no information was available. Among the nonpoor, a tailor had the sewing machine registered in her name, while the sweet oil seller did not think it necessary to have the business listed in her name. The woman who sold bangles did not register her business, while the fourth respondent did not provide any information. Thus, the assertion of female autonomy through ownership of assets was largely missing.

Except for the widow, all other poor female respondents confirmed that the income accrued to the family. Among the nonpoor, the sweet oil seller asserted her control over the income. Two others responded that it belonged to the family; the fourth did not comment. This is yet another sphere of decision making in which a limited role for women could not be ruled out.

More often than not, repayment obligations were the woman's responsibility, since she was the borrower. An exceptional case was that of the frame maker, as repayment of the loan was a joint responsibility of the mother and son. Among the nonpoor too, joint responsibility for repayment of loans was uncommon.

Four out of the eight female beneficiaries (two poor and two nonpoor) confirmed a more than moderate increase in their household incomes and an improvement in their standard of living. Moreover, since they were seen as possessing skills that could contribute to household incomes, their views were accorded greater importance in household decision making. Male attitudes to income-enhancing activities of women changed significantly from indifference or skepticism to active support within a few years. As some of the income-enhancing activities (e.g., selling snacks and bangles, tailoring) could be easily combined with household chores, the latter were not neglected or delayed. Although this meant extra work for the women, they accepted it willingly.[28] No cases of violence against them were reported, as none defaulted. Some positive externalities of greater self-confidence and assertiveness were exhibited among them. A few women became more actively involved in the panchayats, some fought for banning liquor consumption, and yet another group advocated for better hygiene and sanitation.[29]

AWARENESS, SKILLS, AND TRAINING

Attention has already been drawn to the lack of awareness among the poorest about procedural requirements for individual as well as group loans—a key factor in their exclusion from the MRCP. By contrast, all except one nonpoor participant were well informed. Among the representatives of SHGs (FS), who were mostly nonpoor, all except one were well informed about individual and group borrowing options, repayment obligations, and other procedural requirements.[30]

Except for a few respondents, who belonged to the control group for individual beneficiaries, most of the others were unenthusiastic about skill acquisition through specialized training. Some had prior experience or benefit of family involvement in the same activity (barber); a few others (a tailor and a bangle seller) were trained by NGOs; and some acquired necessary skills through their own initiative (a snack seller who quickly learned how to prepare the snacks). Those (three poor female [individual] beneficiaries) who underwent residential training programs organized by MCED found two modules particularly useful—accounting and marketing procedures/skills.[31]

NGOs were sponsored by MAVIM or by the participating banks to nurture SHGs in the initial stages.[32] Sahyoginis were the link between NGOs and SHGs. Sahyoginis were trained by MAVIM (as well as by NGOs). They perform a wide range of functions such as awareness building, attending group and block meetings, maintaining records, training women in nonfarm activities, and arranging trips to banks. Much of awareness building/training was imparted through group discussions and demonstrations, street plays, and cultural programs. The duration was usually short, one to three days. Despite the short duration, however, the usefulness of the training was borne out by some respondents. Whether this type of training in its present form was appropriate for the poorest segment (e.g., tribals) was doubtful, given their special requirements.

MCED had a different focus. It trained entrepreneurs and assisted them through market surveys, documentation, and management counseling. Usually, the training was given in groups of 30 to 60 persons. Training was residential and lasted 12 to 13 days. MCED had no previous experience of working with the rural poor (the MRCP being a new activity in its portfolio); it had limited ability to design suitable training modules for the very poor.

MITCON's major function was to provide quality consultancy for small and medium enterprises at affordable fees. Specifically, project reports in Marathi, prepared by its staff, were supplied to potential entrepreneurs. After a project was selected, MITCON was supposed to monitor its progress. Shortage of field staff was a constraint. Slow growth of nonfarm activities was attributed to illiteracy, lack of entrepreneurial skills and aptitude, and low marketability of products/services.

Residential entrepreneurial development programs offered by MCED and project reports prepared by MITCON were not just too demanding but also inappropriate for large sections of the rural poor—especially women—lacking literacy, numeracy, and exposure to markets. The training provided by NGOs/sahyoginis was, however, more useful because of its emphasis on problem-solving skills in a village environment. However, the sahyoginis were overburdened as they covered a large number of SHGs. Also, since the remuneration was not linked to performance, no incentive was provided for them to induce the poorest to participate in the scheme in question.

CONCLUSIONS

Two new village level institutions, the village development assembly (VDA) and village development council (VDC)—especially the latter—were given key roles in the selection of beneficiaries and monitoring of loans. In most cases the returns were sufficiently high to cover the repayment requirements. Average returns would be higher for group activities if the skills of group members were complementary and there were economies of scale in production/marketing. Moreover, whatever the variability of returns, the sharing of risks would limit welfare losses.

Appropriateness of the size of SHGs and their regulation are related issues. If an SHG has more than 20 members, it will attract the provisions of Section 11(2) of the Companies Act of 1956.[33] As a result, SHGs usually consist of 15 to 20 members, so whether this is an optimal group size is not self-evident. An option is mandatory registration of SHGs—either as a partnership, a trust, a cooperative society, or a company with built-in reporting requirements, as described in Table 10.1. Some modifications of the existing legal provisions would, however, be necessary to preserve the autonomy and flexibility of SHGs. In particular, careful attention must be given to registration and reporting requirements under various forms in line with minimal but effective regulation. Or, alternatively, self-regulatory mechanisms could be strengthened (e.g., an external annual audit could be mandatory).

A concern, however, was whether group cohesiveness was synonymous with homogeneity of social background (e.g., caste affiliation) and/or economic status. The MRCP guidelines (as well as those for the bank linkage scheme of NABARD) emphasized group homogeneity to avoid domination of the poor by the nonpoor. In the present sample, however, mixed groups performed well. This suggests that under certain conditions social heterogeneity was not a barrier to group cohesiveness. An issue therefore was whether the village environment—i.e., whether there was a great deal of intermingling of different castes or groups in village events such as a major festival—had a role in it. In this context, of particular significance was the formation of other SHGs induced by the existence of a few successful SHGs in the same or a neighboring village over a period of time (or, in other words, dynamic group network externalities). The potential of such dynamic externalities for enhancing the cost effectiveness of microcredit interventions along the lines of the MRCP may well be substantial.

In Phase I, a few commercial banks participated in the MRCP. Their limited presence in rural areas (one-person branches in selected villages) impeded direct contact with the poor. Moreover, the absence of competition inherent in the service area approach meant that few incentives existed for the banks to perform better.[34] In Phase II, however, regional rural banks

TABLE 10.1. Registration options for SHGs.

Options	Advantages	Disadvantages
As trust		Members of a trust cannot become staff of the trust or in any way derive income from the trust's activities.
As partnership firm		A firm with more than 20 members will attract the provisions of Section 11(2) of the Companies Act.
As society	Membership in the society can be open to all those who subscribe to the business of the society.	Society cannot declare any dividend or distribute profits to its members. As and when the society is dissolved, the common property cannot be distributed among its members. The SHG may thus lose the savings kept in the corporate fund.
As cooperative society	Registration in accordance with Cooperative Societies Act of concerned state. Cooperatives provide for equal rights to members for participation in decision making.	Control by Registrar of Cooperative Societies and Government Regulation over functional autonomy.
As company		Registration procedures are lengthy and time-consuming.

Source: Based on Karmakar (1999).

(RRBs) and district cooperative credit banks (DCCBs) were allowed to participate.[35] Although competition among financial institutions following the participation of many more banks in the MRCP was likely to result in better coverage and more efficient financial operations, some specific concerns remain.

It is not obvious why the interest rates charged by SHGs are so high—especially because the transaction costs of lending within the group are so much lower. Even if the interest rates are expected to fall, as they have with the maturing of SHGs in Tamil Nadu, they might still be an impediment if the returns on investment on nonfarm activities fall somewhat rapidly over time.

A major concern is whether the financial infrastructure created is sustainable. Once linked to a bank, the SHGs may continue to borrow, as transaction costs are low and repayment rates are high. However, the poorest SHGs, which did not get linked, might face difficulties in the absence of NGO support.

Chapter 11

Small Enterprises Development Project in Bangladesh

Rushidan Islam Rahman

INTRODUCTION

The microfinance institutions (MFIs) in Bangladesh have adopted the strategy of lending mainly for the rural nonfarm (RNF) sectors. Most MFIs target only the landless households. The absorptive capacity of such households is small and so are the loan sizes. The upper limit of "general" loans for most MFIs is less than 20,000 taka. Therefore there is a niche between MFIs and the commercial banks (who provide large loans, usually above 5 lac taka [$9,000]), which remains unused. This part of the market may be utilized by specialized financial institutions. The Small Enterprises Development Project (SEDP) has come forward to develop the RNF sector through the provision of financial and other services to rural households who are better off than the households targeted by MFIs but do not have access to commercial banks. SEDP provides credit to both landless and landowning households. In that respect it compares with types of activities pursued by households with different initial conditions.

The SEDP study is based mainly on primary data collected through a survey of SEDP entrepreneurs as well as on discussions with the project director of SEDP and a few field-level officers. In addition, secondary data from the reports of SEDP and other available studies have been used.

The questionnaire-based survey was conducted among the SEDP entrepreneurs in Trishal thana of the Mymensingh district. Trishal is not included in a municipality, which is usually defined as an urban area. Fifteen percent of the borrowers were randomly selected, giving 43 RNF entrepreneurs. They lived in eight villages around the thana center, at various distances up to 12 kilometers. The thana center is connected with the district headquarters by a hard surface road which also connects it with the capital city. The villages in which the entrepreneurs live are connected with the thana center by all-season roads.

BASIC FEATURES, TARGETS, AND ACHIEVEMENT

The objective of SEDP is to develop entrepreneurship and to support small-scale, labor-intensive nonfarm enterprises. SEDP credit targets a special group of existing as well as new entrepreneurs in rural and semiurban areas in a region in Bangladesh, the Faridpur and Mymensingh districts. The enterprises are expected to use local resources, technology, and manpower. Women entrepreneurs' participation is especially encouraged. SEDP provides credit and training. Credit use is closely monitored to ensure the success of entrepreneurship building and timely loan repayment.

Enterprises financed by SEDP are located in semiurban areas and in market centers. The borrowers live in semiurban areas and villages surrounding market centers. One reason for choosing the SEDP clients from the semiurban market centers is that it is easy to locate them and to monitor their activities.

SEDP has been initiated in the Faridpur and Mymensingh districts with quite different characteristics. Mymensingh is close to the capital city, with good communications; the other, Faridpur, is far from the metropolitan areas and does not have a well-developed physical infrastructure.

Since the SEDP clients desire loans larger than those offered by MFIs, and smaller than the sizes provided by commercial banks, the minimum and maximum loan sizes were fixed at Tk 5,000 and 500,000, respectively. SEDP credit is channeled through local branches of Agrani Bank (one of the largest nationalized commercial banks). Fifty percent of the fund was provided by the donor and the rest by Agrani Bank.[1]

SEDP's organizational structure consists of a head office with the project director and two district offices. Although the head office is responsible for overall implementation of the program, the loan processing and disbursement takes place at the district offices and local branches of Agrani Bank. At the district SEDP offices, the district project coordinator (DPC), with assistance from the field officers (FOs), identifies, assesses, and approves the selection of borrowers. The local Agrani Bank branch then examines the loan applications and approves and disburses the loans.

The branch manager of the bank usually accepts the SEDP office's recommendation. The FO initially identifies and assesses the credit worthiness of the borrowers and this then is finally examined and approved by the DPC. In rare cases when branch managers do not have full confidence in the SEDP's field officers, they may make their own assessment of borrowers. Some duplication of effort can occur. After loan disbursement, FOs monitor the use of the loan and the SEDP office uses its influence to ensure a timely loan repayment. Some risk occurs in the division of responsibility, but so far this has not been a problem.

LOAN DISBURSEMENT, COST, AND PERIOD OF LOANS

SEDP has set goals for the number and amount of loans to be disbursed during a period.

The target number of loan disbursements has been exceeded every year. Achievement has gradually increased over the years. In contrast, the total amount disbursed did not achieve the target in 1998 and exceeded the target in 1999. The amount of loans disbursed increased from Tk 56 million in the first year to Tk 174 million in the fourth year, an increase of 46 percent over the four-year period. The cumulative target number of disbursements for the five-year period was 11,160, which was already exceeded within the first four-year period.

The overfulfillment of the loan target number indicates that sufficient enthusiasm exists among the small entrepreneurs. No record is available to document the number of applicants from which these are selected, but according to the thana project officer and the district project coordinators, the enthusiasm is high. It should be mentioned that there is no specific requirement about the length of experience of enterprises for obtaining SEDP loans.

The overachievement of the target number of entrepreneurs by SEDP should be cautiously evaluated in relation to loans disbursed vis-à-vis target. Regarding entrepreneur development prospects, the following observations can be made on the basis of Tables 11.1 and 11.2.

The larger the targeted number of loans, the smaller the average loan size, which is less than the targeted average size. Thus the number of loans must be increased to achieve the target of the total disbursement.

Achievement of the total amount of targeted disbursement through a larger number of loans implies

- higher management cost of disbursement per taka, and
- expansion of existing industries and fewer new enterprises that are likely to require larger investment.

TABLE 11.1. A comparison of target and achievement numbers and amounts of SEDP loans.

Year	Number/amount	Target	Achievement
1999	Number	2,280	4,369
1999	Amount (000 Tk)	168,000	174,556
1998	Number	2,400	4,032
1998	Amount (000 Tk)	168,000	150,162

Source: Annual Report, SEDP (various years).

TABLE 11.2. Progress of SEDP loan size: 1996-1997 to 1999-2000.

Year	Targeted average loan (Tk)	Average amount of loan per borrower (Tk)	Percent increase in average loan size over previous year
1996-1997	62,000	26,034	–
1997-1998	–	32,120	23.38
1998-1999	71,000	37,202	15.82
1999-2000	78,000	39,663	6.62

Source: Calculated from SEDP (2000) and Midterm Review, SEDP.

Throughout the existence of SEDP, the average loan size has been much smaller than the target, and is at least partly due to the lack of realistic market assessment. No systematic market assessment has been made before targets are set.

A small loan in relation to the total amount of fixed and working capital of an enterprise involves a smaller risk. The same is true for loans to existing enterprises, which are generating profits and have excess capacity. This in turn implies that loan repayment performance will be better for such small loans. Good repayment performance is one of the most important criterions for assessing the efficiency of the credit program. Therefore, it is not surprising that the program managers prefer small average loans and existing enterprises as clients. Borrowers also prefer to make small loans.

The larger loans involve higher rates of interest. The rate of interest is 10 percent for loans Tk 5,000 to 50,000; 12 percent for loans Tk 50,000 to 250,000; and 14 percent for loans Tk 250,000 to Tk 500,000.

Larger loans involve complicated administrative procedures. Those above Tk 25,000 require approval by the project coordinator and the zonal manager of Agrani Bank. Loans above Tk 250,000 require collateral and insurance, thus making the loan sanction process lengthier and costlier.

The borrowers obviously prefer small loans in view of lower cost per taka borrowed. Even if they require a larger amount for investment, they may prefer to take two or more small loans consecutively for short durations (the sanctioning process is short).

It should also be noted that the average loan size is not substantially higher than the upper limit of loans from MFIs. In view of this, the need for increasing the lower as well as the upper ceilings of SEDP loans is being recognized and is expected to be implemented in the next phase of the project.

There has been no specific target for the gender composition. The gender distribution of loans is shown in Table 11.3. In 1999, about 25 percent of the total borrowers were women, but only 17 percent of the loans were given to women, thus giving a much smaller average size of loan. This figure contrasts with the gender distribution of loans from MFIs, in which 80 to 90 percent of the borrowers are women.

Initially a few women have come forward to apply for SEDP loans. Since SEDP loans are larger, they are not suitable for the poor and landless women who borrow from NGOs, but in nonpoor households, women's status and the impact of such loans on their status are sensitive issues. Therefore, the thana project officer and field officers frequently try to convince the family enterprises in which women provide inputs to borrow in the name of women. No direct incentive is provided to women borrowers (i.e., lower rate of interest) or to the SEDP/Agrani Bank staff to provide loans to women.

The survey has revealed the following specific problems that act as constraints on female borrowers.

In many families, it is not socially acceptable for women to borrow money.

When a large loan is needed, collateral must be provided, and land is the most common form of collateral. Land is usually owned by males and therefore women cannot provide collateral.

In many cases, money borrowed by women is handed over to male family members who may not make timely repayments. SEDP officials cannot hold the men responsible for repayment because they are not the legal borrowers. Similar situations arise in the case of the female borrowers of MFIs, in which women constitute the majority of borrowers. In the case of MFIs, the female borrowers often repay from other sources of their earnings. This is not possible with SEDP where the loan size is much larger.

SEDP has a fixed interest rate policy, specifying varying rates for three different loan sizes. Few loans above Tk 250,000 are sanctioned and most loans are between Tk 25,000 and Tk 50,000. Therefore, the weighted average interest rate is close to 10 percent. This is a highly subsidized rate in comparison to commercial bank rates and those of MFIs. Commercial bank

TABLE 11.3. Gender distribution of borrowers of SEDP from 1999 to 2000.

Gender	Total loan amount (Tk 000)	Percent	Number	Percent
Female	29,587	16.95	1,147	26.09
Male	144,969	83.05	3,222	73.91
Total	174,556	100.00	4,369	100.00

Source: SEDP (2000).

loans usually bear interest rates in the range of 14 to 18 percent. Effective interest rates charged by the large MFIs range from 20 to 25 percent.[2]

Lower interest rates on smaller loans from SEDP may be justified on the grounds that new entrepreneurs are likely to receive these loans. New enterprises may not generate a high return immediately. Moreover, if the loans are used for activities requiring fixed investment, the profitability may be low during the initial years. However, the background information on the enterprises receiving loans, presented in the following chapter, indicates that few loans went to new entrepreneurs.

Interest rates are low, and no collateral is required for loans up to Tk 250,000. SEDP rules for loan processing are also easier compared to the normal rules of commercial banks. About 20 days are required for the loan-sanctioning process. No additional costs such as extra legal payments are required. During informal discussions, clients expressed the view that there is no corruption in SEDP. This aspect is also emphasized by the project director.

For the borrowers surveyed in the present study, the average cost of processing a loan from SEDP was Tk 877, which is 2.16 percent of the average amount of the loan (Table 11.4). Larger loans involve slightly longer processing times, but the cost of processing as percent of the loan varies with the loan size. Lower cost is due mainly to less paperwork; there is no need for repeated visits to the bank branch because SEDP's FO performs the role of a link person.

Thus, the SEDP borrowers are privileged in comparison with the clients of commercial banks or MFIs. Questions may be raised about the appropriateness of such a highly subsidized rate of interest charged by SEDP. A higher rate of interest may provide more rational guidelines for channeling funds to the RNF. Low interest rates may generate large demands for funds, which may be channeled to inefficient uses and may also result in funds' diversion for purposes other than investment. Higher interest rates are primarily used by MFIs to meet high administration costs and social mobilization

TABLE 11.4. Cost of obtaining an SEDP loan by the size of loan.

Last loan amount from SEDP	Amount spent for preparing papers	Days required	Average loan	Cost as percent of loan
1-25,000	537.33	17.33	22,733	2.36
25,001-75,000	840.77	21.69	40,692	2.07
75,001+	3,900.00	20.00	172,500	2.26
Total	877.21	20.09	40,558	2.16

Source: Author survey.

as well as to earn enough income to achieve financial stability over time. For SEDP, financial viability of the program is not a major concern, because it is viewed as an experiment to derive ways to develop entrepreneurs.[3]

The fund cost is low for SEDP because 50 percent of its resources is provided by a grant from a donor and the average fund cost turns out to be around 4 to 5 percent. However, the administration cost is high because of the need for close supervision. Though the estimated operational cost per taka is not available, one may reasonably guess that it is not likely to be much lower than the MFIs. Thus, for the SEDP loans without collateral, interest rates that are intermediate between rates charged by MFIs, on the one hand, and by commercial banks, on the other, may be appropriate.

The SEDP loans at higher interest rates should still be more attractive than loans from commercial banks because they have other convenient features, including quick sanctioning, no collateral requirement, and low processing cost. For the larger loans, which require collateral and insurance, the interest rates charged by the commercial banks provide the effective ceiling. A discount on the interest rate may be offered to new enterprises/entrepreneurs to encourage them to invest. Even at currently prevailing low rates, only a few new entrepreneurs came forward to borrow. Therefore the interest rates for new borrowers may be lower than those charged to existing enterprises.

In SEDP, the loan period is not fixed and ranges from one to seven years. The grace period after which repayment begins also varies depending on the activities' gestation period. The repayment schedule also varies and is usually made in monthly installments. In practice, variations are not large. In the present sample, loans for two- and three-year periods have been observed (Table 11.5). The two-year term is used for service sector loans. For other nonfarm activities, the period is three years. The shorter loan duration for the service sector is based on the hypothesis that the return rate for such activities is high. The repayment period of three to four years is not conducive for fixed capital investment. To encourage such investment, the repay-

TABLE 11.5. Repayment system for SEDP loans.

Last loan amount from SEDP	Months till repayment begins	Period of repayment in months
1-25,000	1.00	35.20
25,001-75,000	1.00	36.00
75,001+	1.00	36.00
Total	1.00	35.72

Source: Author survey.

ment period may be extended up to seven years, which is already acceptable under SEDP's regulations.

Financial stability has not been adopted as an SEDP objective. This concern has not been built into the program design, and the program is dependent on donor funds. MTR (2000) provides some useful recommendations on this subject and suggests that SEDP should aim at cost reduction without increasing the loan size so that it does not lose the focus on small enterprise development. The repayment performance, which now stands at about 94 percent, should improve (SEDP, 1999-2000).

BACKGROUND OF BORROWERS

SEDP was started for the purpose of targeting rural entrepreneurs who do not have access to the prevailing sources of institutional finance. Has SEDP succeeded in reaching this special group?

SEDP's borrowers cannot simultaneously borrow from other organizations that have branches in these areas. The borrowers must provide a certificate to SEDP stating that they do not have outstanding loans with those organizations. This rule is strictly adhered to, but clients of other organizations can repay loans and still borrow from SEDP. If a large-scale shift in the affiliation of borrowers from NGOs or other commercial banks to SEDP takes place, then SEDP may be considered as duplicating efforts.

Survey data reveal that in only 4.5 percent of cases were SEDP borrowers previously members of other NGOs. (One person was a member of BRAC; another was a member of a small local NGO.) Loans from the NGOs were the source of start-up capital for these persons. Information on their level of education, age, literacy, and numeracy is presented in Tables 11.6 and 11.7.

Most of the SEDP borrowers have an average of seven years of schooling. This is substantially higher than the schooling for MFI clients. Eighty-six percent of the borrowers are literate and numerate, much higher than the national level of literacy.

TABLE 11.6. Age and education of SEDP borrowers.

Sex of borrowers	Years of education	Age
Male	7.03	39.82
Female	4.60	34.89
Total	6.74	39.23

Source: Author survey.

TABLE 11.7. Literacy and numeracy among the borrowers.

Sex of borrowers	Possess basic reading and writing skills (%)			Can perform small calculations (%)		
	Yes	No	Total	Yes	No	Total
Male	86.8	13.2	100.0	86.8	13.2	100.0
Female	80.0	20.0	100.0	80.0	20.0	100.0
Total	86.0	14.0	100.0	86.0	14.0	100.0

Source: Author survey.

The clients served by MFIs usually consist of landless and marginal landowning households. In contrast, there is no land-based targeting of SEDP borrowers. Information on the ownership distribution of arable land of SEDP borrowers is presented in Table 11.8. Among the SEDP borrowers, only 34 percent belong to the marginal landowning group. The average size of cultivated land owned by the borrowers is 1.99 acres. Owners of larger size lands received larger loans. About 30 percent of the borrowers are owners of large- or medium-sized lands. Another notable feature is the difference in the landownership of male and female borrowers. Most of the female borrowers belong to landless or marginal landowning groups.[4]

Loan Size Determinants

Previously it was stated that smaller than target size loans are due to cautionary practices exercised by the lending bank and the SEDP management. It may be also partly due to the borrowers' lack of absorptive capacity. However, the following evidence casts doubt on the limited absorptive capacity of the borrowers:

- The rate of return on capital is high in most enterprises. In all cases, it is substantially higher than the rate of interest.
- When the current borrowers are asked whether they would like to borrow more than the sanctioned amount, about 95 percent responded positively.

Data on the replies to this question are presented in Table 11.9. Only 5 percent of the borrowers who have obtained three or more loans from SEDP have given negative replies. It may also be noted that the subsequent loans are larger in size in most cases.

TABLE 11.8. Land distribution of SEDP borrowers.

Landholding (decimal)	Percent of borrowers		
	Male plus female	Male	Female
0-50	32.5	26.3	80.0
51-250	37.2	39.5	20.0
251-500	25.6	28.9	–
501+	4.7	5.3	–
All	100.0	100.0	100.0

Source: Author survey.

TABLE 11.9. Larger SEDP loan survey.

Last loan from SEDP (takas)	Percentage of those wanting a larger loan		
	Yes	No	Total
1-25,000	93.3	6.7	100.0
25,001-75,000	96.2	3.8	100.0
75,001+	100.0	–	100.0

Source: Author survey.

The borrowers' background data show that they are mostly landowning households. Most are literate and have obtained some schooling. Tables 11.10 and 11.11 show that those who obtain more loans and larger loans are better educated and own more land.

Which factors have a significant influence on the amount or size of loan?[5] The results of a regression analysis are presented in Table 11.12. Among the independent variables, age and landownership have significant positive impacts on loan amounts. Other variables include an education dummy for male borrowers and a number of sector dummies. Education has a positive coefficient, but none of these variables are significant.

The older, educated, and larger landowning households obtain bigger loans due to a number of considerations. Experience, reflected in age, enhances the absorptive capacity of the borrower and also increases lender confidence in the borrower. Landownership makes the borrower more confident about ability to repay the loan from alternative sources of income and encourages him or her to borrow a large amount. Moreover, land is the collateral provided by those who borrow an amount above the collateral-free ceiling.

Most of the SEDP-financed enterprises are based on family labor. Among the enterprises surveyed, the number of hired employees is larger than the

TABLE 11.10. Background of borrowers by the amount of the latest loan.

Last loan amount from SEDP	Education	Age	Total arable plus homestead land	Education of wife/husband
1-25,000	5.33	38.20	102.53	3.87
25,001-75,000	7.50	39.35	273.71	6.22
75,001+	7.50	45.50	475.00	4.50
Total	6.74	39.23	223.36	5.25

Source: Author survey.

TABLE 11.11. Background of borrowers by number of loans taken.

No. of loans taken from SEDP	Education	Age	Total arable plus homestead land
1	6.17	38.17	143.00
2	6.96	39.57	238.37
3	7.00	39.88	300.75
Total	6.74	39.23	223.36

Source: Author survey.

TABLE 11.12. Determinants of the amount of latest loan obtained from SEDP: Results of OLS regression.

Explanatory variables	Dependent variable: Last loan amount from SEDP		
	B	T	Sig.
Constant	−14,634.90	−0.47	0.64
Age	1,184.45	1.72	0.09
Education	163.30	0.11	0.91
Cultivable land in decimal before SEDP membership	60.70	2.50	0.12
Sector dummy 1	−17,479.80	−0.75	0.46
Sector dummy 2	−7,798.89	−0.78	0.44
N	43.00		
Value of F	2.41		0.05
R - bar square	0.14		

Source: Survey of SEDP borrowers, February 2001.

number of family workers in 35 percent of cases. If applying the criterion that enterprises hiring more than five regular employees are hire labor-based enterprises, then only 4 percent of the enterprises fall into this category. All of the enterprises use less than ten workers (including both hired and family labor).

Training

SEDP provides two types of training to borrowers: (1) entrepreneurship development training (EDT) and (2) skill development training (SDT). EDT is a prerequisite for obtaining loans and all borrowers participate in this three-day training program at the project offices at Mymensingh and Faridpur. The course includes discussions on the role of bank loans, keeping accounts, and issues of management, etc. The same training method is not suitable for both literate and illiterate (or semiliterate) borrowers. Therefore, separate training sessions are conducted for the two categories of borrowers, even if the basic content is the same.

SDT is not attended by all borrowers. It is attended only by those who are interested or engaged in the activities covered by SDT. The topics covered by SDT include tailoring, dairy, fish culture, poultry, etc. The project director seems to be aware of training the inadequate and the need to provide training in other suitable nonfarm enterprises. Skill training requires specialized resource persons and training materials. The present staff numbers are insufficient for organizing such training. In the future, SDT might be implemented through the involvement of other specialized training organizations.

EDT is viewed as a loan-qualifying process and almost all borrowers attend the course. In the present sample, all the borrowers have attended EDT. In the midterm review, 90 percent of the borrowers attended EDT. In replying to a question about the usefulness of the training, 63 percent of the borrowers affirmed that they found it very useful and 37 percent found it somewhat useful. All women who have borrowed from SEDP found the training very useful.

Most borrowers found that the issues discussed in the training session were familiar and they did not learn much. However, training may still be useful because it gives them a feeling that they are undertaking a larger responsibility. One practical reason for considering the training as unattractive is that it is held at the district town and borrowers must travel a considerable distance and find it difficult to arrange for accommodations. As a result, in most cases the SEDP offices complete the training course within two days, even if it is planned for three days.

Women are more serious about attending the EDT training course. Among all those who have received training, 69 percent are women, whereas among the borrowers only 27 percent are women. The finding on skill training is rather different. Only 13 percent of borrowers (midterm review) have received such training. No borrowers in the present sample have received such training.

Moreover, EDT is not likely to generate sufficient enthusiasm because most of the borrowers are experienced entrepreneurs. EDT may be more useful for borrowers who wish to set up new enterprises. Therefore, SEDP's strategy for its training program will depend on whether the program wants to finance the new or concentrate on the experienced entrepreneurs.

NONFARM ENTERPRISES FINANCED BY SEDP

SEDP has consciously adopted the policy of financing mainly noncrop agricultural and nonagricultural activities (Table 11.13).[7] This table does not provide a representative sectoral breakdown of all loans over the four-and-a-half-year period. From 1998, a deliberate policy was adopted for "not granting loans for trading activity," except to the female borrowers. When SEDP adopted this strategy in 1998, entrepreneurs engaged in trading accounted for

TABLE 11.13. Sectoral distribution of loans from SEDP.

Sector	Percent of total loan	
	1998	1999
Noncrop agriculture and agroprocessing	71.5	74.4
Textile/garments	11.3	10.1
Forestry products	4.4	4.4
Cottage industry	1.0	1.3
Chemicals and pharmaceuticals	2.5	1.2
Engineering enterprises	4.4	3.0
Rural transport	1.1	1.1
Trading	1.3	2.3
Others including service[a]	2.4	2.2
Total	100.0	100.0

Source: SEDP (1999-2000).

[a]Service sectors include hairdressers, training (computer), etc.

35 percent of the borrowers (MTR, 2000). SEDP's decision to not lend to those involved in pure trading activity is based on two considerations:

1. This sector is more or less saturated; high profitability of such activity may simply mean that SEDP clients are taking a share of the existing market away from others.
2. SEDP intends to encourage rural industries as an objective of the program.

SEDP decided to increase the number of loans to the entrepreneurs engaged in production and service. Even after adoption of such guidelines, the extent of sectoral diversification of loans was rather limited.

In the official reports of SEDP, noncrop agriculture and agro-processing loans have been grouped together. Agro-processing includes rice milling, production of poultry feed, etc. In the present survey, agro-processing has been kept separate. This survey shows that, among the nonfarm production activities, two manufacturing subsectors dominate both in terms of the number and amount of loans. These are rice milling[8] and garments/tailoring: first, due to the ease of access to raw material, i.e., paddy; and, second, due to local consumption demand. It is doubtful whether access to SEDP loans has led to growth in the capacity for paddy processing and rice milling, i.e., whether investment has occurred in making a net addition to fixed capital. If the borrowed funds are used as working capital (for example, for purchasing the paddy), this may actually imply that the business is being captured by the borrowers at the cost of the mills that do not have access to such credit. Some of these concerns may also apply to the garment industry, though in this sector a greater capacity has been added or a greater utilization of existing capacity has led to a growth of aggregate production.

The other notable feature is the relative unimportance of the transport sector. This contrasts with MFI's lending in which rural transport is an important area of investment. The nature of investment required in this sector may explain the difference. Most of the microcredit recipients investing in the transport sector are practically owner-renters of rickshaws or are owner–rickshaw pullers. The total investment required for a new rickshaw ranges from Tk 3,000 to 8,000. These are less than the effective floor of the SEDP loan size. Moreover, the borrowers of SEDP do not belong to the group that is engaged in physical labor such as rickshaw pulling, thus ruling out investment in rickshaws.

Mechanized transport may be a useful area of investment of SEDP. However, this would require a large loan size close to the approved ceiling or higher, and possibly would form an investment by new entrepreneurs. SEDP management may be hesitant to provide such large loans to new enterprises.

New enterprises provide an impetus to the growth of the rural economy as well as avoid a concentration in the hands of the existing entrepreneurs. Data from the present survey reveal that SEDP finances few new entrepreneurs. Out of 45 RNF enterprises financed by SEDP, only one (less than 3 percent) is a new project, a photocopy and computer service enterprise. The midterm review by SEDP observes that 7 percent of the SEDP loan recipients are new entrepreneurs. The reason for this discrepancy between the current survey and the midterm review is due to the inclusion of noncrop agricultural activities such as poultry, livestock, fish culture, and power tiller purchases in midterm review, etc., which were excluded from the present survey.

A number of reasons are cited for preference given to the existing enterprises. These are:

It is easier to identify the promising enterprises when these are actually in operation. Whether the entrepreneur is able to use additional finance for increasing the activity scale and/or using excess capacity can be identified through actual observation of the enterprise.

There is less (or no) fear of business failure. Even if the marginal return from the additional investment is low, a loan may be repaid from current earnings.

When substantial equity capital is available and an enterprise has been in business for a number of years, it is very likely that the entrepreneur will not abandon the business or change the location. With new entrepreneurs, lending without collateral is considered somewhat risky (loans up to Tk 250,000 are collateral free).

In the subsectors that have a number of enterprises already in operation, there are informal barriers to entry for the new enterprises. For activities requiring shops/factories in the existing rural/semiurban market centers, a separate operational location must be found as well as a space or structure for selling goods and services. It may also be difficult to secure a space either for purchase or for rent. In some thana centers, open spaces owned by the government are leased out by a "market committee" composed of local elites, formed by local administration. These spaces are rented cheaply to temporary hawkers/sellers on weekly market *(haat)* days. This may not be a suitable arrangement for larger businesses that are financed by SEDP. Similarly, small temporary shops (if not authorized on public *[khas]* lands beside public roads) may provide space for a nonfarm enterprise. This, however, requires connections or influence with the officials in charge.

For home-based enterprises, entry may be easier. That is why most of the new enterprises financed by SEDP concentrate on home-based poultry or livestock raising and fish culture. Home-based paddy processing and trading, without ownership of a rice mill or shop, is another such activity that new enterprises may enter.

The bank and the project personnel consider the fulfillment of the loan disbursement target and the timely repayment of loan as the most important criteria for the program's success. Providing loans to new entrepreneurs is thus given secondary importance. Use of the loan by new entrepreneurs also requires more rigorous skill training and management training on their part. SEDP is currently not in a position to provide such an effective training program. Moreover, to identify new entrepreneurs and to encourage them to invest requires a great deal of initiative and risk taking by project personnel, which is not usually expected from temporary project officers. An easier option for them is to provide loans to existing enterprises because this enables them to fulfill the targets.

Since all the enterprises in the sample except one were in business before obtaining an SEDP loan, they were asked about the source of start-up capital. In 88 percent of the cases, their own savings were the source of capital. In 49 percent of the cases the "father" was the source of capital; in 14 percent of the cases the "father-in-law" was the source of capital. Moneylenders are also an important source of capital. Only one person obtained start-up capital from an NGO and none borrowed from a commercial bank.

Most of the SEDP borrowers were above the poverty threshold. Precise estimates of pre- and post-SEDP household income data were difficult to obtain. The purpose of an SEDP loan is not to lend directly to the poor and to have an indirect impact on their poverty. The loans are made mostly to the nonpoor, who provide employment to the poor in their enterprises. Only 5 percent of the borrowing households were at or below the poverty threshold.

USE OF AND RETURNS ON SEDP LOANS

Table 11.14 indicates the distribution of loans among various uses. The overwhelming proportions of loans, about 90 percent, were used for meeting working capital requirements. In a small number of cases, part of the loan was diverted to house building. None of the borrowers used it for consumption purposes.[9] However, without new investment in expanding productive capacity, the absorption of working capital is likely to reach an up-

TABLE 11.14. Distribution of SEDP loans by type of use.

Use of loan	Percent of borrowers[a]	Average amount (for all borrowers)	Percentage of loan spent for each purpose
Fixed capital	16.3	3,884	9.58
Current capital	90.7	35,163	86.70
House construction/ repair	9.3	930	2.29
Others	4.7	581	1.43
Total	–	40,558	100.00

Source: Survey of SEDP borrowers, February 2001.

[a]Some borrowers used the loan for more than one purpose, therefore the total will be larger than 100 percent.

per limit within a short period. Therefore, ways must be found to accelerate fixed capital formation to expand lending. Using the current data set, an OLS regression equation was estimated to identify some of the factors contributing to fixed investment (Table 11.15).

The equation uses the total loan amount used for fixed capital as the dependent variable. Explanatory variables include age and education of the borrower; area of cultivable land owned by his or her household; number of loans taken from SEDP; dummy for gender (male = 1); and two dummies for type of sector (one for all manufacturing and the other for service).

Variables, which turned out to be significant and positive, are education and the dummy for the manufacturing sector. The coefficients of the dummy for the service sector and for male borrowers are negative and significant. Other variables have insignificant coefficients. The negative coefficient of ownership of cultivable land deserves attention. The large landowning groups, as expected, are not interested in investing in nonfarm capital. Expansion of the nonfarm enterprises through investment in fixed capital takes place only when entrepreneurs are educated or are engaged in the manufacturing sector. Educated entrepreneurs make new investments in fixed capital; this may be due to their better managerial ability.

The return per unit of loan capital obtained from SEDP, after paying the costs of labor and other inputs (including family labor valued at average wage rate in the RNF activities) should be higher than the rate of interest paid (10 percent per year in 96 percent of cases). However, return to the entire (borrowed and owned) capital may not accrue at this rate. A much smaller rate of return from the fund supplied by the entrepreneur may be acceptable if this comes from the entrepreneur's savings or family sources,

TABLE 11.15. Determinants of the amount of SEDP loan used as fixed capital.

Explanatory variables	Dependent variable: Loan used as fixed capital		
	B	T	Sig.
Constant	885.95	.12	.91
Age	118.13	.71	.49
Education	844.10	2.44	.02
No. of loans taken from SEDP	622.66	.53	.60
Cultivable land in decimal before SEDP membership	−3.40	−.57	.57
Sector dummy 1	27,569.56	4.97	.00
Sector dummy 2	−4,668.54	−1.99	.06
Male	−7,945.72	−2.09	.04
N	43.00		
Value of F	5.84		.00
R - bar square	.45		

Source: Survey of SEDP borrowers, February 2001.

because the alternative use of such savings is not easy in view of limited saving instruments available for them. An estimate of the cost of entrepreneur's own capital, which is the same as the rate of return on the alternative use of his own savings, can be obtained. This may be assumed to be equal to the average of (a) the rate of borrowing from SEDP and (b) rate of interest on saving deposits of commercial banks. The weights for calculating the average are relative proportions of SEDP loans and entrepreneur's own savings.

The appropriate wage rate that should be used for calculating the returns to family labor enterprise can never be satisfactorily answered when the market for unskilled labor is characterized by unemployment, underemployment, and seasonally surplus labor on the one hand, and a scarcity of entrepreneurs on the other. Moreover, many rural women workers, especially from well-off households, prefer self-employment and may not participate in the labor market at any wage rate close to the current wage. In this case, any positive return to labor from self-employment may appear worthwhile for the female workers, especially in the households that own some land and/or live above the poverty threshold. The SEDP-financed entrepreneurs come mostly from these households. Therefore, it should be pointed out that the returns to capital obtained after calculating market wage rate to family labor results in underestimation of the returns.

Several interesting features emerge from the frequency distribution of return rates as given in Table 11.16. Profitability is high in most of the enter-

TABLE 11.16. Average return from SEDP activities by subsector.

Sector	Rate of return (percent)
Agricultural-based processing industry	125
Food industry	147
Textile/garment	102
Cottage industry	414
Chemical industry	100
Other industry	110
Service	80
Big shop/trading	75
Medium shop/trading	111
Others	215
Total (average)	128*

Source: Survey of SEDP borrowers, February 2001.

*Weighted average of returns from all sectors.

TABLE 11.17. Average return from activity financed by SEDP loan by the number of loans.

No. of loans	Return (%)
1	166.91
2	112.46
3+	115.14
Total	128.15

Source: Survey of SEDP borrowers, February 2001.

prises. The average rate of return is about 128 percent. The lowest return is 75 percent. Table 11.17 compares the average return for first-time borrowers and older borrowers. Those who borrowed once have a higher return compared to those who obtained two or more loans.

With the assumption of a lower opportunity family labor cost (based on the current wage rate and a probability of unemployment)[10] the rate of return to capital is slightly higher than the rate of return presented previously. The differences between the rates of return calculated on the basis of the two types of opportunity cost of labor are small.

The rate of return from most activities is much higher than the rate of interest charged by SEDP. Highest returns accrue from cottage industries and restaurants, followed by the food industry. In contrast, the service sector and large trading yield low returns (Table 11.16). The question then is whether an excess demand exists for such a loan. Also, as is seen in Table 11.18, most borrowers have applied for a larger loan, which conforms to the findings on high rates of return.

CONCLUSIONS

The SEDP experience apparently has proved to be a success. Loan targets and amounts have been overfulfilled. The loan recovery rate is about 93 to 94 percent and is thus satisfactory. Loan use is closely monitored, and as a result, most of the loans have been utilized in productive activities.

Most of the loans are used in existing enterprises. If only rural nonfarm activity (RNA) is considered, less than 5 percent of SEDP-financed enterprises are new. For all borrowers, new enterprises constitute 7 percent of the total.

About 75 percent of the loans are given for (1) noncrop agriculture and (2) agro-processing. If half of them are assumed to have been invested in noncrop agriculture, then less than 65 percent of all loans have gone to RNA (if agro-processing is included in RNA, but noncrop agriculture is excluded).

Only a small percentage of loans are invested in fixed capital. About 87 percent is invested in capacity utilization in existing enterprises and 10 percent is devoted to fixed capital formation.

The rate of return on capital in SEDP-financed enterprises is much higher than the rate of interest on SEDP loans. This indicates that other constraints act as barriers to entry. Lack of technical and managerial skill may be important reasons that explain the lack of growth by new enterprises. In addition, there may be insufficient demand for nontraditional products.

Given the high return rates from most activities, one may ask why the number and average size of loans for similar activities do not increase. At present, SEDP is not expanding the loan size fast, nor are the new enterprises borrowing to any large extent. SEDP personnel believe the main constraints lie with both the borrowers and SEDP. On the one hand, the pool of

TABLE 11.18. Survey of borrowers who wanted larger SEDP loans.

| No. of loans from SEDP | Percentage of borrowers desiring larger loans | | |
	Yes	No	Total
1	100.0	—	100.0
2	100.0	—	100.0
3+	75.0	25.0	100.0
Total	95.3	4.7	100.0

Source: Survey of SEDP borrowers, February 2001.

the prospective entrepreneurs with technical and managerial skills has been exhausted. The EDT and SDT programs are not able to develop new entrepreneurs. However, the market for most of the activities described previously is rather limited.[11] The declining profitability of additional investment is demonstrated by the lower return for a larger number of loans (Table 11.17). In fact, to overcome the drawbacks of a limited market in the next phase, SEDP is likely to start functioning in new locations rather than expanding activities in the existing locations.

Chapter 12

A Study on Grameen Uddog, MEDU, and Kishoreganj Projects

Sajjad Zohir

INTRODUCTION

A number of microcredit programs have been tested within Bangladesh. Broadly, they may be identified as follows:

1. NGO/MFIs disbursing credit from own/donor funds and group savings to support nonfarm income and employment-generating activities at the beneficiary level.
2. NGO/MFIs engaging in commercial activities and establishing business relations with their beneficiary groups, eventually leading to expansion of rural nonfarm activities.
3. Link programs between grassroots-level organizations (formed under a project or by an NGO/MFI or otherwise) and commercial banks to facilitate rural banking for promoting nonfarm activities.
4. Public-sector banks opening up special credit lines to support microenterprises that are expected to promote rural nonfarm activities.

The first of these programs has been widely dealt with in the literature. The current study is therefore confined to programs 2, 3, and 4. First is the specialized case of Grameen Check, which represents the second variant. We choose the Kishoreganj Sadar Thana (KST) project of the United Nations Development Programme's (UNDP) South Asian Poverty Alleviation Programme (SAPAP), as an example of a community-based organization linked with Bangladesh Krishi Bank (BKB), and providing credit for rural nonfarm employment activities as well. Finally, the International Fund for Agricultural Development (IFAD) support program, along with the Agrani Bank (commercial bank) to promote microenterprises is an example of the fourth variant. This section describes the three selected projects highlighting the differences in the practices of promoting rural nonfarm employment for poverty alleviation.

GRAMEEN UDDOG: A COMMERCIAL VENTURE TO PROMOTE RURAL NONFARM EMPLOYMENT

Origin of Grameen Uddog

Handlooms account for almost a quarter of total rural nonfarm employment in Bangladesh. The 1989 handloom census puts the employment figure at 1.03 million persons, and the estimated value addition is in the range of Tk 10 billion.[1] The sector has faced persistent problems due to increasing competition from the garment industry and from illegal imports from India. Although clothes are woven on handlooms throughout the country, traditional weaving communities in certain parts have weavers whose inherited skill and whose livelihood depends on the sustenance of the handloom sector. Various nongovernment organizations (NGOs) and microfinance institutions (MFIs) have often encountered problems in assisting communities engaged in specific trade (e.g., fishermen, weavers, potters, etc.). Grameen Uddog's initiative provides an alternative approach to alleviate poverty by integrating the poor communities with the market.

Grameen Bank, established in 1983, has led the way in banking with rural poor by giving collateral-free loans. Over the past decade, a number of sister concerns emerged out of the Grameen Bank (GB) initiative, each addressing a specific area of activity. These include Grameen Krishi (Agriculture) Foundation, Grameen Motsho (Fishery) Foundation, Grameen Telecom, Grameen Cybernet, Grameen Communication, Grameen Sakti (Energy/technology) Foundation, Grameen Uddog, Grameen Samogree (products), Grameen Fund, Grameen Kallyan (welfare), and Grameen Trust.[2] However, except in the cases of telecom, cybernet, and communication, these institutions emerged to meet specific needs of the beneficiary groups of the Grameen Bank, and their activities largely center around these beneficiary groups.[3]

There are two stories about the origin of Grameen Uddog and Grameen Bank Family's (GBF) involvement in the production of Grameen Check. One story suggests that the success of Madras Check in India was brought to the notice of Professor Yunus, who then initiated the process of implementing a similar project within the umbrella of GBF.[4] The other story emphasizes the initiatives of field staffs of Grameen Bank in the Sirajganj district, who were looking for ways to help the weaver community in the area. We assume that there are elements of truth in both the stories. It recognized that the weavers had to incur losses due to their dependence on informal sources of credit for the purchase of raw materials, and had to bear all the risk for marketing the produce.[5] The route sought by the GBF was to act as a marketing agent and organize production through a putting-out system,[6] involving a network of weavers.

Operations of GU

The GU venture started with an initial borrowed capital of Tk 50 million, availed from several commercial banks. By 2001, such borrowed capital stood at Tk 250 million. Major types of fabrics produced were check, stripe, flannel, sheetings, and dobbie. During the early years, it was only the weavers in some selected areas of Sirajganj district who participated in production. The coverage later expanded to Araihazar thana under the Narayanganj district. Production in Sirajganj is organized through six handloom-based units and one powerloom-based unit; one production unit is located in Araihazar.[7]

It is the responsibility of the head office to contact buyers, translate the demand into appropriate designs, and procure yarn and dye materials. Since timely delivery is essential, feedback is received from unit offices (equivalent to production units) via the regional/area offices on ways the work may be shared.[8] The area office coordinates dissemination of technical knowledge, dyeing of yarn and its distribution through unit offices, and procurement of final output. The unit offices are responsible for identifying the enterprises through weaving managers,[9] distributing yarn, supervising work to ensure quality, and procuring the final output.

GU relies on local markets for all the raw materials. It procures yarn from local mills upon receiving work orders. It also buys dye from local markets. During the early years of its inception, GU organized small private dyeing groups in the handloom belt to dye the yarn under GU supervision. Later in 1998, with a view to derive economies from vertical integration, a textile mill was set up jointly with the Gano Sasthya Kendra.[10] The mill was set up to handle three specialized activities: dyeing, washing and finishing, and flannel brushing. Unfortunately, mechanical dyeing proved to be quite costly, and this particular operation was discontinued in the mill. Instead, GU set up its own dyeing unit in 1999, resorting to traditional technology, and since then it organizes the dyeing activity with its own manpower at the regional/area level.

Upon receiving a work order, GU claims to follow a set of criteria in distributing the work at two levels. Since GU activities are largely confined within the Sirajganj district, work distribution across districts is not yet a

major concern. In distributing the work among various unit offices, GU officials take into account the following factors:

1. location-specific skill,
2. location-specific wage rates,
3. idle production capacity,
4. current workload, and
5. past performance of a unit.

In distributing work among various weaver-managers within a unit, priority
is allegedly given to marginal and poor weaver-managers who own five
looms or less. Other important factors considered include skill, current
workload, and past performance in terms of timely delivery or quality output. Information recorded in monthly statements helps ensure equity in
work distribution among the weavers spread across different units.[11]

Weavers' failure to deliver the products on time seriously affects the
time-bound delivery to a buyer. This substantially reduces the chance of
getting a subsequent work order from the respective buyer, which is a threat
to the sustainability of the organization. Sometimes, because of concern
with financial viability, GU must take into consideration the previous record of a unit in supplying the product in time. Such concerns, quite
expectedly, often conflict with the broad objective of poverty alleviation.
This is also reflected in GU's gradual move toward powerlooms, which are
found to be more efficient compared to the handlooms, but are less labor intensive.[12]

Unit-level GU staff are mainly assigned the responsibility of identifying
the weavers who would produce a particular work order. Weavers who own
a number of looms and run these looms with family and/or hired loom
workers are called "weaver-managers" by GU. The loom workers are often
identified as "ordinary weavers," in order to distinguish them from weaver-
managers. GU delegates the work of producing a particular volume of GU
product within a specified period to weaver-managers. Weaver-managers
are responsible for delivering the products to the GU's unit office.

The output market is rather thin on the domestic front. Generally, GU obtains work orders from buying houses located in the country and the garment industry. Currently, 95 percent of GU products are exported, while the
rest are sold through GU's own retail stores.[13] Germany has been the major
destination of GU products, which also acts as the hub for the European
market. Lately, GU products have made considerable inroads to the U.S.
market.

GU's Health Service: Combining Social Concern with Commercial Initiative

Grameen Uddog offers a health insurance plan for its member weavers, including the loom workers. Each participating household is required to pay Tk 60 for health coverage over a period of six months. This enables the members of the respective household to visit a physician at a nominal fee of Tk 5 per visit. Health centers are open twice a week, and a trained physician attends the patients. While GU management claims that the offer is open to all, during our field survey we found only the participants making use of the offer.

It is important to note that GU does not provide any additional benefits to the weavers who work for it, either in the form of training or general education. However, an on-the-job transfer of knowledge occurs in the process of adapting new designs.

Survey Design

In addition to the handloom and powerloom units, one dyeing unit is located in Sirajganj, which has been in operation since 1999. A total of 2,095 weaver-managers have been engaged so far by different production units located only in Sirajganj to produce the stipulated output.[14] As a mirror image of production, Syedabad-Sirajganj and Gopalpur-Belkuchi units have engaged the highest and lowest number of weavers, which is 1,960 and 745, respectively. For all other units, the total number of weavers engaged varies from 1,500 to 1,600. To assess the GU achievements, information was sought at the weavers' level from a group of GU weavers as well as from the non–GU weavers.[15]

The weavers can be classified into three major groups according to their socioeconomic and entrepreneurial status. One group of microentrepreneurs has more than 20 looms and runs a home-based factory-type production unit. These entrepreneurs do not participate in the actual production process and produce Grameen Check or traditional items by employing wageworkers. Another type of home-based production unit has more than one loom and is run mainly by family members along with a very small number of hired weavers. The third group of weavers owns only one or two looms and engages family labor in the production process. These three groups of weavers were included in our survey (Table 12.1).

Since GU's activities are demand driven, its success in generating employment opportunities depends on the total demand for Grameen Check, both in domestic and international markets. Again, GU, being predomi-

TABLE 12.1. Distribution of surveyed weavers, Grameen Uddog.

Thana	Unit	Nature of participation		
		GU weavers	Non–GU weavers	Total
Belkuchi	Chala-Belkuchi	20	5	25
Belkuchi	Bawra-Belkuchi	20	5	25
Shahajadpur	Control area	0	10	10
Total		40	20	60

Source: Author survey.

nantly an exporter, is also affected by different financial policies of the government.[16] GU supplies yarn and other inputs to weavers by buying it from different private sources. GU's failure to collect yarn on time not only seriously affects the time-bound delivery to a buyer, but also reduces the chance of getting the next work order from this buyer.[17]

Findings

There are a significant number of large loom owners among participants, and many small loom owners among the nonparticipants. When asked, the GU officials would persistently assert that GU involves marginal and poor weavers, with rarely anyone owning more than five looms. They note that skill, idle loom capacity, and commitment to timely delivery are emphasized while choosing weaver-managers from the pool of marginal and poor weavers. The GU officials also argue that their practices in respect of input delivery and output collection, which require the weaver-managers to visit unit offices and the regular supervision by GU staff, discourage the middle-class and rich loom owners from joining in their network. Our field survey does not lend much support to this assertion. Although the very rich loom owners do not join GU, a large percentage of those owning 7 to 19 looms participate in the GU program (Table 12.2). The finding is comparable to those generally observed in the field of microfinance: the hardcore poor are left out and the relatively poor segment of the nontarget population joins.

From the GU perspective, there are two important considerations: (1) low prices to be paid for the work (which depends on wage rates) and (2) timely delivery of quality output. The low-wage consideration leads GU to choose areas with depressed wage rates, which normally reflect higher incidences of poverty in the area. Thus, GU's commercial objective matches its objective on poverty alleviation. The second consideration, however, leads GU to choose weaver-managers who are dependable and can ensure timely prod-

TABLE 12.2. Distribution of respondent households by loom ownership groups (in percent).

Number of looms owned	Program GU participants	Program non-participants	Control non-participants
1 to 3	32.5	50.0	20.0
4 to 6	32.5	30.0	50.0
7 to 9	17.5	0	0
10 to 19	17.5	10.0	20.0
20 and above	0	10.0	10.0

Source: Author survey.

uct delivery. These qualities are expected among loom owners who have command over some minimum resources, and for whom weaving is the main source of earning. The dual objectives of the GU are likely to draw marginal and middle-sized loom owners, which is discussed in the data.

However, not all of the aforementioned loom owners are willing to join the GU. About 30 percent of the nonparticipants in the program area have reported their unwillingness to join GU because year-round work is not available with GU and the profit margins are low compared to some other competing agents.[18]

Contribution to Nonfarm Employment and Income

The GU has introduced a product that has not displaced other domestically produced goods; due to idle loom and labor capacity, GU has been able to create additional employment and income for the weaving community. Since in the past, output and employment of the handloom-based weaving industry, in general, has been declining, a positive contribution of GU may show up in decreasing the decline rate.

GU originated the product popularly known as Grameen Check, which is alleged to have imitated the Madras Check. From the Bangladesh perspective, the product partially replaced the imported fabrics, without having any significant substitution effect on other locally produced textile products.[19] In the raw materials market, GU relies on the open market, in which production and imports are organized under the private sector. Although dye materials (chemicals) are imported, GU production has strong backward links for the yarn-producing (spinning mills) sector. It is important to note that the weavers in the GU net are free to produce anything they want, and for anyone they want. GU identifies a subset of existing weavers in a locality, and depending on the size of the work order it receives, gets the work done by some of these weavers, who are identified as "GU participants." If timely delivery of the product was not an important consideration (which may impose substitution during the peak period of activities), in the presence of idle loom capacity and idle loom workers, the entire increase in production will imply a net addition to employment and income of the loom owners and workers.

The survey findings support the previous conclusions. Capacity utilization of looms owned by program participants is higher than both types of nonparticipants (Table 12.3). Also, the GU participants enjoy a longer period of peak months compared to the other weavers (Table 12.4). On average, a weaver-manager enjoys a peak period of 6.4 months in a year, whereas it is less than four months with the nonparticipants. The trend in production (Table 12.5) is consistent with the general observation that handloom production has been on the decline. However, the annual rate of decline in production during 1993-1999 was only 2.3 percent for program participants, as opposed to 9.9 percent decline for program nonparticipants and 5.2 percent decline for control villages. Net income from weaving, re-

TABLE 12.3. Average number of looms owned and operated (per enterprise).

Number of looms	Program participants	Program non- participants	Control non- participants	All surveyed households
In operation	4.68	3.50	4.60	4.47
Not in operation	1.45	3.10	2.60	1.92
Total looms owned	6.13	6.60	7.20	6.38
Capacity utilization (% in operation)	76.35	53.03	63.89	70.06

TABLE 12.4. Average length of peak and slack periods (in months).

Average number of	Program participants	Program non- participants	Control non- participants	All surveyed households
Peak months	6.4	3.7	3.6	5.6
Slack months	5.5	8.3	8.4	6.3

Source: Author survey.

Note: Months during which the respondents can keep most of the looms in operation (due to sufficient work order) are identified as peak months.

ported in Table 12.6, largely reflects the trend in production. While it has declined for all sample groups, the rate of decline is lower for program participants. Furthermore, share of net income accrued from fulfilling the GU orders has persistently increased for the participants, and currently stands at 96 percent. This implies that GU has gradually settled down with a subset of weaver-managers, who also depend almost solely on working for GU.

Although minimum skill is a prerequisite for being a loom worker, the quality of labor is not homogenous. It is therefore expected that the urge to deliver products within a short time[20] may often lead to driving up the wage rates. Such an increase was not reported in the program area. One plausible reason is that production for GU is carried out over a period (see Table 12.7), most of which is traditionally identified as a slack period (June to October). During the months of November and December, when the former

TABLE 12.5. Average annual production per enterprise (figures in yard).

Year	Program participants	Program nonparticipants	Control nonparticipants
1992-93	14,135	9,700	14,920
1993-94	14,705	8,622	13,350
1994-95	15,180	6,973	11,200
1997-98	12,427	5,733	9,922
1998-99	12,151	4,739	10,289
% decline/year	2.34	9.91	5.17

Source: Enterprise survey.

Note: Nonparticipants are not year specific, but participants are.

TABLE 12.6. Annual net income per enterprise from weaving and contribution of Grameen Check.

	Program participants		Program	Control
Year	Net income	% share of GC in net income	nonparticipants	nonparticipants
1992-1993	34,838	33.57	38,823	54,706
1993-1994	33,597	81.36	34,067	47,254
1994-1995	27,433	90.44	22,356	32,769
1997-1998	17,174	94.77	145,554	27,415
1998-1999	17,986	96.02	12,193	26,046

Source: Author survey.

Note: Measured in taka/year per enterprise.

TABLE 12.7. Months of peak and slack activities in handloom weaving (modal values).

Item	Program GU participants	Program nonparticipants	Control nonparticipants
Beginning, peak	June-July	October-November	November
End, peak	December	November-December	January
Beginning, slack	January	December-January	February
End, slack	June	September	October

Source: Author survey.

Note: The first row identifies months when the peak period of activity begins, while those in second row identify the months when the peak seasons end. Similarly, the third and the fourth rows are for slack seasons, respectively.

overlaps with the traditional peak period, obtaining skilled loom workers is often difficult. Since it is the responsibility of the weaver-managers to deliver the products, they may do so on time if they are able to contract these workers ahead of time and engage them over a longer period. The problem is well recognized by the GU management. Since the size of the export market is rather limited, it is not possible to guarantee year-round utilization of all the looms owned by the GU weaver-managers. Engaging a limited number of weaver-managers through long-term contracts in accordance with the volume of order is not feasible for the following two reasons:

1. the volume of work orders fluctuates substantially, and
2. since GU provides fabrics for summer wears, the production period rarely extends beyond six months in a year.

It is, therefore, quite expected that GU relies upon the existing pool of weavers to fill the work orders it receives. However, absence of long-term contracts often makes it difficult for GU to ensure timely delivery of quality products. It is quite likely that the more-skilled workers and the weaver-managers (who are better able to organize production) are employed elsewhere. This concern led GU temporarily to engage in production not backed by orders from the export market. The experiment turned out to be a failure, since inventory was built up and financial losses had to be incurred. GU has therefore decided to fill work orders only, relying on a limited number of weavers who are able to engage workers for a longer period. This is reflected in the increasing share of GU's order in total net income of the participant weavers.

Nature of Contract

Handloom weaving is organized under an intricate set of arrangements in the input, output, and labor markets. Prior to the GU intervention, the handloom owners generally bought the yarn and dye with cash or on credit, dyed the yarn and wove the fabric with family labor and/or hired labor, then sold the output mostly on credit. The laborers engaged in all processes, washing, dyeing, drying, reeling, warping, and weaving, and were paid on a piece-rate basis. Under the old system, most risks in both production and marketing were borne by the loom owners. It was also their responsibility to finance all stages from input procurement to final sale of the products. Occasional work orders were received from specific buyers, but the predominant arrangement followed the previous description. With the emergence of GU, this arrangement had changed significantly for the participants. Even though the GU weaver-managers were free to work for others, they eventually ended up by running their looms mostly to fill GU work orders. Since new products such as Shahajadpur Check came about on private initiatives, some of the nonparticipants worked under the output system as well (Table 12.8).

In 1995, an average enterprise would supply its output to three or more buyers. Currently, all the GU participants covered by the survey are engaged in delivering products for GU only. Although this has reduced the risk they bore previously, it has also transformed them into the status of "workers," dependent on a single employer. Since GU has no contractual obligation to ensure regular work orders, one expects that these participants are more vulnerable due to their dependence on a single buyer/employer. For example, if the single agent (GU) fails to provide work, many of the participants could face problems in switching back to the non–GU market. The responses to the question of whether they would face problems in switching

TABLE 12.8. Current modes of input procurement.

Mode of input procurement	Program participants	Program nonparticipants	Control nonparticipants
Procured on credit—repaid in cash	0.3	34.0	54.5
Cash purchase	4.8	46.0	35.5
Buyer of product provides inputs	95.0	10.0	0.0

Source: Author survey.

Note: Percentages of total value of inputs purchased under each procurement mode.

to new products or to new buyers do not support this fear. The loom workers, who have acquired greater skill in producing GU products, are likely to find jobs elsewhere in the handloom industry. Similarly, the entrepreneurs feel that they may switch back to producing lungi and other materials. Such optimism prevails when no sudden decline occurs in the overall demand. In the event of such declines, a long-term contract with a single buyer may be costly for the entrepreneurs.[21]

Household-Level Impact

Findings on current school enrollment rates for children five to nine years old (Table 12.9), changes in assets owned by households (Table 12.10), and changes in financial and food status of the weaver households (Table 12.11) clearly indicate that the program participants are better off than the nonparticipants.

GU's health assistance program has been noted previously. The survey reveals that the incidence of disease is lower among members of the participant households (Table 12.12). In spite of it, labor days lost by these sick persons (over a recall period of 15 days) is relatively higher among participant households, which may reflect the fact that they are more regularly employed than the members of nonparticipant households. Most important, average medical expenditures per respondent household are significantly lower than those incurred by the nonparticipants.

Institutional Aspects

In comprehending the kind of decision-making problems that GU confronts, several questions may be posed. First, why did GU select Sirajganj and Araihazar as their areas of operation? Second, how does GU distribute

TABLE 12.9. Gross enrollment rates at primary education, by sex and sample groups.

Sex	Program participants	Program nonparticipants	Control nonparticipants
Boys	72.73	50.0	50.0
Girls	86.96	60.0	25.0

Source: Author survey.

Note: Gross enrollment rate = (five- to nine-year-old children enrolled) × 100 / (total number of five- to nine-year-old children).

TABLE 12.10. Percentage of weavers reporting increases in assets over two years preceding the survey.

Assets	Program participants	Program nonparticipants	Control nonparticipants
Number of looms	32.5	9.09	22.22
Vehicle for commercial use	5.0	0	0
Bullocks	15.0	9.09	2.22
Cassette player	17.5	0	0
Cycle	7.5	9.09	0
House	47.5	18.18	11.11
Ornaments	17.5	0	22.22
Radio/transistor	12.5	9.09	0
Trees	12.5	9.09	0
Television	2.5	0	0
Wristwatch/clock	25.0	18.18	0
Shop	10.0	0	11.11

Source: Author survey.

TABLE 12.11. Changes in financial situation and food status over two years preceding the survey.

Percentage of respondents reporting	Program participants	Program nonparticipants	Control nonparticipants
Working capital increased	27.5	9.09	22.22
Financial status improved	77.5	18.18	22.22
Volume of business credit increased[a]	2.5	45.45	0
Volume of business debt increased[b]	10.0	54.55	88.89
Personal credit increased[a]	10.0	18.18	0
Personal debt increased[b]	35.0	54.55	44.44
Food status of the house-hold improved	77.5	18.18	22.22

Source: Author survey.

[a]credit = payments to be received from the market.

[b]debt = payments to be made to other agents in the market.

· TABLE 12.12. Incidence of sickness and medical expenses.

Item	Program participants	Program nonparticipants	Control nonparticipants
Weavers suffering from diseases during past 15 days (%)	37.5	45.45	77.78
Sick persons whose work was affected during past 15 days (%)	85.71	60.00	42.86
Average number of work days lost by the affected persons during past 15 days	4.41	3.5	4.33
Average medical expenditure per respondent house hold during past 15 days (in Tk)	174	350	277
Respondents suffering from diseases during past six months (%)	60	72.73	66.67
Sick persons whose work was affected during past six months (%)	66.67	87.5	50
Average number of work days lost by the affected persons during past six months	9.75	18.42	11.75
Average medical expenditure per respondent household during past six months (in Tk)	258	378	420

Source: Author survey.

its work among various units? Third, how does a unit distribute its work among the weaver-managers? Last, how does GU sustain its operation?

GU entered the market with Grameen Check, whose design can be better reproduced by weavers who are familiar with using yarns of different colors, both in the warp and weft. Lungi, a traditional cloth worn by males in Bangladesh, is often woven with checkered designs. Weavers of Sirajganj and Narsingdi have an established reputation in the production of good quality lungi. This initial concentration of skill induced GU to commence their operation in Sirajganj. Another reason for the choice of the area, as claimed by GU officials, is the initiation of a proposal to promote employment for the weaving community from Grameen Bank staff in Sirajganj. A final reason may lie in the fact that Sirajganj had long been identified as a resource-poor area, characterized by river erosion and low wage rates, and the anticipation that the Bangabandhu Bridge over the Jamuna River would soon be completed.[22]

The GU setup has a head office in Dhaka, where the production and marketing department is the most important one. This department supervises the one area office in Sirajganj and another emerging area office in Araihazar. There are six unit offices under the Sirajganj area office, each headed by a unit officer whose salary is in the range of Tk 8,000 per month. Normally, eight to ten villages are covered by each unit office. On average, about 350 weaver-managers are under each unit, and the average number of looms operated by each weaver-manager is 4.5. Thus, each unit office during the peak activity period needs to supervise production in 1,500 to 1,600 looms. The day-to-day supervision work is done by GU employees known as *taant karmi*, whose monthly salary ranges between Tk 2,000 to Tk 4,000. Each taant karmi is expected to supervise 200 looms, even though they are currently underutilized due to low work volume. During the field survey, each taant karmi supervised approximately 125 looms.

A technology specialist is attached to the area office whose responsibility is to train the unit officers and taant karmis whenever a new design is introduced. Such events are normally associated with each work order. The taant karmi is responsible for ensuring that GU purposes are served at the field level—that is, quality products are produced by the weaver-managers and are delivered on time. For this, the taant karmis are expected to visit the enterprises of contract weaver-managers. In an ideal situation, a taant karmi is expected to cover 200 looms. If the average number of looms per enterprise was five, this would imply visits to 40 enterprises. It is expected that such visits be made at least once every two days. Prior to the field survey, we anticipated that the taant karmis might have incentive to choose large-sized enterprises so that the number of enterprises to visit would be lower. Although we have observed a significant number of weaver-managers owning more than ten looms to participate in GU programs, apparently this was not due to the urge to avoid work. In reality, due to limited workload, a taant karmi ends up attending to only 125 looms. The observed presence of large enterprises among GU participants is largely due to the urge to ensure timely supply. As noted elsewhere, this concern has led to both avoidance of very large enterprises as well as very small ones.

It is important to note that the taant karmis employed by the GU are normally matriculates, and are residents of the locality where they work. Thus, the salaries offered to them, in spite of being low by urban standards, are able to ensure their loyalty to GU initiative. Training is provided to these workers in the unit and area offices. However, such training is very specific to the job to be performed, and the skill is not easily tradable. We have not found sufficient evidence to conclude that they are being lured by other private-sector enterprises in similar trade.[23] Other than the taant karmis, all the GU staffs at the field and head office levels are university graduates. Since they expect GU operations to expand in the future, they aspire to upward mobility.

Summary: What Is the Uniqueness of the GU Approach?

The salient features of the GU approach are summarized as follows:

1. The marketing agent (GU), who also coordinates production by a large number of small producers, must bear all financial risk. Since the size of the market is limited and there is a great deal of competition, the agent must be innovative in designing new products and be aggressive in marketing strategy.[24] An important question in this regard is what are GU's advantages in undertaking the initiative, compared to either a private entrepreneur or a public sector agency such as the Bangladesh Small and Cottage Industries Corporation (BSCIC)? Ideally, BSCIC, with its network of 64 district-level offices, is well placed to identify products and establish liaison with potential producers. It also has the manpower to design products. Unfortunately, marketing of products as a commercial venture is not its tradition. Moreover, its institutional design and incentive structures are not conducive to undertaking commercial activities. However, undertaking investment in a network to organize production in rural areas is a risky proposal. GU was preceded by Grameen Bank's credit programs, which reduced their initial cost of experimentation. Moreover, the GBF could undertake the initial costs for the initiative before it groomed into a separate entity.[25]

2. Hypothetically, as a marketing agent, GU has the choice of relying on market supply of its product or on intermediaries to coordinate with the producers. Examples of the latter include private seed traders' negotiation with NGOs to coordinate production of soybeans by numerous farmers. In such instances, the respective NGOs are paid a commission and are supplied with the right kind of seeds and technical knowledge. Another example following GU is BRAC's unsuccessful attempt to produce BRAC Check through intermediaries.[26] GU's approach involves direct linkage with the weaving community, bypassing all intermediaries. One may question if it is better to rely on direct supervision of production through the weaver-managers or to depend on the market, including intermediary agents. When production re-

quires very little or no supervision, and when cost due to shortfall in quality is not borne by the marketing agent,[27] it is cost-effective on the part of the latter to depend on the market. Dependence on intermediaries at levels higher than the weaver managers runs the risk of nonfulfillment of contracts. This is particularly true when markets for such services (as those provided by the "intermediaries") are not well developed, and there is an absence of competition, which facilitates accountability in order to ensure repeat contracts. Since the GU activity is essentially similar to that of the readymade garment industry, the products to be delivered are prespecified, and they must be delivered within a prespecified time. Under such circumstances, relying on the market for supply is not a feasible option. While the other two options are feasible, experience with GU and BRAC suggests that some degree of direct control over the production process is crucial in running the business.

3. Due to the uncertainty in the market, it is not possible for GU to negotiate long-term contract with the producers; the weaver-managers are also unable to secure the better loom workers in long-term contracts. At the same time, it is important for GU to ensure dependable weaver-managers to enable it to keep its commitment on work orders received. GU has gradually settled down with a selected group of weaver-managers, with whom short-term contracts are repeated on a regular basis.

4. It is necessary to ensure that the field-level GU workers (taant karmi) provide information truthfully and diligently supervise the production activities undertaken by weaver-managers. However, all production contracts are over short periods and the performance indicators of taant karmi are easily verifiable. For example, quality of output is measured in terms of (1) the number of threads in warp and in weft and (2) the quality of yarn (measured in terms of counts), which are observables. An appropriate remedial action may be taken before the next work order.

5. GU provides a good example of how rural poverty at the microlevel may be addressed by a microfinance institution. Such a venture, however, may be fragile due to severe competition in the product market. It is important to acknowledge that technologies and institutions, which were once pro-poor, may at one stage of their development change in ways that may make them less pro-poor. It was already noted that GU has undertaken the production of gray cloth in powerlooms, which are less labor intensive and are likely to adversely affect the owners of handlooms. This appears to be GU's major dilemma in future planning.

EMPLOYMENT GENERATION PROJECT FOR THE RURAL POOR: AN AGRANI BANK INITIATIVE

In a bid to generate employment in the rural areas, especially for the landless and women, and to promote potential microentrepreneurs, Agrani Bank, one of the largest nationalized banks in Bangladesh, initiated a project of advancing loans to a targeted group. This program is known as Employment Generation Project for the Rural Poor. The microenterprises financed by the project include agro-processing such as production of puffed rice and paddy threshing; manufacturing such as jute goods and other handicrafts (pottery, handloom, metal products, and fabrication); and services such as restaurants, light engineering repair workshops, tailoring, and block and screen printing.

A new wing, named Microenterprise Development Unit (MEDU), has been set up within Agrani Bank for implementation, monitoring, and reporting the project activities. The donor assistance (IFAD) provides the overhead support such as maintaining a project office with a minimum number of staff, transportation at the branch and head office levels, office equipment, expenses for training, and other allowances. Two types of approaches have been adopted to disburse loans: (1) direct lending through selected branches of Agrani Bank and (2) indirect lending through different nongovernment organizations (NGOs).[28]

The target group of the project includes the following:

1. Owners of existing microenterprises who have proven managerial and technical skills in self-employment and who wish to expand, upgrade, or diversify their business (direct lending).
2. Persons who have acquired skills and capacity to operate microenterprises, undergoing an apprenticeship or working as skilled laborers and who want to start their own business (indirect lending).
3. Persons capable of "graduating" from ongoing poverty alleviation programs, such as those supported by NGOs (BRAC, Proshika, ASA, and Grameen Bank) in related sectors and who are interested in expanding their ongoing small income-generating activities (direct lending through the advisory services of NGOs or indirect lending).
4. The unemployed landless, who fall into two categories:
   • Persons such as graduates from the universities or schools of higher education who have general education levels that assist them in running a business (indirect lending).
   • Those seeking self-employment in preference to wage employment (indirect lending).

In the case of direct lending, the branches are selected first before the potential borrowers are identified within the perimeter of each branch's operation. In the case of indirect lending, the partner-NGOs are identified first, who subsequently lend to members.

Branch and Client Selection Under Direct Lending

The zonal offices are assigned with the responsibility of proposing a list of branches that may potentially be involved in direct lending. Guidelines followed in this initial selection are as follows:

- The selected branches should have most of their activities in the rural area. The branch may be located in the municipal/urban area, but its main clients have to be rural.
- The branch area should have dormant or manifested prospect for nonfarm activities.
- The selected branches must have an adequate workforce to run extra credit operations without adversely affecting ongoing activities.[29]

The final list of branches is prepared by the head office, while the initial lists of branches are provided by the zonal office. In doing so, the head office assesses the capability of a branch and reevaluates its location, previous record, and the prospect of nonfarm activities within the area of that branch. The final authority for involving any branch in direct lending belongs to the board of directors.

A potential MEDU borrower needs to meet the following criteria:

1. *Residence:* The intending borrower must be a permanent resident of the project area or must have tangible assets (owned or leased-in land, constructed house/building, installed machinery and equipment, etc.) in the project area.
2. *Age:* He or she must be within 18 to 45 years of age.
3. *Education:* He or she should have a minimum education of primary level.
4. *Experience:* He or she should have a minimum of six months' experience in running an enterprise.[30]
5. *Income:* His or her annual net income from all sources must be within the range of Tk 10,000 to Tk 30,000.
6. *Current liability:* Individuals having any outstanding liability with any credit institution other than the project (or a defaulter of any credit) are excluded.

7. *Type of enterprises:* Prospective borrowers should be involved in running enterprises that primarily fall under any of the following three broad categories: agro-processing, manufacturing, and services. However, Agrani Bank branches involved in direct lending may finance other enterprises that are locally suitable and profitable.

Under direct lending, MEDU has established targets in terms of the number of enterprises and total disbursement for every branch selected.

Selection of NGOs

Attracting the NGOs to borrow from the MEDU fund has been quite difficult, especially since alternative sources of finance are available to them at lower cost.[31] However, faced with increasing credit constraints, and encouraged by the prospect of lending to the microenterprise sector, some NGOs have expressed interest in borrowing from MEDU. While applying for loans, NGOs are expected to provide details on their program and performance, as well as specific nonfarm activities that they intend to support. Upon receiving the MEDU loan, a successful applicant (NGO), is expected to lend to its members who have "graduated" to undertake larger ventures. Even though the criteria for client selection at this level is not explicit, one would expect the client to have received previous loans from the respective NGO, which also implies membership with the NGO over a substantial period.

Loan Application and Sanctioning Procedures
Under Direct Lending

The microenterprise customer relations officer (MECRO) of the participating branches of Agrani Bank is mainly responsible for seeking promising entrepreneurs from the eligible target group. The eligible applicants may seek the help of the MECRO for preparing their loan applications. With loans exceeding Tk 50,000, an applicant needs to produce evidence of ownership and possession of collateral to be offered for the loan before the loan application is processed.[32] Owned land, buildings, and other immovable properties whose sale value is at least equivalent to the sanctioned loan amount are taken as collateral security.

The MECRO/branch manger visits the proposed site of the business/enterprise to verify the genuineness of the information provided by the applicant. After reviewing the professional and technical capacity of the applicant and the financial viability of the project for which the loan is sought, the branch manager and MECRO jointly certify in writing whether an applicant belongs to the target group and meets all the eligibility criteria.

Although a branch manager can sanction a loan of up to Tk 50,000 on the basis of the recommendation of the MECRO, loans exceeding Tk 50,000 and up to Tk 250,000 (inclusive of both fixed and working capital loans) are sanctioned by the zonal heads on the basis of recommendations of the branch manager, if necessary by verifying the viability of the project and the creditworthiness of the borrower, sometimes through on-the-spot verification of the proposed undertaking.

Interest Rates and Loan Repayment Schedule

MEDU/Agrani Bank obtains from the Bangladesh Bank (the central bank of the country) funds at an annual interest rate of 7 percent. Agrani Bank generally charges 13 to 15 percent interest rates for its normal loans; it charges 12.5 percent for fixed capital and 14.5 percent for working capital.[33] In the case of indirect lending, the partner–NGO borrows from MEDU at 3 percent less than those quoted for direct lending. However, there is a provision under the MEDU project for a subsidy equivalent to 3 percent of the disbursement under the project to MEDU. Actual transfer of this subsidy is yet to be realized, even though it is recorded on paper.[34]

Approvals on working capital are made on an annual basis, and a successful applicant opens a cash credit account, which is closed at the end of the year.[35] MEDU lending on fixed capital is normally for a period of two to five years and repayment commences six months after loan disbursement. The repayment period and the installments are open for negotiation between the borrower and the bank officials. Indirect loans through NGOs have a repayment period on working capital of 12 to 18 months; the maximum repayment period for fixed capital is three years. These periods apply to the NGOs repaying the bank, and these are normally applied by the respective NGOs to their clients as well.

Monitoring and Provision of Incentives for Bank Officials

The MECRO and the branch manager are expected to closely supervise a borrower to ensure that he or she is making the best use of the borrowed fund. Normally, a MECRO is expected to contact each borrower and visit each of the concerned enterprises at least once a month. If the total number of borrowers in a branch exceeds 50, the MECRO contacts each borrower once every two months. The branch manager personally contacts each borrower once every three months.

The zonal head is also involved in monitoring. During his or her routine visits to a branch every month, he or she checks on the progress of implementation and recovery, and randomly visits one or two enterprises that have borrowed from a MEDU account. Among the MEDU officials, the project director and project coordinator visit the zonal and branch offices once every two months and once a month respectively. The technical and financial advisor visits branch offices and the field level enterprises visit more frequently to check the implementation of the project activities.

To ensure target disbursement and high recovery rates, the project has a number of incentives. A branch is rewarded in accordance with the schedule described in Table 12.13. Although there is no precise module to distribute the total incentive money among the branch officials, the branch manager and the MECRO get the larger share of the amount. The zonal officials, particularly the zonal head and concerned officer of the zonal office together get 10 percent of the total incentive amount earned by concerned branches of the zone, but not exceeding Tk 10,000 for a particular accounting year. At present, incentive is no longer contingent upon the achievement of the disbursement target. Maintaining a floor recovery rate of 90 percent is sufficient to qualify a branch for incentive.

Apart from these pecuniary incentives, some nonpecuniary incentives are also provided to the bank officials. Provision is made for training bank officials who are involved in direct lending. This on-the-job training is believed to strengthen their candidacy for future promotion. Every branch involved in direct lending gets a motorcycle, which the branch manager and the MECRO can share. For a midlevel bank officer, such as the MECRO, owning a motorcycle may be quite prestigious in a rural and/or semiurban social setting. On average, the branch manager and the MECRO attend at least two workshops per year. In these workshops, they report recovery and other performances of the project activities in front of very top-level bank officials. Rewards are also distributed

TABLE 12.13. Incentives to Agrani Bank branches under MEDU.

Sl. no.	Minimum disbursement as % of target	Recovery rate as % of total recoverable amount	Incentive amount as % of total loan recovered
1	75	90	0.2
2	75	91	0.3
3	75	92	0.4
4	75	93	0.5
5	75	94	0.6
6	75	95	0.7
7	75	96	0.8
8	75	97	0.9
9	75	98-100	1.0

Source: Banking plan MEDU, p. 13.

during the workshops. The workshops provide the branch-level bank officials with opportunities for reporting their performance to the top-level management. This encourages the manager and the MECRO to work hard.

At the end of 1998-1999, a total of 89 branches of Agrani Bank, spread over 31 districts of the country, were involved in the implementation of the project. As of June 1999, cumulative loan disbursement stood at Tk 288.60 million including repeat loans of Tk 91.064 million under the direct lending approach of the project. A total of 4,071 loans were procured by 2,981 micro-entrepreneurs during the same period. Approximately 42 percent of the total loan amount was disbursed to service enterprises, followed by 32 percent to agro-processing and 26 percent to manufacturing enterprises. Most loans (92 percent) were within a range of Tk 20,000 to Tk 50,000, which did not require any collateral security. The highest percentage share of collateral free loans was observed in the service sector (96 percent), followed by the manufacturing sector (93 percent). In agro-processing, the percentage share of collateral free loans was 86 percent. Detail on per activity group lending is shown in Table 12.14.

Lending through the NGOs commenced in 1998. Till the end of FY1998-1999, only five NGOs were involved in indirect lending: ASA, Grameen Fund, Grameen Uddog, Grameen Samaj Kendra, and Uttara Development Programme.

TABLE 12.14. Allocation of MEDU lending by sectors (percentage of total yearly disbursement in each activity group).

Sector	1996	1997	1998	1999
Fishery	0.00	0.00	0.28	1.16
Flour mill	0.00	0.00	0.84	0.00
Horticulture	0.00	1.60	1.92	2.57
Oil mill	19.77	5.74	5.84	7.54
Others	9.88	11.67	8.71	8.02
Poultry/dairy	3.74	22.19	17.53	23.24
Puffed rice (Ohira Muri)	0.87	3.33	2.62	1.78
Rice mill/threshing	64.07	41.13	52.30	48.05
Sweet preparation	1.87	14.36	9.95	7.63
Total agro-processing	100.00	100.00	100.00	100.00
	(1.34)	(21.52)	(28.15)	(26.93)
Aluminum product	0.00	3.00	4.48	1.41
Bakery	18.13	9.83	7.12	8.67
Candle making	0.00	0.67	1.40	0.00
Carpentry	0.00	0.67	0.30	0.56
Chemical products	0.00	2.56	1.56	3.86

TABLE 12.14 *(continued)*

Sector	1996	1997	1998	1999
Dyeing & finishing	0.00	0.00	0.00	0.56
Food products	0.00	2.60	2.23	5.85
Garments	0.49	6.48	13.56	8.81
Handicrafts	0.00	0.00	0.30	1.41
Handloom products	0.00	4.24	4.03	3.44
Metallic spares	0.00	0.67	0.88	1.93
Others	29.70	24.95	20.94	15.45
Plastic products	0.00	1.90	0.36	0.88
Pottery	0.00	0.00	0.58	0.21
Printing	0.00	3.91	1.43	3.83
Saw mill	14.56	2.22	4.49	7.09
Spare parts	0.00	0.60	0.75	0.56
Steel furniture	7.28	7.55	6.70	5.53
Tiles making	0.00	3.06	2.00	1.77
Wood furniture	29.83	25.13	26.87	28.17
Total manufacturing	100.00	100.00	100.00	100.00
	(19.52)	(26.61)	(26.80)	(24.81)
Grocery	0.00	1.58	1.51	1.75
Laundry	0.00	0.51	0.33	0.36
Light engineering	23.77	51.97	15.71	15.11
Others	13.48	2.05	31.78	33.67
Power tiller	0.00	0.54	1.07	1.75
Repairing and maintenance	11.19	13.00	15.66	15.41
Restaurant	8.81	6.16	8.17	7.43
Tailoring	42.75	23.93	25.60	24.06
Veterinary	0.00	0.27	0.17	0.46
Total service	100.00	100.00	100.00	100.00
	(79.08)	(51.87)	(45.04)	(24.81)

Source: MEDU data.

Note: Figures in parentheses are percentages of total annual disbursement to each subsector. "Others" is not defined in the MEDU database. Given our observations from the survey, much of this may be in trade.

Study Design

The survey for the current study had to be limited to only two operation areas due to time and financial constraints. In order to capture both direct and indirect lending practices of the MEDU operations, one area under each was selected for the purpose. Of the 89 branches of Agrani Bank, only 19 had been involved in direct lending since 1996. Keshabpur under the Jessore district, which primarily serves rural populations, was chosen for direct lending. In cases of indirect lending, Grameen Samaj Kendra was selected, whose activities were confined to Kapasia (in Gazipur district), which is largely a rural area.

In drawing the sample (borrowers/entrepreneurs), a list of all MEDU borrowers was obtained from the specific branch and from the respective NGO. In examples of the bank borrowers, 20 entrepreneurs were selected, taking proportionate representation from three broad categories of activities for which the loans were taken. The three activities are agro-processing, manufacturing, and services. Only 17 borrowers were involved in indirect lending, all of whom were included in the survey. Choosing of a control group in both the areas was more problematic. In the absence of a local chamber group (association of entrepreneurs), we obtained information on clients from other commercial banks and prepared lists of enterprises that were visible in the vicinity. Operationally, the task was made easy due to the presence of similar enterprises in the same vicinity. This enabled us to pick up a non–MEDU enterprise engaged in an activity similar to that of our program respondent. A total of ten respondents in the control group were chosen this way from each of the areas.

Besides the questionnaire survey, MEDU data on branch-level disbursement and recovery were used. Information was also sought through the interviews of MEDU officials at the head office in Dhaka and at branch levels, and by conducting focused group discussions with the beneficiary and nonbeneficiary entrepreneurs.

The survey findings are summarized under three broad groups: entrepreneurs who have not borrowed from a MEDU source (nonparticipants), those who have borrowed under direct lending programs of MEDU (participants of direct lending), and those who have borrowed from MEDU's partner–NGOs (participants of indirect lending).

Targeting: Selection of Branches

The branch selection following the guidelines mentioned previously is predominantly a subjective exercise, and here lies the danger of making errors. A wrong selection may arise due to wrong assessment of the appropriateness of the branch or of the prospect of nonfarm activities in the respective branch's area of operation. Zonal offices as well as branch offices receive some pecuniary incentives for achieving a given disbursement target with a floor recovery rate. One way of increasing the total amount of financial awards for a zonal office is to involve as many branches as possible in direct lending on the expectation that they would be able to achieve the given disbursement target and the floor recovery rate. This may lead to the selection of branches that do not have an adequate workforce to run the direct lending activities. It may also lead to the selection of branches in areas where prospects of nonfarm activities are not bright.

Erroneous branch selection, although quite few in number, has seriously affected the overall performances of the project activities. If adverse selection is avoided, the current recovery rate[36] could be much better. The MEDU data show that 59 percent of total overdue loans are concentrated among only 13 branches. The problem of underperformance has become so incurable that project activities have been closed in four of these branches. Two of them, Laksam and Tomsombridge, are plagued by inefficient bank officials. In two other branches, Sapahar and Purba Dhala, wrong assessment of the prospect of nonfarm activities is at the root of the problem (see Exhibit 12.1).

Targeting Sectors: Has MEDU Lending Made Any Difference?

The sectoral distribution of total loan amounts disbursed under MEDU has been as follows: half the amount goes to the service sector, and the rest

EXHIBIT 12.1. Fallouts of wrong branch selection.

Laksam branch of Agrani Bank under Comilla was selected for direct lending in January 1996. By January 1997, a total of Tk 675,000 was disbursed to 15 enterprises, which were engaged mainly in service activities. The recovery rate was well below the overall recovery rate of the project. The head office blames the bank officials, the branch officer, and the MECRO, who were in charge of looking after the project activities, for this underperformance. Rather than replacing the incompetent officials, mainly because of fear that they would also jeopardize the project activities of other branches, the MEDU activities of the branch were closed. At the time of the branch closure, a total of Tk 180,000 was overdue, which was about 27 percent of the total disbursement.

The Tomsombridge branch of Agrani Bank, under the Comilla region, distributed Tk 590,000 under direct lending among 16 enterprises involved in agro-processing, manufacture, and service activities. The program faced the same problems, and for the same reason. The Sapahar branch of Agrani Bank under the Nagaon region, which commenced direct lending in July 1997, could find only one suitable entrepreneur to disburse loans before its exclusion from the project activities. Purba Dhala branch in Netrokona district even failed to find a single qualified enterprise to disburse loans.

is split equally between agro-processing and manufacturing (Table 12.14). There seems to be a bias in favor of the service and agro-processing sectors. The sectoral distribution of microenterprises in the two areas as represented in the sample surveyed in the present study suggests few differences across the two groups of MEDU participants. Direct lending through Agrani Bank appears to be biased in favor of the service sector, while the borrowers under indirect lending (through the NGOs) appear to be investing in agro-processing and in some of the new activities that have emerged due to linkage effects. The two cited sets of evidence clearly indicate a MEDU bias against manufacturing (as distinct from agro-processing). One possible reason for such bias lies in the fact that MEDU operations are located primarily in semirural areas where manufacturing activities are less prevalent than in the urban and semiurban areas.[37] Moreover, in general, investments in both agro-processing and manufacturing sectors are perceived to involve higher risks than those in the service sector. Conventional lending by the formal risk-averse banking sector tends to be biased toward the service sector. Direct lending under MEDU does not appear to have made a significant breakaway from this bias. In contrast, the NGOs appear to be able to pick their borrowers, who venture into areas of higher long-term growth potential.

Targeting of Borrowers

Several other characteristics of the sample enterprises are summarized in Tables 12.15 to 12.19. Even though no significant difference is found in the distribution of enterprises by modes of acquisition (Table 12.16), a relatively higher percentage of the indirect borrowers (72 percent) established the firms themselves than the direct borrowers (68 percent) and the nonparticipants (55 percent). As one would expect, family businesses are better supported under indirect lending (16.7 percent) than under direct lending (4.8 percent) (Table 12.17). The reason largely lies in all the legalities involved in collateral-based borrowing from the banks, which require a borrower to define ownership in legal terms. The requirements are less stringent in cases of indirect lending or for those who borrow from NGO and informal sources.

Direct lending requires a borrower to operate his or her business for a period of at least six months and, therefore, the enterprises have already established their standing prior to borrowing. Table 12.18 shows that about 68 percent of the borrowers under direct lending established their firms prior to 1995, while 55 percent of those under indirect lending and 50 percent of the nonparticipants had done so. Finally, the borrowers under indirect lending

TABLE 12.15. Activities undertaken by sample enterprises.

Activity	Nonparticipants	Participants, direct lending	Participants, indirect lending
Agroprocessing	(25.0)	(8.7)	(41.2)
Poultry/hatchery	2		5
Oil mills	1		
Rice milling	2	1	
Fish drying			2
Flour mills		1	
Services	(35.0)	(60.9)	(11.8)
Tailoring	1	5	1
Restaurant		2	
Fast food	1	1	
Fertilizer/cement shop[a]	1		
Selling footwear[a]	1		
Photo binding		1	
Computer training and art		1	
Shop for selling poultry feed and fish feed[a]	1	1	
Sell and repair of electronics		1	
Selling broilers[a]		1	1
Rickshaw rental		1	
Barbershop	2		
Manufacturing	(40.0)	(30.4)	(47.0)
Bakery	2		
Bangles	1	1	
Bobbin manufacture		1	
Engineering work	1	1	
Ice and ice cream	1	2	
Wooden furniture		1	4
Sanitary slabs	1		
Jewelry	1		
Body of rickshaw/vans			1
Iron grills			2
Dry cell battery			1
Roof tiles, brick	1	1	
Total number of enterprises	20	23	17

Source: Author survey, 1999-2000.

Note: Some of the activities, marked by ([a]), are essentially trade and should not have been financed as a service sector activity. For a distribution of MEDU disbursement by activity, see Table 4.2.

TABLE 12.16. Distribution of enterprises by mode of acquisition (column %).

| Mode of acquisition | Participation in MEDU | | |
	Nonparticipants	Participants of direct lending	Participants of indirect lending
Inherited	35.00	31.82	27.78
Self-established	55.00	68.18	72.22
Purchased	10.00	0.00	0.00

Source: Author survey.

TABLE 12.17. Distribution of enterprises by type of ownership (column %).

| Type of ownership | Participation in MEDU | | |
	Nonparticipants	Participants of direct lending	Participants of indirect lending
Single	80.00	90.48	77.78
Partnership	5.00	4.76	5.56
Family business	15.00	4.76	16.67

Source: Author survey.

TABLE 12.18. Distribution of enterprises by period of their establishment (column %).

| Year of establishment | Participation in MEDU | | |
	Nonparticipants	Participants in direct lending	Participants in indirect lending
Up to 1979	15.00	4.55	11.11
1980-1989	15.00	31.82	22.22
1990-1994	20.00	31.82	22.22
1995-1999	50.00	31.82	44.44

Source: Author survey.

were found to be more educated than the borrowers from other sources (Table 12.19). The fact that they had already received collateral-free loans and that they subsequently had been introduced to formal sources of loans under this project through the NGOs lends support to the conjecture that NGOs are possibly better able to identify entrepreneurs. This is also indicated by the responses to queries about sources of information for different types of

loans (Table 12.20). NGO workers are a significant factor in getting potential borrowers in touch with the banks, even in the case of direct lending. Table 12.19 (education level of entrepreneurs) suggests that the requirement
of a minimum level of primary education (i.e., above class 5) is not always
fulfilled in the case of direct lending; this is suggestive of other considerations in the selection of borrowers.

The survey started out with two broad hypotheses. First, the borrowers
under direct lending are possibly the past borrowers of Agrani Bank, and
the transfer exists only on paper, without the project being able to contact a
new group of entrepreneurs. The second hypothesis relates to the borrowers
under indirect lending; these borrowers have not really graduated from the
NGO's microcredit program. We find the first hypothesis to be true in the
initial year—two out of three borrowers under direct lending are previous
borrowers of the Agrani Bank (Table 12.21). However, this changes drasti-

TABLE 12.19. Distribution of entrepreneurs by their level of education.

	Participation in MEDU		
Level of education	Nonparticipants	Participants of direct lending	Participants of indirect lending through NGO
Class 1-5	20.00	26.09	0.00
Class 6-10	20.00	26.09	27.78
SSC and above	60.00	47.83	72.22
Average age (years)	35.40	35.00	34.50

Source: Author survey.

Note: Owners, who look after the day-to-day management of their enterprises,
are identified here as entrepreneurs.

TABLE 12.20. Source of information about different sources of loan.

	% of respondents across sources of loan benefited from this source		
Source of information	MEDU direct lending	MEDU indirect lending	Others
Banker	32.5	0	53.4
Other borrowers	17.5	0	0
NGO workers	40	82.4	37.1
Neighbors	10	5.8	9.5

Source: Author survey.

TABLE 12.21. Number of MEDU borrowers in the sample and their reliance on MEDU source.

Items	1996	1997	1998
Number of new MEDU borrowers	3.0	10.0	10.0 (17)
Percentage who borrowed previously from Agrani Bank (NGO)	66.7	10.0	0.0 (11.8)
Percentage who continued to borrow from MEDU in subsequent years	66.7	100.0	100.0 (n.ap)

Source: Author survey.

Note: Figures in parentheses refer to indirect lending through NGOs.

cally after 1997, and all new borrowers under direct lending have no previous standing with Agrani Bank, nor have they borrowed from other formal sources in the previous years. The second hypothesis is found to be true. Fifteen out of 17 indirect borrowers have no previous association with the respective NGO. This finding suggests that there may not be too many graduates of the microcredit program who would like to borrow for investment in microenterprises. However, as credit retailers, NGOs are well suited to identify potential entrepreneurs because of their standing in and regular interaction with the local community.

Borrowers' Preference

Ninety-six percent of the MEDU borrowers reported that MEDU loans offered a better deal, while the other 4 percent have not found much difference from other sources of loans. Even though the noninterest cost of borrowing is similar between direct lending and regular operation of Agrani Bank, the respondents are explicit in expressing their preference for MEDU loans.[38] Reasons for such preference are summarized in Table 12.22. Most borrowers under the direct lending program find that the MEDU loans carry lower interest rates and are more easily accessible than other sources of loans. This is also true for the borrowers under indirect lending. Some others find the collateral-free nature of the MEDU loans and easy repayment installments to be attractive features of the loan program. While only 10 and 5 percent of the respondents under direct and indirect lending programs have reported problems, the corresponding figure is as high as 27 percent for the nonparticipants. Entrepreneurs who borrow from non-MEDU sources mention high interest rates, inadequate loan size, and misguidance from bank officials to be important elements that characterize non-MEDU loans. In contrast, only a few of the MEDU borrowers have reported facing difficulty in getting the MEDU loan on time (Table 12.23).

MEDU Performance: Targets, Achievements, and Recovery Rates

At the planning stage, MEDU has fixed targets regarding the number of enterprises to be supported and the amount of loans to be disbursed. The two targets also imply prior planning regarding average loan size. These targets are fixed for each branch in cases of direct lending, and for each NGO/MFI location in cases of indirect lending. A summary of statistics on targets and achievements is presented in Table 12.24, which reveals some in-

TABLE 12.22. Reasons for identifying MEDU as a better source of loan (% of participants).

	Direct lending		Indirect lending	
	1st reason	2nd reason	1st reason	2nd reason
Easy access	31.82	23.81	100	
Easy installment	9.09	38.10		
Lower interest rate	50.00	14.29		100
Mortgage/collateral free	4.55	14.29		
Other	4.55	9.52		

Source: Author survey.

TABLE 12.23. Problems faced by different borrowers.

	Nonparticipants	Participants of direct lending	Participants of indirect lending
Percentage of borrowers facing problems	26.7	10.0	5.0
Of those facing problems, percentages reporting			
Insufficient amount of loan	70.0		
Misguidance by bank officials	10.0		
High interest rate	20.0		
Not getting loan in time		100.0	100.0

Source: Author survey.

TABLE 12.24. Summary statistics on targets and achievements, end April 2000.

Item	Indirect/credit retailer	Direct lending
Targets		
Number of enterprises	1,112.0 (113)	2,835.0 (45)
Loan disbursement (000 Tk)	247,050.0 (1000)	396,726.0 (8090)
Average loan size (000 Tk)	222.2 (8.85)	139.9 (179.8)
Achievements		
Number of enterprises	68.0 (17)	6,196.0 (156)
Loan disbursement (000 Tk)	216,475.0 (1000)	449,172.2 (9590)
Average loan size (000 Tk)	3,183.5 (58.8)	72.5 (61.5)
Achievement as percentage of target		
Number of enterprises	6.1 (15.0)	218.6 (346.7)
Loan disbursement (000 Tk)	87.6 (100.0)	113.2 (118.5)
Average loan size (000 Tk)	1,432.9 (664.4)	51.8 (34.2)

Source: Author survey.

Note: Figures in parentheses are for the two study areas, Kapasia under indirect lending and Keshabpur under direct lending. Our estimates on average borrowing are substantially lower for Kapasia.

teresting dimensions of MEDU operations. Information specific to the two study areas is also included in parentheses for comparison with the survey findings. The important conclusions are discussed below:

1. It has been noted previously that MEDU could commence its indirect lending program only at a later stage of the project due to lack of interest on the part of NGOs. It is therefore no wonder that only 6 percent of the target enterprises has been reached through this program. Interestingly, however, almost 88 percent of the target disbursement under the program has been achieved, with a consequent high average loan size figure (more than 14 times that of the target). This raises doubt about the actual uses of the MEDU resources by the participating NGO/MFIs. Even after adjusting for the average number of loans taken, actual disbursements to the borrowers (by the NGOs) fall far short of what has been reported. It suggests that a large share of the MEDU funds are spent by the NGOs for purposes other than those intended by MEDU. It is quite possible that these NGOs use a substantial part of the MEDU funds to support their normal microfinance programs.

2. An exactly opposite scenario emerges in cases of direct lending, i.e., the number of enterprises is more than double the initial target, and the average loan size is consequently half the initial target.[39] MEDU appears to have been quite conservative in its lending under the direct program; given what is known about the fixed capital requirement for microenterprises, actual disbursement figures are low. It is therefore expected that much of the MEDU direct lending program has financed working capital requirements of existing enterprises.

Regarding the recovery rate of MEDU lending, a comparison of MEDU recovery rates is made with those of Grameen Bank—the highly reputed successful microcredit pioneer (Table 12.25). The recovery rates are roughly comparable for Bangladesh as a whole—there are differences between the two in regard to regions of Bangladesh. In both cases, the recovery rates have declined during the late 1990s. Compared to the norms in nationalized commercial banks (where the recovery rates have been around 30 percent), MEDU's performance has so far been satisfactory. It should be noted that MEDU's target group, relatively the richer entrepreneurs in the rural and semirural areas, is perceived to have contributed to large-scale defaults on the commercial bank loans in the past. Therefore, compared with the commercial banks, MEDU's performance is to be commended.

TABLE 12.25. Recovery rates: A comparison.

District/division	MEDU direct lending		Grameen Bank	
	1999-April 2000	1995-2000	December 2000	December 1997
Chittagong	85	91	96.78	95.17
Dhaka 1	81	88	86.52	89.41
Dhaka 2	50	76		
Khulna	84	92	96.88	97.68
Rajshahi	77	87	91.28	95.84
All Bangladesh	79	88	80.36	90.55

Source: MEDU and Grameen Bank's MIS.

Note: (1) The second column under MEDU is on cumulative basis over the whole project period. When the figures in the first column are less than those in the second column, they suggest declines in recovery rates; (2) Grameen Bank reports on the percentage of loans overdue for more than a year, from which the recovery rates have been calculated (by deducting from 100); (3) For MEDU, the label column describes regions, although they denote new districts for Grameen Bank—thus, the units are not comparable.

TABLE 12.26. Average size of loan, number of loans, and default rate across participants and nonparticipants during 1995-1999.

Items	Non-participants	Participants—direct lending	Participants—indirect lending
Average number of loans	2.8	2.74	1.4
Average number of MEDU loans	0	2.35	1.4
Average loan size (taka)			
Other NCBs	13,750	–	–
Agrani Bank	12,542	53,333	–
MEDU–Agrani Bank	–	40,785	–
MEDU-NGO	–	–	15,854
NGO	4,000	–	–
Others	3,357	–	–

Source: Author survey.

Note: For participants of NGOs' lending, the recovery period was not yet over.

TABLE 12.27. Aggregate investment by different groups of sample enterprises (investment figures are in thousand taka).

Participation/items	1995	1996	1997	1998	1999
Nonparticipants					
Working capital	1,064	1,775	2,349	2,714	3,421.7
Fixed capital	4	84	13.5	122	504
Number reporting fixed capital	2	5	5	7	7
Participants in direct lending					
Working capital	865	985.5	1,284	1,824.5	2,270
Fixed capital	107	315	681	694.9	662.5
Number reporting fixed capital	3	5	9	15	11
Participants in indirect lending					
Working capital	612	970	1,190	1,634	2,235
Fixed capital	0	0	13	93	270.5
Number reporting fixed capital	0	0	2	5	10

Source: Author survey.

Performance Across Enterprises

Information on loan sizes, investments, and their sources, sale proceeds, capacity utilization, and employment are summarized in Tables 12.26 to 12.30.

As expected, average loan sizes are higher under direct lending, and these are comparable with loans availed previously by the same group from Agrani Bank (Table 12.26). Average loan sizes are smaller under indirect lending and are comparable with those contracted by the nonparticipants from other nationalized commercial bank (NCB) sources. Average investments on fixed

TABLE 12.28. Annual gross sale proceeds from main products (000 Tk/enterprise per year).

Participation status	Pre-MEDU	Post-MEDU
Nonparticipants	1,900	1,871
Participants, direct lending	993	1,136
Participants, indirect lending	548	1,585

Source: Author survey.

Note: Enterprises that reported positive sales in all five years have been included.

TABLE 12.29. Average employment and wage bill during pre- and post-MEDU periods.

Items	Nonparticipants	Participants of direct lending	Participants of indirect lending
Pre-MEDU period			
Employed person per enterprise	6.75	6.03	4.68
Days of employment/ person/year	285	290	280
Annual wage bill per person (Tk)	12,410	12,276	13,324
Post-MEDU period			
Employed person per enterprise	5.96	6.95	5.33
Days of employment/ person/year	279	290	273
Annual wage bill per person (Tk)	10,564	12,419	15,164

Source: Author survey.

Note: Enterprises that were in operation for the entire five-year period for which information was collected have been included in calculating the averages.

TABLE 12.30. Number of installments by purpose.

Purpose of loan	Total number of installments
Poultry	4
Fishery	4
Dairy cow	50
Beef fattening	1[a]
Rickshaw	50
Mortgage	50
Nursery	4
Banana and papaya plantation	2
Hand tube well	50

Source: Author survey.

[a]Interest is deducted during disbursement.

capital are smaller under indirect lending, compared to those under direct lending (Table 12.27). Over the program period, the MEDU participants have progressed well compared with the nonparticipants in terms of increases in sale proceeds, capacity utilization, and employment (Tables 12.28 and 12.29).

Financial Sustainability of MEDU Operations

Given that the project has an extensive monitoring schedule, the overall sustainability hinges upon two factors: (1) whether the relevant bank officials have sufficient incentives and (2) whether the project is financially viable.

As described earlier, the project has some pecuniary and nonpecuniary incentives for the bank officials participating in the MEDU program. Apart from the provision of incentives for good performers, there are also some disincentives, albeit in indirect form, for bad performers. If a branch performs badly, the concerned branch manager and MECRO are identified as bad performers by the top-level management. This will seriously jeopardize their futures. The bank officials involved in direct lending would prefer to avoid such consequences, and will also have the financial incentive to follow the monitoring schedule and perform better.

The government is committed to provide a subsidy of Tk 3 for a disbursement of Tk 100, regardless of the repayment period. Since the repayment period varies across loans, the weighted sum of this subsidy becomes equivalent to 2.28 percent on disbursement per annum. The project was not financially viable at the initial stage. However, since 1997-1998, the project has started to make a net profit. Although government subsidy and losses during the early years affected the outcome, the project made net profits in 1998-1999 and 1999-2000, even when government subsidy is excluded.

Lessons for MEDU Experience

The important lessons include:

1. The limited evidence on the MEDU experiment clearly shows that group lending or peer monitoring is not a prerequisite for advancing collateral-free loans—often, good banking may encourage good response from the borrowers! Close supervision, incentives to bank officials, flexibility in repayment, and appropriate loan sizes are some of the factors in such banking.

2. It is unclear how long such performance may be sustained, even with the incentive structure that is in place. Until now both pecuniary and nonpecuniary incentives appear to have worked in most branches. However, such positive response to incentives may be short-lived. The recovery rates have been on decline, and currently stand at around 80 percent (for FY 2000), with some branches performing worse than an average non-MEDU branch. Moreover, the reported profit from MEDU operations is shown to be positive because it does not yet consider the overdue loans in the loss portfolio. If the project currently sponsored and financed by an external donor, i.e., IFAD, is not actively owned by the bank's management and if it suffers from the envy of nonbeneficiaries within a large corporate enterprise, the initial success of the project is likely to degenerate. It may also be noted that having a floor recovery rate (90 percent) to generate a positive honorarium to the bank officials may backfire, since there will be a tendency for the recovery rate to decline further if it gets significantly lower than the floor rate.

3. It has been noted that the borrowers from the NGOs, under indirect lending, are not graduates of the NGO's normal program. These clients have also been found to be running enterprises that involve large sums of money. In the absence of further investigation into this issue, two possible explanations are offered. First, some entrepreneurs run their business with their own funds or with loans from informal sources, and yet do not seek loans from the commercial banks due to large hidden costs involved in transacting with the banks. They may feel more comfortable with an NGO as the credit retailer. A second possibility is the likely collusion between erstwhile defaulters of the commercial banks (who have long been unable to seek new loans

from the commercial banks) and the partner NGO. The indirect lending gives a 3 percent commission to the partner NGOs. The borrowers under indirect lending may never decide to borrow under direct lending, even under the best of circumstances, if the respective NGO shares a part of the commission it receives from MEDU.

The survey generally suggests that there is truly a demand for credit size exceeding the current offers made by the MFIs. However, the regular loans from the nationalized commercial banks (NCBs), offered in the rural and semirural areas, can meet such size-specific demand. At the same time, microenterprises, defined by employment size, may vary widely in terms of capital intensity and their demand for credit may vary widely.

Is there a need for special credit for the entrepreneurs in the "missing middle" group? One may identify a number of reasons.

1. A demand exists for a loan size that exceeds the amount being currently offered by the MFIs; the aggregate demand for such loans is less than that offered by the NCBs.
2. Demand for such loan capital is normally associated with microenterprise development, and credit supply for the microenterprises will meet such demand.
3. Regular lending of the NCBs may be biased against the microenterprises because of one or both of the following reasons: (a) profitability of the sector is low and therefore banks find them risky, (b) profitable projects exist, but the entrepreneurs cannot access the NCB loans because they either cannot offer collaterals or for social and cultural reasons the bank officials are inaccessible to them.

Among the three, the last rationale, including the two elements, appears more appropriate. To encourage disbursement to a particular sector, which is otherwise not considered profitable, one may justify giving additional incentives to the banks. Similarly, to induce investment in that sector, clients/potential investors may be offered interest rates lower than those offered elsewhere. All these elements are there in the MEDU project design. Moreover, instituting a separate project office and delegating a few officials as monitors recognizes the difficulties in running the MEDU operations within the current setup and incentive structure. Bank officials are expected to be more accessible when positive financial benefits are attached. Our findings lend support to these two aspects of MEDU—lower interest rates and more accessibility. However, one would expect collateral-free loans to be an additional factor in making bank loans more accessible.

KISHOREGANJ SADAR THANA PROJECT: A COMMUNITY-BASED APPROACH

Introduction

Following the recommendation endorsed in the 1993 Dhaka summit of the South Asian Association for Regional Cooperation (SAARC) Heads of State or Government, the United Nations Development Programme (UNDP) initiated and sponsored a subregional project titled "Institutional Development at the Grassroots for Poverty Alleviation." A demonstration pilot project was started in Kishoreganj Sadar Thana (KST) of Kishoreganj district, in October 1994. The KST approach differs from other approaches to poverty alleviation and rural development in a number of ways. Important among them are its nonexclusion principle and built-in incentives for village organization (VO) managers in running the credit program.

Of the major activities undertaken under the KST Project, the following are noteworthy:

1. Organize villagers under the umbrella of village organizations (VO), to be led by a president and a manager from among the members of the organization.
2. Encourage savings mobilization through VOs.
3. Establish contacts among the VOs and government (and other) agencies engaged in the delivery of services. One important agency with whom contacts have been established to facilitate delivery of credit to VO members is the Bangladesh Krishi Bank (BKB). Others include the thana-level government officials, especially those in the agriculture extension, livestock, and fishery departments.
4. Initiate and administer a credit delivery system that can obtain funds from a number of sources, in which VO managers play a key role.
5. Encourage social mobilization[40] and investment on social infrastructure (e.g., schools) through the VOs.
6. Provide training on matters related to (a) bookkeeping (which is expected to lead to better management of the savings and credit programs) and (b) to activities, agricultural and nonagricultural (which are expected to ensure timely delivery of extension services at the grassroots level through a cadre of social activists).

Organizational Structure

The social organizers (SOs) and the VO managers, including the auditors chosen from these managers, are the key players at the grassroots level. Under the guidance of the project coordinator, the SOs organize groups (VOs) to oversee the election of the manager and president,[41] motivate VO members for social mobilization, supervise programs to propagate the idea of savings mobilization, regularly attend group meetings,[42] monitor records on financial transactions maintained by the VO managers, approve loan applications and forward them to the project coordinator, assist subject specialists in organizing training and select participants for the training as well as liaison with bank officials and other government officials at the thana level.

The grassroots unit through which the KST model operates is the village organization. It is important to note that the VOs in the KST are essentially *para*-based organizations, and there are often three or more VOs in a village.[43] Ample sociological evidence suggests that there are kinship ties among households located within a *para,* even though they may not be at the same economic and financial level. The relative strength of vertical links vis-à-vis horizontal links had once been an important subject of discourse among rural sociologists, and there is no a priori reason to believe that one always supersedes the other. Under the circumstances, it is quite plausible that some forms of activities are better pursued through social institutions with vertical links. Thus, KST-VOs, primarily functioning as conduits of credit delivery, and organized at *para* levels, are likely to be more inclusive than one would expect from village-level community-based approaches.

A VO has a president, normally an elderly person, and a manager who must be educated to do all the record keeping. The manager effectively runs a VO. Since he or she should have a minimum level of education, it is most often the case that the manager hails from a rural rich family. The choice of president is influenced by a host of factors: maintaining harmony between factions within the same *para;* choice of someone from the poor segment; or choice of an elderly person who commands respect from all groups.

The VO managers are responsible for the following:

- organizing VO meetings;
- motivating others to join the VOs;
- maintaining records on resolutions of the meetings as well as records of savings and loan repayments;
- assisting members in filling out loan application forms;
- regularly (normally, once a week) visiting the bank branch (where the account is maintained) to conduct financial transactions;
- maintaining regular contact with the respective SO;
- attending meetings with other VO managers in the ward (called the cluster meeting);

- participating in training and nominating members for training; and
- completing monthly report forms that are submitted to the SOs.

Some of the VO managers have the additional responsibility, at supplementary remuneration of Tk 500 per month, to act as auditors. As auditors, they are expected to audit all records on financial transactions of the VOs in a given ward; some of these reports are to be submitted once per month.[44]

The current structure has evolved over time. It is important to note that the structure and the activities of the project largely center around savings mobilization and credit operation. The intended link with government departments is yet quite loose and social activists are expected to play an important role in ensuring extension services. In the case of financial services, the Bangladesh Krishi Bank is the key outside organization involved; funds are made available from UNDP sources as well as from the savings of VO members.

Main Features of the KST Credit Program

Unlike many other MFIs, the KST credit program does not have a target group that constitutes an observable segment of the rural society. This saves them from any additional effort in monitoring targeting. At the same time, declared inclusion of the rural rich widens the market for credit, which is crucial for enhancing the financial viability of grassroots organizations (VOs) delivering such credit. Third, the structure of incentives provided to different agents involved in the KST approach differs significantly from those in other MFIs. In the case of a conventional MFI, the beneficiaries (i.e., group members) are distinctly separate from the MFI staffs who deliver credit and monitor repayment. The service charges normally accrue to the respective MFI fund, out of which salaries are paid to the staff. In contrast, the KST-VO managers get a significant share of the service charges, which is often enough to make a decent living for a rural youth.

A fourth departure of the KST credit program is in the mix of funds available for loan disbursement. Most MFIs have their own seed capital and they also draw from members' savings. Some of the MFIs also borrow from such institutions as the Palli Karma Sahayak Foundation (PKSF) and lend to their beneficiaries.[45] The KST credit program, in contrast, has three distinct sources of funds: individual savings, commercial and specialized banks, and supplementary funds provided by the UNDP. Since rural rich (VO members) are free to contribute to savings, the latter has seen phenomenal increases in size.[46] Interest payments received from lending out of individual savings accrue to the members, and the UNDP funds provided are also

VO-specific so that among VO members there is a sense of owning the funds.

A final difference lies in the way the groups are organized and made to function. The MFIs, targeting the poor, organize groups whose size ranges from 5 to 30; sometimes, several small groups are federated into one large one. In the case of the KST project, the minimum size of a VO is around 35, and the size often is as high as 100. Since the VO managers bear the primary responsibility for ensuring repayment, with which their incentives are tagged, they have found it convenient to regroup the members into smaller subgroups, with their respective leaders.[47] It is also important to note that conventional MFIs normally target the women, even though some of them have lately organized male groups. In contrast, the KST model has targeted both male and female groups from the very beginning.[48]

Modality of Credit Disbursement, Repayment, and Incentives

The loan disbursement process is initiated in the weekly meeting of the VO. The meeting decides the allocation of loans among competing applicants, primarily on the basis of the following factors:

1. the amount of savings of the applicant;
2. the repayment ability of the applicant;
3. the purpose of the loan and the ability of the applicant to carry out the stated activity;
4. whether the applicant has taken any training on the proposed activity;
5. the need for capital of the applicant, evaluated on the basis of his or her socioeconomic background; and
6. other members' opinions about the repayment behavior of the applicant.

The applications of the selected members, signed by the VO manager and approved by the SO, are sent to the project officer for final approval, while Bangladesh Krishi Bank disburses loans to the enlisted members. In the case of individual savings and UNDP funds, the members need not go to the bank after they have been approved. The manager withdraws the money on their behalf and disburses it among them. In case of BKB loans, every borrower must go to the bank to receive his or her loan.

The total number of installments for the repayment of loans varies, depending upon the main purposes of the loan as presented in Table 12.30. The variation in the total number of installments is permissible in cases of the VO's own savings; any loan taken from UNDP sources irrespective of loan purposes must to be paid in 50 weekly installments. Every borrower is required to make repayments to the manager in the weekly meetings. Ideally, a manager is required to deposit the collected money to the respective bank on the next day. However, due to absence of regular monitoring, managers often hold on to the collected money.

The practice of group lending is not strictly enforced in the KST model. However, since the remuneration to the VO manager is linked to recovery, the latter has an incentive to put pressure on the defaulters, including visits to the defaulter's house with some other members to put social pressure on the borrowers to repay. Since the VO manager more often than not comes from a relatively rich family, he or she has an extra leverage to pressure a defaulter.

All types of loans, regardless of the source of funds and the repayment process, charge an interest of 15 percent per year. Since both principal and interest are paid together from the second week, the effective interest rate stands around 34 percent per year. Total interest receipts from loans granted from the UNDP fund are distributed between the manager and all the members. Managers also get 5 percent of the 15 percent interest charged on loans financed by BKB. Total interest received from VO's own savings is distributed among the VO president, all VO members, the risk fund, the manager, and the organizational fund (Table 12.31).

Auditing was introduced in late 1995 to crosscheck total weekly collections and bank deposits. Most of the managers are the managers of VOs. An auditor must attend cluster, ward, and union meetings. On average, an auditor is required to audit 35 VOs each month.

An Appraisal of the KST Project: Its Achievements and Prospects

Broadly, three sets of issues call for attention in the particular case of the KST approach. The first pertains to the role of VOs as models for rural development. The second relates to the efficacy of the institutional framework

TABLE 12.31. Sources of credit and distribution of interest receipt (in %).

| Sources | Distribution | | | | | |
	Managers	President	Member	Group fund	Risk fund	BKB
Own savings	3.75	2.25	5.25	2.25	1.5	0
BKB	5	0	0	0	0	10
UNDP[a]	5	0	0	10	0	0

[a]UNDP refers only to UNDP1 fund.

in mobilizing rural savings and administering credit operations as distinguished from the conventional target-group-based approach of the MFIs. The third set of issues has a narrower focus on the role of the KST model in promoting rural nonfarm employment. The first set of issues is not discussed in this study. However, the second and third sets of issues are discussed since (as in the case of MEDU) credit delivery is instrumental in promoting economic activities, including that of the nonfarm sector.

The following analysis is based on primary and secondary information as well as on primary data collected using three different instruments: interviews, *focused group discussions,* and structured questionnaires. Secondary information on disbursement and nonfarm activities was collected from the project office.

In administering the structured questionnaire, samples have been drawn in two stages. It has been observed that nonfarm activities promoted by KST are concentrated mainly in the areas adjacent to Kishoregonj district towns. It is expected that the problems in promoting nonfarm activities in remote rural areas (away from district towns) are different. To study the differences, we selected two male VOs, Brahman Kachuri North and Barakapon, which are in close proximity to Kishoregonj district town and two (one male and one female) VOs from Hazirgol, which is a remote village. Hazirgol has a poor infrastructure. Initially, a census of all members in the selected VOs was taken. A structured questionnaire was later administered to a selected number of VO members, who had borrowed for nonfarm activities. Apart from this, all types of nonfarm activities (which have been promoted by KST but not found in these two areas) were also covered.

The national project director and relevant project officials, e.g., credit and nonfarm activity specialists, were interviewed. Focused group discussions were arranged with four VOs, namely Brahman Kachuri North (male), Hazirgol (male), Hazirgol Middle (female), and Barakapon (male).

Social Mobilization versus Credit Retailing: Identifying KST Activities

During the inception of the project and the early years of its operation, social mobilization was identified as the major objective of the project. This was to be achieved by forming village organizations, motivating the VO members to save and participate in social actions, and by training some of them as extension agents to access the government services in a more effective manner. The VOs formed during the first two years (1994-1995 and 1995-1996) were also encouraged to develop "village plans" to promote their villages. Ironically, this was not a saleable idea. When the participants came to know of the "no credit" package offered by the UNDP project, participation in the VOs did not appear very lucrative. Only 112 female and 59 male VOs were formed during the 1995-1996 fiscal year.[49] More important, a decline occurred in average size of the VOs as well as a decline in the total number of VOs in certain unions during this period. This early experience suggested that social mobilization of the rural people in Bangladesh was not possible unless and until it was attached to resource delivery of some sort.

The KST management recognized the problem, and credit from savings funds soon commenced. This did not prove sufficient, and BKB loans were introduced during 1996-1997. Along with this came the UNDP fund, which was allocated to individual VOs out of which credit was disbursed. No looking back has occurred since then, and all evidence cited so far in different studies suggests that the primary function of the VOs and the project office centered around savings mobilization and credit disbursement. Shares of different sources of credit funds are presented in Table 12.32.

Reaching the Poor Through Community-Based Organizations

The KST model targets the whole village population. It is commonly perceived that village organizations quickly turn into "clubs of the rich and for the rich." In general, this is not true in the case of KST-VOs. A census of all members in a selected number of VOs undertaken by the present study reveals that 50 percent of all members are functionally landless, while another 21 percent originate from households owning 0.5 to 1.0 acres of land. More important, a higher percentage of land-poor households than that of land-rich households are found to receive credit (Table 12.33). Information

TABLE 12.32. Shares of different sources of KST credit funds.

Source	1996	1997	1998	1999
Savings	50.53	58.13	44.72	10.87
BKB	43.99	36.82	30.64	35.84
UNDP[a]	5.48	5.05	24.65	12.57
UNDP[b]	0	0	0	40.72

Source: Author survey.

Note: Both UNDP[a] and UNDP[b] refer to funds provided by the UNDP to the VOs through KST. UNDP[a] refers to the first allocation received from UNDP in late 1996; UNDP[b] refers to the allocation received during 1999. UNDP[b] is often identified as BGD/UNOPS. While distribution of interest earned from UNDP[a] lending is finalized, till the time of our survey, no decision was made with regard to the decision on interest receipts from UNDP[b].

in terms of self-perceived poverty ranking also suggests that the poor are not left out of the VO membership. Differences in the average number of loans taken between developed and backward villages (Table 12.34) are not large enough to indicate any pronounced bias in favor of the rich.

Access to credit is generally believed to positively affect the well-being of the recipient households. A comparison across VO members, members of other MFIs, and nonmembers (in Tables 12.35 and 12.36) suggests that a

TABLE 12.33. Percentage distribution of VO members, borrowers, and average size of current loan by land groups and socioeconomic condition.

Land groups	% of VO members	% of respondents who borrowed from KST	% of borrowers for nonfarm activities	Average size of loan for nonfarm activities
Up to 50	50.47	50.90	58.90	6,906
51-100	21.23	22.16	20.55	9,533
101 and above	28.30	26.95	20.55	13,266
Poverty ranking				
Rich	16.04	14.37	14.89	14,100
Middle	40.09	39.52	35.11	9,484
Poor	34.91	36.53	41.49	6,772
Very poor	8.96	9.58	8.51	4,500

Source: VO Census and Field Survey, Kishoregonj, 2000.

TABLE 12.34. Average number of loans received by period of joining.

Village organization	Period of joining the VO			
	1994-1996	1997-1998	1999	All periods
Developed village				
Brahman Kachuri, male	3.35	1.11	0.07	2.12
Barakapon, male	2.14	1.11	0.20	1.61
Backward, Hazirgol village				
Male	1.91	1.44		1.70
Female		0.76		0.76
All VOs	2.93	1.00	0.09	1.78

Source: Author survey.

Note: The male group in Hazirgol was formed during the latter part of 1994-1996.

greater percentage of the VO members have increased their savings and have invested in agricultural machinery. In regard to asset accumulation, the VO members are found to have a bias in favor of such unproductive assets as television, watch/clock, and ornaments.

TABLE 12.35. Household level changes during 1997-1999, selected villages in Kishoreganj Sadar Thana.

	Percentage of sample households in each group			
Situation improved/ increased	VO member	Dual membership in VO and other MFIs	Member of other MFIs	Not member of either VO or other MFI
Financial	63	89	50	80
Food status	47	50	17	50
Law and order	42	44	50	60
Savings	100	83	50	40
Investment on increase				
Fishery	47	67	50	40
Poultry	5	0	17	0
Agricultural machinery	11	11	0	0
Commercial transport	11	0	17	0

Source: Table A.3.2 in Zohir et al. (2000).

TABLE 12.36. Percentage of households with net accumulation of assets over 1997-1999.

	Percentage of sample households in each group			
Type of assets	VO member	Dual membership in VO and other MFIs	Member of other MFIs	Not member of either VO or other MFI
Land	21	22	16	20
House	21	33	17	20
Bullocks	5	5	16	30
Ornaments	37	17	0	30
Bicycle	−5	17	0	10
Watch/clock	32	11	17	−10
Television	16	6	0	0

Source: Table A.3.3 in Zohir et al. (2000).

The VO Managers and the Performance of the Credit Program

Although evidence in the recent past suggests all-inclusiveness, there are fears that the institutional dynamics will eventually lead the VOs to be overtaken by the richer sections of the rural community. Since the VO managers are identified as the key individuals to determine the future direction of the program, an outline of their profile is discussed in Table 12.37.

TABLE 12.37. Gender-wise percentage distribution of VO managers across different occupational groups, different levels of education, and age.

	Percentage share of managers		
Factor	All VOs	Male VOs	Female VOs
Level of education			
Just read and write	1.53	1.55	1.50
Up to class five	6.34	5.31	7.50
Between classes seven and ten	42.02	39.38	45.00
SSC or equivalent	31.46	32.08	30.75
HSC or equivalent	12.32	13.72	10.75
Graduate	6.33	7.96	4.50
Total	100.00	100.00	100.00
Occupations			
Agriculture	19.74	30.07	8.06
Service	11.70	13.14	10.08
Small business	24.59	37.42	10.08
Housewife	23.05	1.11	47.86
Village doctor	1.77	2.45	1.01
Student	16.08	11.36	21.41
Pisciculture	0.24	0.45	0.00
Tailoring	0.59	0.45	0.76
Technical works	0.71	1.11	0.25
Others	1.53	2.44	0.49
Total	100.00	100.00	100.00
Age group (in years)			
15-25	38.01	29.57	47.55
26-35	40.56	42.55	38.32
36-45	16.71	21.15	11.68
46-55	4.72	6.73	2.45
Total	100.00	100.00	100.00

Source: Zohir (2000).

About two-thirds of the VO managers have secondary levels of education. The type of work that a manager must do requires some expertise in accounting and bookkeeping. Very few persons with only primary education have succeeded in acquiring this expertise to assume the responsibility of a VO manager. Only about 6 percent of the VO managers are graduates.

In respect to their occupation, the highest percentage of male managers is engaged in small business, followed by agriculture. More than two-thirds of the total male managers belong to these two occupational groups. Almost half of the managers of the female VOs are housewives. More than 20 percent of the female managers are students, mostly at the level of secondary education. These two groups together constitute more than the two-thirds of all the total managers of the female VOs.

The average age of a VO manager is found to be as high as 29 years; the average age of the managers of the male VOs is about 31 and for female VOs only 28 years. A larger percentage of managers of female VOs belongs to the under-35 age group.

Most of the factors taken into consideration for disbursing loans are subjective and dependent on the discretion of the VO managers. No threshold level of savings needs to be achieved as a prerequisite for getting a loan. The level of training or education of the borrowers is not seriously taken into account in deciding the eligibility of a borrower or the amount of loan. Evidence suggests that a VO member who has been trained in fishery has received a loan for a nursery. Lack of well-defined rules, coupled with the subjectivity involved in the selection process, paves the way for the VO managers to exercise their discretionary power. The manager, being the only VO level signatory to a loan application, has emerged as the key figure in the whole selection process. Some managers even perceive that it is their prerogative to decide who will receive credit and how much, since they are mostly responsible for maintaining a good recovery rate. Members are considered to be unruly if they make queries regarding the total inflow and outflow of the organization's funds. When asked about the loan delivery process, invariably everyone replies in the first person saying "I will not give any credit to an unruly member." Some managers are quite explicit about their autocratic attitude and make such statements as "I will quickly expel an unruly member because his presence will create distrust and raise doubt about my honesty among other members, which is bad for the organization."

During the field survey, credit program monitoring was found to be very weak. It is generally the responsibility of the auditors, who are a subset of VO managers, to monitor the financial transactions. Given the workload, an auditor is unable to audit every VO in every month. Since an auditor cannot check all the accounts and since they are themselves the VO managers, the financial irregularities perpetrated by managers are not reported. Upon verifying the bank register and comparing the financial transactions of a number of VOs with those reported in audit reports, Zohir (2000) found gross discrepancies between the two. The incidence of misreporting in the audit reports has been increasing over time.

In spite of the previously mentioned deficiencies in the KST credit program, there are reasons to believe that most of the VO managers would like to see the program sustained. Estimated recovery rates are comparable with those of many MFIs (Table 12.38). The benefits that a VO manager receives from the program have been mentioned. Other than those associated with training, networking, and easy access to credit, Zohir (2000) estimated an annual average income of more than Tk 9,000 for a VO manager during 1999. Given their academic background and opportunities elsewhere it is argued that, with appropriate monitoring in place, the KST credit program has a good prospect for success.

Nonfarm Sector Within a KST Framework

The statement of objectives of KST makes no explicit mention of the promotion of nonfarm rural employment. It does, however, mention the overall development of the villages through organization and utilization of local savings and external resources. Since credit has so far been the main vehicle of promoting rural development, its contribution to the development of the nonfarm sector is discussed in the following text.

Since computerized data on sector-specific allocation of the KST do not seem to reflect the uses or purposes of loans as stated by the clients, information has been collected from all the members of a selected number of VOs. About 39 percent of total loans disbursed to the members of these VOs is reported to have been used for nonfarm activities. Tables 12.39 and 12.40

TABLE 12.38 Recovery rates under the KST, by sources of fund.

Source of fund	1998	1999
Group savings	93.00 (67.53)	62.48 (63.36)
BKB	89.48 (71.19)	95.14 (61.59)
UNDP1	95.22 (75.94)	90.09 (62.14)
All sources	91.47 (70.41)	88.86 (62.36)

Source: Author survey.

Note: Figures in parentheses include only past due in its calculation.

TABLE 12.39. Information on borrowings by male VO members, by year of joining KST.

Year of joining KST	% of respondents	% of all borrowers (% who borrowed in each group)	Average number of loans	% of loan amount taken for nonfarm	Average loan size (Tk) for nonfarm	Average loan size (Tk) for farm activities
1994	9.5	12.0 (100.0)	2.29	15.8	5,500	9,000
1995	36.3	45.1 (98.5)	3.38	30.1	14,353	12,326
1996	10.6	13.4 (100.0)	2.16	56.2	6,833	9,143
1997	9.5	12.0 (100.0)	1.65	81.2	8,643	9,333
1998	16.2	15.5 (75.9)	1.09	45.2	8,889	7,462
1999	17.3	2.1 (9.7)	1.00	50.0	5,000	10,000
2000	0.6	0.0 (0)	–	–	–	–

Source: Author survey.

TABLE 12.40. Information on borrowing by male VO members, by financial status.

Self-perceived financial status	% of respondents	% who borrowed	% of loan amount taken for nonfarm	Average loan size (Tk) for nonfarm	Average loan size (Tk) for farm activities
Rich	13.97	12.7 (72.0)	37.4	22,000	15,333
Middle	39.66	39.4 (78.9)	47.8	10,429	11,393
Poor	36.87	38.0 (81.8)	27.1	6,889	9,278
Very poor	9.50	9.9 (82.4)	41.8	4,714	6,571
All	100.00	100.0 (79.3)	38.8	9,638	10,639

Source: Author survey.

indicate the purposes of borrowing by different groups of borrowers, defined in terms of year of joining, land ownership, and self-perceived financial status. A number of observations are worth noting.

1. Members from all groups have borrowed from KST.
2. Generally, the poorer households tend to spend a larger share of their loans on nonfarm activities.
3. For rich borrowers, the average loan size for nonfarm activities is larger than that for farm activities.
4. The opposite is true for the poor borrowers, whose loan size for nonfarm activities is smaller.

Most productive nonfarm activities are of low productivity, which attracts mostly the poor households, especially in villages not well endowed with infrastructure. Villages that are farther away from the thana headquarters or towns (or main roads), are offered opportunities for such low productivity occupations as vendors and other petty trades (Table 12.41).

About 17 percent of loans are used in agro-processing, which primarily includes rice mills (Table 12.42).[50] All the others have gone to the service sector, mainly shops and trade. In spite of this pattern of lending, the KST program with its large infusion of credit into rural Kishoreganj has brought significant changes in the occupational structure. Table 12.43 indicates the change in the primary occupation of the VO members. Almost one-fourth of former agriculturalists and one-fifth of the former laborers have stated "business" as their primary occupation. The number of unemployed persons has been drastically reduced and some of the rickshaw pullers have found it lucrative to move into other activities.

The previous findings are based on a sample survey and do not adequately capture the other nonfarm activities that were promoted by the KST. One important activity was poultry farming. Initially, some of the investors incurred losses due to a lack of adequate knowledge and an absence of market links. This has gradually been overcome; a project office, with access to information through the network of VOs, has enabled the farmers to establish links with outside markets. This indicates the potential of the KST framework to develop producers' cooperatives in areas where nearby markets are not readily available.

Another important element in the KST framework can be used to promote nonfarm activities. The proposed model has the provision for training village-level extension workers (volunteers), who can then be linked up with the thana-level government agencies through ward and union level coordination. These activities have related traditionally to agricultural exten-

TABLE 12.41. Information on borrowing by male VO members, by developed and backward region.

Development status	Village	% who borrowed	% of loan amount taken for nonfarm	Average loan size (Tk) for nonfarm	Average loan size (Tk) for farm activities
Developed	Brahman Kachuri	74.0	34.9	12,724	11,279
	Barakapon	86.1	36.0	7,833	8,789
Backward	Hazirgol	100.0	77.4	5,647	9,333

Source: Author survey.

TABLE 12.42. Borrowing by the nonfarm sector, by activities.

Nonfarm activity	Percent of borrowers	% of current loan amount	Average size of current loan (taka)
Rice mill/trade	9.5	15.80	15,000
Fish processing	1.4	1.70	11,000
Agroprocessing	10.9	17.50	14,611
Cloth trade	4.1	4.20	9,333
Grocery shop	12.2	16.00	11,777
Timber trade	8.1	4.80	5,333
Rickshaw purchase	14.9	8.60	5,181
Hawker	2.7	2.30	7,500
Hotel/sweetshop	8.1	7.40	8,166
Trade—medium	6.8	6.00	8,000
Trade—small	8.1	5.90	6,500
Rickshaw garage	2.7	2.30	7,500
Mason	1.4	2.30	15,000
Shoe shop	1.4	2.30	15,000
Sugar trade	1.4	2.00	13,000
Cosmetic sale	1.4	2.30	15,000
Furniture shop	1.4	1.50	10,000
Rickshaw driver	2.7	1.80	6,000
Bookshop	1.4	1.50	10,000
Medicine business	5.4	6.20	10,250
Sand business	1.4	1.50	10,000
Others	4.2	3.90	8,333
Trade and services total	89.1	82.50	8,268
Total	100.0	100.00	8,959

Source: Census of VO members.

Note: Nonfarm sector accounted for 39 percent of total current loan.

sion. The KST experiment shows that the training of village assistants (extension agents) generates positive benefits, but it has not been possible to link this network of extension agents with the thana-level government services in an effective manner. The reason partly lies in the project design, which has not involved the local government officials from its inception and has possibly alienated them. However, having a technology specialist in the project office who is linked with the entrepreneurs through the SOs and the VO managers has the potential to promote nonfarm activities.

TABLE 12.43. Change in occupation of male VO members.

Before KST \ After KST	House work	Stu-dent	Teacher	Busi-ness	Unem-ployed	Agri-culture	Doc-tor	Ser-vice	Rice mill	Rick-shaw puller	Day labor	Car-penter	Hotel busi-ness	Fisher-man	Fish culti-vation	Guard	VO	Tailor	Row total
Housework	2																		2
Student		2	1	1		3											2		10
Business				30															30
Unemployed	1			3	1			1	1								1		8
Work abroad				1															1
Agriculture				16		47		1							1		1		66
Doctor							4												4
Service								8					1						9
Rickshaw puller				3		3				19				1	1				26
Day labor	1			4		2					10								18
Carpenter												2							2
Hotel business																1			1
Fisherman				1															1
Tailor																		1	1
Column total	4	2	1	59	1	55	4	10	1	19	10	2	2	1	2	1	4	1	179

Source: Author survey.

CONCLUSIONS: LESSONS FOR THE FUTURE

Although MFI's success has largely been identified in the areas of poverty alleviation and collateral-free banking with the poor, emergence of institutions with potentials beyond microcredit has too often been neglected. Most of these institutions have access to donor grants, which help in drawing together a large body of activists and organizing groups that deliver services to rural areas. Along with it, especially due to an urge for self-reliance, there has been an accumulation of both tangible and financial assets. The microcredit program provides two additional assets (1) access to financial resources generated through members' savings and individual accumulation (on account of service charges), and more important, (2) a network of groups of rural people who may aptly be called producers-cum-consumers. All these factors pave the way for some of the large NGO/MFIs to undertake marketing and production activities on a commercial basis. Compared with big private business houses, the latter group of corporate NGOs has comparative advantages in several areas. First, as noted, they may access financial resources at a relatively lower cost. Second, presence of multiple programs under the same umbrella allows them to derive economies of scope (i.e., reduces cost per program or activity undertaken).[51] These factors enable an NGO/MFI corporate body to have considerable potential in undertaking activities spread over the rural areas of Bangladesh.

They are also able to promote links between the urban and/or the international market with rural producers. One example is the United States Agency for International Development's (USAID) Agriculture Technology Development Project (ATDP), which provided credit and technology support to the agro-processing industry (mostly located in urban centers),[52] and linked them with rural growers of agricultural produce—either directly or through the network of NGOs.[53] While such an initiative promoted the value addition in agriculture through the development of an urban-based agro-processing industry, one may think of a similar approach in promoting rural nonfarm employment and income. The sustainability of the project and its activities depended on the continuation of foreign assistance for a subsequent phase. Unfortunately, the experience of ATDP suggests that the activities were not owned and pursued by any of the local institutions.[54] A different example is that of the Hortex Foundation, formed with World Bank support to promote production and export horticulture products. The foundation identifies buyers in the international market, provides technical assistance to local NGOs, and enters contracts with NGOs to undertake extension and marketing services. The work of the foundation is still at an initial stage; no assessment of its achievements is yet available. It is suspected that the overhead expenses might be too large for this kind of initiative.

More important, in both cases the key element lacking is a domestic interest group or an institution that will sustain the activities in the long run.

A number of lessons may be drawn from the Bangladesh case studies:

1. The three Bangladesh models are not strictly comparable; the promotion of rural nonfarm activities through one route does not necessarily preclude the others.

2. A well-managed credit delivery program holds good prospects in providing the rural poor access to credit for enabling them to engage in nonfarm employment. Moreover, there are options to package credit programs in ways that are self-targeting, both in terms of clients as well as activities. The GU-type initiative, if replicated in the case of appropriate or suitable products, holds a good chance of being sustainable. The case for MEDU in achieving self-sustainability is not self-evident.

3. The KST model, as practiced, has been largely supported by subsidy/ grant from the UNDP. Although the model with its unique feature of community-based approach and VO managers as credit retailers has the potential for mobilizing rural savings and for administering credit delivery to a wider segment of the rural population, there are several limitations. First, the ownership (and therefore accountability) is less clearly defined in the case of the KST model. Instituting a project office manned by in-service government officers at the helm of management is not a sustainable option. Second, it is unclear as to who really will own the large amount of UNDP funds that came in as grants at the expiration of the project. The active VO members may feel that it is theirs; some VO managers may provide for future access to resources through inappropriate sanctioning of loans.

   A few modifications in the current KST model may be suggested: (a) the loan size should remain small at the VO level; (b) a consortium of VOs (or their representatives) may command a pooled fund and take joint responsibility of funding larger projects; (c) a local NGO in an area may be entrusted with the responsibility of organizing groups and supervising the activities leading to formation of VOs and their consortium body; (d) during all this time, a trustee body may oversee the NGO's activities and monitor the progress—this body will also legally own the seed money to be provided by the government.

4. Why will a commercial bank not contract individuals (in place of a VO manager) to act as its retailing agent in return of a commission? One may argue that with the prevailing state of poor governance in the commercial banks, a market-based approach to promoting credit re-

tailers may be futile. It is important to undertake a review of the banking sector reforms, with a view to making the branches more client-oriented and officials accountable for any mismanagement. To create a special project office like MEDU that becomes alienated from the rest of the bank may not be conducive to harmony and support for new initiatives. There should be a special focus on rural nonfarm activities and on incentives for officials linked to good performance.

5. The GU model is open for replication by any private entrepreneur. What is needed is the choice of an appropriate product and technology and the availability of seed money to compensate for risks during the initial years. Ideally, any product that substitutes imports or finds new markets and is produced in the rural areas creates net rural employment and income. The initiatives can be taken by either the private sector or the NGO/MFI sector to institutionalize the support services; lessons may be drawn from ATDP of USAID. A trustee board may be created to establish a network of entrepreneurs who are engaged in (or are willing to engage in) marketing nonfarm products with rural origin. This board may provide support to these entrepreneurs in the form of market research, establishing links with potential producer groups, policy advocacy, and providing seed capital (or undertaking the risk of their initial investment).

Chapter 13

A Review of the Employment Guarantee Scheme in Maharashtra

Raghav Gaiha

INTRODUCTION

The Employment Guarantee Scheme (EGS) in the Indian state of Maharashtra is an innovative antipoverty intervention. A variant, the Employment Assurance Scheme (EAS), for example, was launched on a large scale in some of the most backward regions a few years ago.[1] The present study reviews the performance of EGS, drawing upon existing studies as well as a small survey designed to complement them.

As the analysis is based on three different data sets, a brief description of their salient features is given in the following text.

District-wise data on participation in the EGS, type of EGS works, and expenditure were obtained from a document prepared by the Planning Department, Government of Maharashtra. Two sets of results based on household data are reported here: one set is based on the International Crops Research Institute for the Semi-Arid Tropics (ICRISAT) panel survey that covers the semiarid tract in Maharashtra and Andhra Pradesh, and another on a survey designed specifically for this study and conducted in two villages in Ahmadnagar district in Maharashtra. These surveys are reviewed in the following sections.

ICRISAT SURVEY

Three representative districts were selected on rainfall, soil, and cropping criteria. Next, typical talukas (i.e., smaller administrative units) within these districts were selected, followed by the selection of six representative villages within these talukas. Finally, a random stratified sample of 40 households was selected in each village. This comprised a sample of 30 cultivator and ten landless labor households. To ensure equal representation of different farm-size groups, the cultivating households were first divided

253

into three strata, each having an equal number of households. A random sample of ten households was drawn from each of the three strata. Ten land-less labor households were also randomly selected. Landless labor house-holds were defined as those operating less than half an acre (0.2 hectare) and whose main source of income was agricultural wage earnings.

Survey data on time allocation, especially time spent on public employ-ment programs, were collected continuously for three villages from the sample households for the period 1979 to 1984. Of these, two (Shirapur and Kanzara) are located in Maharashtra where the EGS operates. Given the sample design, 12 households were selected randomly from each stratum in the sample villages. The present analysis is based on the samples for these two villages.

Briefly, the contrast between Sholapur and Akola villages in terms of rainfall, soil quality, cropping patterns, and technological advancement is striking. On the other hand, the shares of laborers, cultivators, and others as well as average farm sizes of small, medium, and large cultivators are largely similar. Thus, a small subsample of two villages is unlikely to be a serious constraint.

AHMADNAGAR SURVEY

To supplement the analysis based on the ICRISAT panel survey, a special survey was designed for this study. Although some overlap exists between the two surveys, the latter focuses more on the *process* of implementation of EGS. Accordingly, a small sample of participating and nonparticipating households, and representatives of official agencies at the state, district, tehsil (a block of villages) block, and village levels were interviewed.

It was decided to conduct the fieldwork in Ahmadnagar district as the level of EGS activity was reported to be high there. Another consideration that favored the selection of this district was the high concentration of the tribal population in it. In this district, two tehsils, Akole and Sangamner, were selected, with Akole recording the highest number of registered labor-ers and Sangamner the next highest (for the EGS). From each tehsil, one vil-lage was selected in consultation with tehsil officials. From the tribal zone in Akole, Padoshi was chosen, and, from the dry region in Sangamner, Panodi. Hereafter, this is sometimes referred to as the Ahmadnagar sample. It was decided to restrict the sample to 20 EGS participants. In consultation with the chairperson of the village panchayat, a list of participants in the EGS was prepared. A random sample was then selected, making sure that SC/ST, agricultural laborers, and women were represented in it. From the two sample villages, ten nonparticipants were randomly selected, making sure that these belonged to relatively poor households (i.e., SC/ST/other backward classes [OBC]).

Several officials involved in the implementation of the EGS at different levels were interviewed. These comprised officials in the planning department, district collector, tahsildar,[2] and village-level worker (VLW, or gram sevak).

SALIENT FEATURES

The scheme guarantees that every adult who wants a job in rural areas will be given one, provided that the person is willing to do unskilled manual work on a piece-rate basis.[3] Self-selection of the poor is built into EGS. First, no choice of work is offered. Second, until 1988, the wage rate was usually below the agricultural wage rate.[4] Third, as the guarantee holds at the district level, a person may be required to travel a long distance for a few days of temporary work.

The employment seeker must register his or her name under this program with the registering authority of the village (e.g., the VLW) by filling out a form. Thereafter, a formal request for employment is made to the samiti officer (i.e., the tahsildar) by completing another form. The tahsildar is obliged to provide work within 15 days of receiving the "demand for work." The employment seeker is required to work for a minimum of 30 days on the site assigned by the tahsildar. The person must present himself or herself for work within seven days after the letter is issued by the tahsildar. Failure to provide employment within 15 days entitles the person to an unemployment allowance (of Rs 2 per day).[5] Ex gratia payment up to Rs 10,000 is admissible in the case of death or disablement of a worker on the site. Some amenities provided on the site include potable water, crèches, a resting place, and first aid.

The scheme operates through identification of projects, which must satisfy two criteria; they must be labor intensive and must create productive assets. The labor-intensity criterion is defined rather strictly—the ratio of cost of unskilled labor to equipment, materials, supervision charges, and so on must be 51:49 or higher.[6-7] Productive works are, however, somewhat loosely defined as those that directly or indirectly lead to an increase in production or which, if not undertaken, would cause production to decline. With a view toward minimizing the recurrence of droughts, priority is given to moisture conservation and water conservation works (e.g., percolation and storage tanks). Other priorities are soil conservation and land development works, afforestation, roads, and flood protection schemes. It is mandated that work under the EGS should be so organized that it does not interfere with normal agricultural activities. Also, this scheme is not activated when work is available on other public work projects in progress.[8]

A three-tier setup, comprising committees for planning, direction, and coordination, exists at the state, district, and panchayat levels. At the state level, overall responsibility is vested in the planning department, at the district level in the collector, and at the panchayat samiti level in the tahsildar.

Weekly and monthly progress reports are sent by the implementing agencies to the collectors for onward transmission to the planning department of the state government. To minimize the malpractices, a high level vigilance committee under the chairmanship of the revenue secretary of the state government was formed. Three squads were formed at different levels. The workers were given identity cards-cum-wage books in which their attendance and wages are shown.

The scheme is financed through taxes levied specifically for it and a matching contribution from the state government. The former include

1. a tax on professions and trades;
2. an additional tax on motor vehicles;
3. a surcharge on sales tax;
4. a surcharge on land revenue; and
5. a tax on nonresidential urban land and buildings.

RECENT DEVELOPMENTS

SSGV

A subscheme of the EGS—Shram Shaktidware Gram Vikas (SSGV)—was launched in June 1989. It was designed to take up all development activities in a village in an integrated manner. More specifically, backward and forward links were considered in selecting them. The construction of a percolation tank, for example, was linked to construction of wells and pump installation. Once a village was selected for this program, its implementation continued until all works undertaken were completed. Individual beneficiary schemes such as wells, horticulture, and farm forestry on land owned by marginal and small farmers are financed by the government.[9] All other beneficiaries bear 50 percent of the expenditure. The participation of a village in this scheme is conditional upon an undertaking by the gram sabha (the village assembly) that 50 laborers will be available per day, and two (land-owning) workers will provide free labor for one day a month (or alternatively will bear the labor cost involved). Also, there will be stall-feeding of cattle and a ban on cutting trees illegally. Villages with surface irrigation exceeding 20 percent of the land are excluded from this program. Beyond a radius of five kilometers of selected villages, the usual EGS works will continue.

The following serious doubts have emerged about a few features of this program:

- Would the project selection be guided by the preferences of the community?
- If there is a binding budgetary constraint, what would be the project selection criteria?
- Would all selected projects be subject to the labor intensity norm of EGS?

Some of these difficulties may be causally linked to the decline in the relative importance of this scheme in the overall EGS outlay during the early 1990s (the share fell from 12 to 6 percent during 1990 to 1996).

Horticulture Program

A horticulture program linked to EGS was introduced in 1990, highly labor intensive and remunerative under certain conditions. Its salient features are as follows:

1. It is not restricted to any group.
2. The entire cost of extending it to the lands of SC/ST/small farmers will be borne by the government while others will bear part of it.
3. Plantations are allowed on landholdings between 0.2 and 4 hectares.

Some prerequisites for the success of horticulture are availability of irrigation, transportation, and marketing. Even if these requirements are met, a few other concerns remain. In particular, although some benefits of additional employment will accrue to agricultural laborers, the bulk of the gain is likely to accrue to the landowners. Since the scheme is not restricted to a specific group of landowners, participation of even large landowners is not ruled out. Moreover, it is not clear whether any additional extension provision enables marginal and smallholders to benefit from it.

Some findings based on fieldwork in Akole tehsil are as follows:

1. The income-enhancing potential of this scheme is substantial—especially in plantations. After four years, for example, one hectare of land devoted to mangos yields an additional income of 8,000 to 12,000

rupees (assuming a survival rate of 60 to 70 percent for mango sap-
lings).
2. Although precise estimates of employment generated were not given,
   high female participation was reported (e.g., in watering the trees).
3. The participation of medium and large farmers was high—accounting
   for about 40 percent of the cases processed annually.

Jawahar Wells Scheme

A subscheme of EGS, Jawahar Wells Scheme was launched in 1988. Initially, it was confined to marginal and small farmers below the poverty line.
In 1991, however, it was extended to all marginal and small farmers. Wells
were dug on their lands at government cost. However, quotas have been
fixed for backward classes by land size.[10] Some findings of an eight-district
survey and fieldwork in Akole are illustrative of the functioning of the
program:

1. Considerable variation occurred in targeting this project. In more than
   a few cases, the proportions of relatively affluent farmers were non-
   negligible. In Akole, for example, about 70 percent of the beneficiaries were marginal or smallholders and the rest relatively affluent.[11]
2. Well digging is often organized by the farmers themselves or assigned
   to contractors.
3. Delays in completion are not uncommon, mostly because of the inability of marginal/smallholders to provide their share of the cost in time
   (in absolute terms, the latter could be as much as Rs 50,000). In Au-
   rangabad, for example, 30 percent of the wells were incomplete.
4. Government subsidy is given in several small installments—in gross
   violation of the guidelines—to extract bribes. In some cases, the
   bribes paid were as much as Rs 5,000.
5. Lack of careful attention to geological features (e.g., whether soil
   strata are muddy or rocky) and their cost implications meant that the
   costs exceeded the estimates by a wide margin, forcing the beneficiaries to incur large debts. Also, in one district (i.e., Dhule), 70 percent
   of the wells were submerged due to heavy floods.
6. Access to a well leads to changes in crop intensity and pattern. In
   Akole, for example, it is feasible to grow an additional rabi crop. Also,
   it is profitable to grow tomatoes, onions, etc.
7. On certain assumptions, an additional rabi crop of wheat could yield
   an income of Rs 5,000 per hectare.

PERFORMANCE OF EGS

Employment

As shown in Figure 13.1, EGS participation declined for the period 1980 to 1997—the person-days of employment (crores) fell from 20.55 to 9.01 crores. The expenditure (at constant prices) also fell over this period, from Rs 30.17 crores to Rs 24.66 crores. Although participation fluctuated, a sharp reduction occurred in 1989, following the hike in the EGS wage rate. Between 1987 and 1989, there was a reduction in person-days of employment of over 5.50 crores, a large part of which was due to rationing.[12] Soon after, EGS participation rose until 1993, followed by a steady decline in subsequent years[13] and its importance as a supplementary source of employment has diminished in recent years.

As shown in Figure 13.2, EGS employment continued to peak during the slack period, i.e., April to June. Although the mean EGS participation (about 2 lakhs [200,000]) remained unchanged over the period 1991 to 1996, the coefficient of variation declined slightly—from 53.24 to 49.11. As may be seen in Figure 13.2, EGS employment was slightly more evenly distributed from its peak in 1996 relative to 1991. Whether in fact this is a manifestation of the changing composition of EGS activities (e.g., a higher

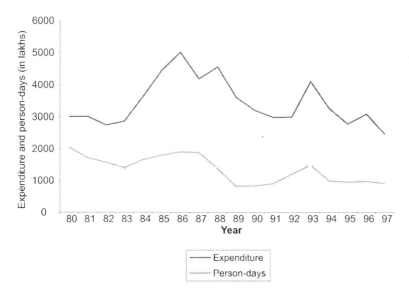

FIGURE 13.1. EGS employment and expenditure during 1980 to 1997.

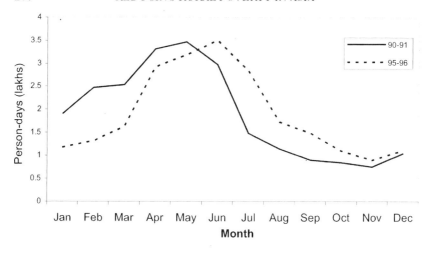

FIGURE 13.2. Monthly EGS attendance in 1991 and 1996. *Source:* Author survey.

share of Jawahar Wells Scheme) cannot be ruled out. Digging of wells is often spread over a few months as the subsidy is released in installments at different stages of completion. Since some small farmers are unable to raise loans to supplement the subsidy quickly, the construction gets delayed. If this trend continues, EGS may have a stronger effect on agricultural wage rate through a spillover of its activities into busier months.

Over the period 1991 to 1996, the share of female participants (i.e., female participants/total participants) ranged between 30 and 39 percent. Although these are high shares, it is arguable that they are lower than expected. In Pune district, for example, the number of females registered for EGS was considerably higher than the number of males registered. Yet the share of females in total EGS employment was much lower.[14]

Another neglected group is tribals, usually confined to isolated settlements lacking basic amenities (e.g., access to drinking water). Possessing few income-earning skills or assets, if any, most of them are condemned to abject poverty. Even EGS has bypassed them—their share in total participants fell from a meager 13.3 percent in 1991 to 8.5 percent in 1996.

Limited participation of deprived groups—women and tribals—raises a concern about the effectiveness of the guarantee. An additional concern is that barely a quarter of those registered under the EGS secured employment within 15 days (of registration) in Pune.[15]

Over the period 1991 to 1996, the composition of EGS expenditure changed significantly. The share of usual EGS projects fell slightly from over 74 to about 70 percent, that of the SSGV nearly halved from about 12 to about 6 percent, while Jawahar Wells rose sharply from more than 13 to nearly 23 percent. Similar patterns are observed in *all* 29 districts of Maharashtra in which EGS operated during this period.

It is plausible that these compositional changes—especially the increasing importance of Jawahar Wells Scheme—impinged on the cost-effectiveness of EGS employment. Some implications of this shift may be serious. Replacement of community assets with individual assets (i.e., wells) could divert the benefits of EGS away from the poorest landless households to the moderately poor or relatively affluent land owners.[16] Moreover, it is not obvious whether financing of such assets through the EGS is preferable to financing through a microcredit scheme. Finally, to the extent that (small) contractors are involved in the construction of wells, it is unlikely that the employment benefit accrues largely to the poorest workers. Guided by profitability considerations, contractors prefer physically strong and dexterous workers, making it harder for the poorest to fend for themselves.[17,18] Besides, the contractors are notorious for some malpractices (e.g., use of inferior material, diversion of resources, etc.).

Some earlier studies have drawn attention to the concentration of EGS activities in a few districts.[19] Almost two-thirds of all EGS employment (averaged over the period 1979 to 1997) was concentrated in one-third of the districts.[20] If, however, the districts are ranked in terms of an index of development, weighted by their shares in the total rural population, a less skewed distribution of EGS employment is observed.[21] The main findings are as follows:

1. The (cumulative) share of the five least-developed districts in total EGS employment was more than 18 percent in 1990 as against about 16 percent in the total rural population. By contrast, in the five most developed districts, the corresponding shares were about 12 and 18 percent, respectively. These figures suggest that, on a rural population weighted basis, the employment benefits of the EGS were slightly greater in the least developed areas.
2. Over the period 1990 to 1995, the spatial distribution of EGS activities changed little. In fact, the shares of both the least and most developed districts in EGS employment rose slightly, i.e., by less than 1 percentage point. Even though the share of the poorest districts continues to be slightly larger, a case for a substantially larger allocation can be made if the findings from the Ahmadnagar sample are replicated in other backward regions with long slack periods.

DIRECT TRANSFER BENEFITS

Earlier studies point to the accurate targeting of EGS. One, for example, reported that 90 percent of the workers in a sample of 1,500 EGS workers in 1978 to 1979 lived below the poverty threshold, as against a headcount index of poverty of 49 percent for rural Maharashtra. More recent studies point to similar conclusions. Of particular interest is a study based on the ICRISAT data over the period 1979 to 1983.[22] It confirms that the EGS was well targeted as days of participation decreased rapidly with increases in wealth, and participation was higher in the more backward of the two villages. As shown in the following, if the ICRISAT data are disaggregated by income class, and analyzed using different tests of targeting accuracy, some strikingly different conclusions follow.

The overall EGS participation rate (i.e., EGS participants as a proportion of total workers) dropped sharply over the period 1979 to 1989 (i.e., from 17.7 to 9.44 percent). In addition, the share of poor participants (i.e., poor EGS participants as a proportion of poor workers) also fell (from 19.3 to 14.29 percent).[23] Thus, not only was the coverage of the poor small but it also reduced over the period in question, implying a worsening of the mistargeting of the EGS.

The distribution of EGS earnings between the poor and nonpoor further corroborates the mistargeting. The poor participants (i.e., those with per capita incomes < Rs 180) accounted for about three-quarters of EGS earnings (75.32 percent) in 1979 and for well over half (56.87 percent) in 1989. Thus, although the poor continued to account for the larger share of EGS earnings, the nonpoor substantially augmented their share. In both years, the mean income of the EGS participants exceeded the poverty threshold and the excess was greater in 1989 than in 1979 (see Table 13.1).

Some evidence indicates a worsening of gender disparity. Female participation was low, reducing from 31 percent in 1979 to about 29 percent in 1989. Also, the average duration of participation of females fell from 83.71 days to 70 days per participant, while participation of males rose from 68.19 days to 114 days per participant. As a result, the male earnings rose more

TABLE 13.1. Descriptive statistics of EGS participation.

Distribution	Mean income (Rs)	Skewness	Kurtosis
EGS participants by income in 1979	188.32	1.28	3.88
EGS participants by income in 1989	226.64	1.63	6.25

Source: Computed from the ICRISAT data at constant prices.

substantially, i.e., from Rs 82.40 to Rs 279.53 compared to Rs 64.81 to Rs 113.30 among the female participants.[24]

Two issues are addressed as follows: (1) the mistargeting of EGS as reflected in a large share of relatively affluent participants in 1979, and (2) its worsening with a rise in the share of the affluent. More specifically, it is examined whether the latter reflected largely the rationing out of the poor consequent to the hike in EGS wage rate in 1988.

Broadly, since alternative employment opportunities exist in rural areas, participation in EGS depends on whether (net) direct transfer benefits exceed those in an alternative activity (e.g., farm or nonfarm employment).[25] Concentrating mainly on the (net) benefits from the EGS, it is arguable that these were higher for the nonpoor, resulting in the exclusion of the poor.

Although the opportunity cost of time is not likely to vary much between the poor and nonpoor, other costs probably do.[26] These relate to (1) long waiting periods between registration and work offers and (2) long distances between the residence and the work site. Some elaboration may be helpful. Several forms must be completed and different officials must be contacted. Further delays occur because projects have to be cleared by technical departments/agencies. After clearance, projects do not get executed promptly due to slow and ad hoc disbursal of funds. Thus, the waiting period is long. It is not uncommon for the poor to bear the brunt of such delays while the relatively affluent manage to avoid them through their influence with the local administration. Since the poor typically lack savings (or, because they are subject to tighter liquidity constraints), other things being equal, the longer the waiting period, the less attractive is the EGS as an employment option. Since the guarantee operates at the district level, participants are required to travel long distances for temporary work.[27] This is particularly disadvantageous to the less energetic poor. Also, since women from poor households combine outside work with exacting domestic chores, the longer the distance, the greater is the disincentive for them to participate in EGS.[28]

On the benefit side was the actual remuneration (and not so much the wage rate) and payment at regular intervals. Given the piece-rate system, it is arguable that two groups of persons would usually opt for it. One is the group of highly skilled persons, i.e., those who can complete a larger number of tasks with the same time effort, and another group comprising persons who lack physical strength and stamina and thus prefer a slower pace of work than under a daily wage contract.[29,30] But piece-rates cause long delays in payment of wages, as assessment of the quantity and quality of work done takes time. Although it is stipulated that wages must be paid once a week or bimonthly at the latest, it is not uncommon for the wages to be delayed for up to four to eight weeks.[31] More seriously, wages paid are often less than the recorded amount. Given the wage rate, the actual remuneration is substantially lower for the poor because they are more likely to be physically less strong and more prone to be victims of cheating because of their illiteracy.[32]

In sum, the disincentives to the poor on the benefit side (e.g., delays in payment of wages) may be just as strong as those associated with long delays in offer of EGS work—especially if they are subject to tight liquidity constraints.

How did the "costs" and "earnings" change for the poor and the nonpoor between 1979 and 1989? First, following the EGS wage hike in 1988, the screening mechanism weakened and EGS became more attractive to the nonpoor. As a consequence, some poor were "crowded out." Besides, the rationing mechanisms further disadvantaged the poor. These took the form of longer delays in responding to the demand for work, and restrictions on opening of new work sites, implying stronger disincentives for the poor. Thus, altogether the wage hike in 1988 resulted in a further reduction in the share of poor participants.

INDIRECT TRANSFER BENEFITS

If EGS provides an effective employment guarantee, it will be reflected in a strong relationship between agricultural and EGS wages. The analysis with the ICRISAT monthly data confirms a strong effect of the EGS on agricultural wages.

Briefly, the procedure used is as follows. Given the interdependence among agricultural, EGS, and nonfarm wages, a Granger-Sims causality test is carried out. After confirming that EGS and nonfarm wages together cause changes in agricultural wages, a dynamic specification is estimated to throw further light on the dependence of agricultural wages on EGS and nonfarm wages.[33] Finally, as there is often a possibility of spurious results with time-series data—specifically if the wage series are random walks— a test was used to rule it out.[34]

Confining to the results for EGS, its short- and long-run effects on agricultural wages are large. Specifically, if EGS wages rise by a rupee, agricultural wages would rise by about 17 paise in the short run, and by about 28 paise in the long run (one rupee equals 100 paise).

A few mechanisms through which the EGS influences agricultural wages are (1) gains in agricultural productivity through the assets created and associated with such gains in a shift in the demand for agricultural labor; and (2) a higher reservation wage as a consequence of a "guaranteed" employment option in slack periods. Besides, since there is some evidence of the contribution of EGS to a sense of collective identity among rural workers, it is plausible that their bargaining position vis-à-vis that of large landholders would strengthen, leading to higher agricultural wages. For all these reasons, effect of the EGS on agricultural wages is likely to be more substantial in the long run.[35]

An earlier survey by the government of Maharashtra and the planning commission indicated that 91 percent of the users of EGS assets were cultivators and 6 percent were agricultural laborers. About 70 percent of the users reported gains in agricultural production, and 40 percent reported changes in cropping patterns. Much of the area that benefited from EGS works belonged to large landholders. Specifically, while 9 percent of the area benefited belonged to small landholders (i.e., with holdings < 5 acres or 2 hectares), 32 percent of the area belonged to large landholders (i.e., with holdings > 25 acres or 10 hectares).

Some recent evidence, based on the ICRISAT panel survey of two villages in Maharashtra, suggests that the poor withdrew from the EGS when a rapid expansion of farm and nonfarm employment occurred. Of those who withdrew from the EGS, a sizeable section was better off. On the other hand, among those who continued to depend on this scheme, there were some who were chronically poor. Whether the latter lacked the motivation and/or skills to augment their incomes or whether they belonged to low castes with few other employment options could not be ascertained. On the whole, therefore, the EGS does not deter income-enhancing choices.

INCOME STABILIZATION

Whether EGS helped stabilize incomes in rural areas could be inferred from a with-and-without comparison of income variability. Although there are agroclimatological differences between the two villages that could have implications for the variability of household incomes, the shares of laborers, cultivators, and others as well as average farm sizes of small, medium, and large cultivators are similar. Since the composition of participants does not differ considerably between the two villages—a comparison of variability of household incomes with and without EGS is of some interest.

First, a comparison of the variance of household incomes in EGS participating and nonparticipating subsamples confirms that income varied much less in the former. This is consistent with the income-stabilizing role of the EGS. Second, based on a variance decomposition procedure, an assessment was carried out of the marginal contribution of variances in the components of household income, namely, income from cultivation, EGS, and nonfarm activities. In each case, the effect is substantial. If, for example, variance in earnings from cultivation is reduced by a given amount, the variance in household incomes falls by a multiple of 1.11. The effect of a reduction in variance in earnings from nonfarm activities is slightly larger (as variance in household incomes falls by a multiple of 1.16). Compared with the effects of these two components, that of a reduction in variance in earnings from EGS is not so large. Since the corresponding multiple is 0.885, a given reduction in variance in earnings from the EGS is associated with a slightly smaller reduction in the variance in household incomes. So, although the marginal contribution of variance of EGS earnings is the lowest, it is still substantial. Much of the variance in EGS earnings reflects differences in the duration of participation, among other reasons. To the extent that the poor are limited by some organizational aspects of the EGS, e.g., time and energy wasted in long journeys to and from work sites, delays in payment of wages, etc., some remedial measures (e.g., greater dispersal of work sites) would help stabilize their household incomes through a reduction in the variance of EGS earnings.

The income (and consumption) smoothing effects of household incomes during the slack period have considerable significance. Failure to do so could result in a reduction in nutrition and possibly deterioration in a short-run health status indicator (weight-for-height) with adverse impact on on-farm labor productivity as well as on income from casual agricultural labor. If these effects are partly offset by the EGS, escape from acute poverty becomes easier—especially if there are elements of a poverty trap.

A more detailed econometric analysis with the ICRISAT panel survey for 1979 to 1984 casts more light on the income-stabilizing role of the EGS.[36] Specifying a system of equations for the determination of agricultural wages, EGS wages, alternative measures of participation in the EGS, and a measure of variability of monthly wage earnings, a reduced form estimation is carried out, using different techniques.[37] Although there is some unevenness in the results, the effect of EGS participation on the variability of labor earnings is negative. Since this measure of variability of labor earnings is constructed from monthly data, and the contribution of the EGS is separated from that of various household (e.g., caste and occupation) and village (e.g., rainfall) characteristics, this is a more convincing demonstration of the income-stabilizing role of the EGS, notwithstanding the fact that the measure of household income variability is a partial one.[38]

SUPPLEMENTARY ANALYSIS

Although the importance of the EGS as a supplementary source of income at the state level appears to have declined in the 1990s, its benefits in some of the poorest areas continue to be substantial, as illustrated below. Indeed, if this analysis has any validity, in the absence of alternative employment opportunities, a scaling down of the EGS could result in severe economic hardships to some sections. The official justification for an across-the-board scaling down is thus contentious.

Profiles of Village and Participants

Located in hilly terrain, Padoshi is a small tribal village containing 225 households and a total population of 1,182. The total area is 1,176 hectares, with a forest cover of 153 hectares and an unirrigated area of 989 hectares. It is a predominantly tribal village as the STs account for 92 percent of the population. The nearest market is five kilometers away. The main (kharif) crop is paddy.

Panodi is a relatively large village containing 411 households with a total population of 2,766. The total area is 2,112 hectares, out of which 708 hectares are irrigated. The SCs/STs account for 22 percent of the village population. The main crops are *jowar* (sorghum) and *bajra* (millet). Although Panodi was a relatively backward village some years ago, it has witnessed some expansion of both farm and nonfarm activities. As a result, farm wages are much higher in this village, as compared with Padoshi. Yet the extent of unemployment/underemployment during a few months of agricultural slack is by no means negligible.

Most respondents participated in EGS as the employment opportunities in the two villages and elsewhere were few and far between during long slack periods of up to seven to eight months in Padoshi. Failure to secure work within the village meant a long and difficult job search in neighboring villages. The few who succeeded traveled one way up to 20 to 35 kilometers daily. Although wages were higher (Rs 60 per day compared to EGS wages ranging from Rs 30 to Rs 50), the net earnings were lower (as travel expenses amounted to Rs 10 to Rs 15 per day). Another reason cited in favor of EGS by some respondents was the chance to earn substantially higher amounts, depending on the quantity and quality of work performed. Finally, the female participants preferred EGS as it allowed them to work with their husbands.[39] Typically, whether a female member would participate in EGS was decided either by the husband or the male head of the household. [40]

The participants became aware of EGS through village meetings organized by a few officials, notices displayed in panchayat offices, neighbors, and initiatives of the mukadam (the assistant supervisor of EGS).[41] There was no formal registration in most cases. The mukadam recorded the names in a diary, usually at the work site. The waiting period was short, varying from 3 to 12 days. The distance between residence and work site varied from one kilometer to five kilometers. No choice of work was offered. The work was essentially unskilled, e.g., digging and carrying mud. Usually, the male participants performed the more strenuous and physically demanding tasks (e.g., digging). Men and their wives worked together on the same site. The duration of participation was moderately high—varying from 30 days to 90 days. Few amenities were provided. Cattle feeding stations were non-existent.[42] But drinking water and first aid (e.g., bandages) were provided.

Considering limited irrigation facilities, much of agricultural employment was seasonal and of short duration (four to seven months). Although the incidence of landlessness was low in both villages—in fact, negligible in the tribal village (Padoshi)—ownership of land was of limited consequence in itself as yields were low. Consequently, the *dispersion* in incomes was low. The distinction between the poor and nonpoor is thus hard to draw. Nevertheless, it is of some interest that out of the 20 participants, 12 were poor. Among the poor, ten were extremely poor.[43] In most cases, acute poverty stemmed from meager agricultural wages during a few months in a year. Although there were eight nonpoor participants, none of them could be classified as affluent as their (per capita) incomes were only slightly higher.[44] Differences in land and other endowments were not marked either. Average land owned among the poor participants was 1.29 acres compared to 2.43 acres among the nonpoor. Moreover, although about 37 percent of the nonpoor had primary or secondary schooling, 25 percent of the poor had similar educational attainments. For the bulk of the poor as well as the nonpoor (75 percent and 87 percent, respectively), the main sources of income were agricultural labor and cultivation and related activities (e.g., dairying). Among the nonpoor, the dependence on the latter (i.e., cultivation and related activities) was slightly higher. Average EGS earnings were substantial for both groups—Rs 3,878 among the poor and Rs 4,312 among the nonpoor (at current prices). The share of EGS earnings in total household income, however, was higher among the poor (30 percent compared to 27 percent among the nonpoor). Thus, not surprisingly, the importance of the EGS as a supplementary source of income was high for both the poor and nonpoor in the two villages.

Wages

Daily wage rates varied over a wide range from Rs 30 to Rs 50 among the poor and from Rs 35 to Rs 80 among the nonpoor. Considering that these are based on piece rates and that some of the nonpoor *are* healthier than the rest, a wider range for the nonpoor is not surprising. The work done was assessed jointly by the mukadam and the overseer or agriculture officer. Most respondents agreed that wages were paid bimonthly, usually on the day before the market day. If a worker was absent when wages were paid, alternative arrangements existed to make the payment soon after.[45] A few respondents, however, experienced delays of up to one month. None protested for fear of termination of employment. Wages were paid after obtaining a signature or thumb impression in the presence of two witnesses.

EGS wages were substantially higher than agricultural wages, especially during the slack period.[46] There were a few options outside the village that offered higher wages but involved expensive search and travel.[47] Net of search and travel costs, the differences between such options and EGS were small, if not negligible. What further reduced the attractiveness of some options outside the village was their shorter duration.[48,49]

Piece Rates versus Time Rates

Piece rates and time rates received a mixed response, with the majority of the respondents favoring piece rates.[50] The preference for piece rates rested on the possibility of higher earnings. Moreover, especially from the point of view of female participants, an advantage was the flexibility in the work schedule.[51] The few who favored time rates did so for two reasons: one was an assured amount at the end of the day; another was that it did not involve elaborate calculations.

Underpayment

A majority of respondents were satisfied that the wages paid were fair and that there was no cheating.[52] Some of the reasons cited in support of this view cannot be rejected outright. One was that the work done was measured and recorded in their presence. Another reason given was the fear of surprise checks by district/state officials. A third reason was that, if a mistake occurred in the assessment of work, it was quickly corrected. Some other responses, however, suggest that it would be somewhat naïve to accept these reasons uncritically. A few participants, for example, were emphatic that, given their illiteracy, they were not in a position to judge whether they were paid appropriately. Moreover, since the records were not in the public domain, it was difficult to detect any cheating or irregularity. Finally, fear of abrupt dismissal following a complaint weighed heavily in their minds.[53] A few anomalies were in fact reported.[54]

In sum, contrary to earlier reports, there were few cases of delays and underpayment of wages.[55] This is not to suggest that disbursement of wages was altogether fair and efficient. Rather, the deficiencies were far from glaring.

Benefits

Using the classification of EGS benefits into transfer and stabilization benefits, a few illustrative estimates are presented in the following text.

Transfer Benefits

EGS earnings were a large share of household incomes of both the poor and nonpoor. Among the former, the share of EGS earnings ranged from 18 to 40 percent, while among the nonpoor it ranged from 18 to 33 percent. However, in order to calculate direct transfer benefits, the opportunity cost of time spent in the EGS must be deducted from EGS earnings. As high job search and travel costs render the option of working in neighboring villages much less attractive than the wage differences imply, a more likely alternative to participating in EGS is farm/nonfarm employment in the same village. Since slack period opportunities are few and far between, the opportunity cost of participating in EGS is taken to be no more than Rs 20 per day. Using this estimate, the direct transfer benefit to one of the poorest households worked out to be Rs 2,400 (i.e., about 60 percent of the EGS earnings).

The effect of the EGS on agricultural wages is a form of indirect transfer benefit. Although agricultural wages rose in the past three years, they were well below those in some neighboring villages. Expansion of irrigation on a small scale made it easier to grow a summer crop. As this crop coincided with peak EGS activity, competition occurred among farmers to hire laborers and agricultural wages rose. However, whether some participants contested the EGS alone made a significant difference. So, given the scale and duration of EGS activities, a small positive effect on agricultural wages is plausible.

Another form of indirect transfer benefit is through the output of EGS assets (e.g., percolation tanks enabling irrigation of farms in neighboring areas). As the percolation tank in Padoshi was located in the foothills and their farms were on top of the hill, several poor EGS participants were deprived of its benefits. Even among those with farms around the tank, the benefits accrued to those with wells. Those who benefited in this way were able to grow another crop. The drinking water facility during the summer, of course, benefited a larger number. The benefits would have been greater if the village panchayat had a role in the selection and location of such assets.[56] Besides, their maintenance by the village community would enhance the flow of benefits. No consideration was given to the maintenance of these assets as a self-interested response from the community in case their ownership was vested in it.

Stabilization Benefits

Without the EGS, the prospects for making ends meet for the majority of the participants (including the nonpoor) were grim. If unable to find employment in the same village, an option would be to seek employment in better irrigated neighboring villages or work in a brickmaking unit elsewhere. Not only does it entail an expensive job search but also long daily journeys—in a few cases of up to 20 to 35 kilometers one way.

Failure to secure employment of any kind, of course, involved grimmer choices: cuts in food expenditure, liquidation of assets, and loans at exorbitant rates of interest. To the extent that EGS facilitated consumption smoothing among poor households and prevented them from making costly adjustments (e.g., sale of livestock) during slack months, the stabilizing benefit was likely to be substantial.

Did the EGS Act As a Disincentive to the Search for Alternative Employments?

The responses were mixed, with a large number of respondents denying the disincentive effects of the EGS.[57] There was a strong desire for economic betterment through self-employment in a nonfarm activity (e.g., brickmaking). What prevented them from engaging in such activities was not the availability of employment under EGS but their lack of access to credit facilities. Whether the latter would do away with the need for anti-poverty interventions such as EGS was disputed by a few on the ground that some protection against market uncertainties would in any case be necessary.[58,59] There, however, was some evidence of a mild disincentive effect of the EGS, discouraging job search in neighboring villages. Availability of work nearer the home and flexibility in work schedules were the underlying considerations, especially for the female participants.[60]

There was no lack of awareness of this program. Their exclusion was largely a result of scaling down EGS activities Work sites had become few and far between and duration of work had shortened considerably. There was little evidence of malpractice in wage disbursement (e.g., delays and underpayment) thus the poor were not discouraged from participating in EGS.

Assessment

In sharp contrast to the evidence reviewed earlier, mostly from the ICRISAT panel survey, the performance in the sample villages (one extremely poor) in Ahmadnagar was quite impressive in some respects. The targeting was good. Besides, the proportion of the poorest among the participants was high. The duration of participation was moderately high too, enabling the participants to substantially supplement their incomes through EGS, especially during slack seasons. For most participants—especially the poorest—it made a significant difference to their economic well-being, as costly adjustments (e.g., liquidation of assets, loans at exorbitant rates of interest) and adverse nutritional effects were avoided. By preventing the decumulation of assets through distress sales over time, households were enabled to withstand idiosyncratic shocks. There was a small positive effect on agricultural wages. However, the benefits of the assets created under EGS (e.g., percolation tanks) were confined largely to households in their immediate vicinity. There was no blatant cheating of EGS workers in the disbursement of wages (e.g., delays and underpayment).

An issue then is whether this new evidence can be reconciled with earlier not-so-favorable assessments. Partly, of course, the superior performance of EGS in the two sample villages in Ahmadnagar reflects their pervasive poverty. In such a context, targeting failures are unlikely. Moreover, to the extent that the awareness of EGS was widespread, blatant violation of the wage schedules would be harder. But whether the district administration also had an important role in ensuring some measure of efficiency in implementing EGS cannot be ruled out.

CONCLUSIONS

From a broad perspective, the following observations can be made.

Overall participation in the EGS fell sharply over the period 1980 to 1997 due to the expansion of farm/nonfarm employment opportunities, decline in total public expenditure, and the rationing that followed the hike in the EGS wage rate.[61] Also responsible was a change in the composition of EGS, i.e., specifically, a substitution of community assets (e.g., soil conservation works) by individual assets (e.g., wells) that involved fewer workers. In some ways, the contrast between the findings from the ICRISAT and Ahmadnagar samples points to the need for a reallocation of EGS outlays. In the ICRISAT sample, a marked deterioration occurred in the targeting of EGS over the period 1979 to 1989, also resulting from a fall in total expenditure and rationing of employment. Improvements in the targeting of EGS with a fixed outlay made little difference in most cases, nor did substantially larger EGS outlays without any improvement in targeting. However, larger outlays with greater participation of the poorest in the EGS yielded unambiguously superior poverty relief outcomes. There was a substantial concentration of the benefits among large landholders. The benefits of the output of EGS assets, as the Ahmadnagar survey reconfirmed, accrued largely to those living in their vicinity—a case in point being farmers located around a percolation tank. As EGS projects were handed down by district authorities (often independently of the proposals put up by village panchayats), this is not surprising.

In striking contrast to some earlier evidence, in the Ahmadnagar survey,

- the registration formalities were minimal and discretionary;
- the waiting period was short;
- men concentrated on physically demanding tasks, whereas women were usually assigned less demanding duties;
- except in a few cases, wages were paid bimonthly; and
- the manipulation of the muster rolls was not a serious problem.

Although it is not easy to reconcile clean and honest implementation of EGS in the Ahmadnagar sample with other evidence, some features of this sample seem pertinent. These are

1. absence of acute economic disparities,
2. widespread awareness of EGS, and
3. small outlays.

So perhaps what the poor lost in terms of small outlays was to some extent compensated for by clean and honest implementation of EGS.

From a broader perspective, it is arguable that the contribution of EGS to political activism and coalition building among the poor is of considerable significance in itself, as it makes the political system more responsive to their interests. A sense of collective identity—despite social and religious differences—is inculcated through close interaction on work sites. Reinforced by a sense of economic security during lean periods, prospects of collective action among the poor are likely to improve.

Some concerns about the design and implementation of Rural Public Works (RPWs) in general and EGS in particular, however, remain:

1. If public investment substitutes for private investment, as in the case of Jawahar Wells Scheme as a component of EGS, the (net) benefit may well be small. It may be more appropriate to promote private investment through easier access to microcredit. A case in point is the Maharashtra Rural Credit Project.

2. Failure to integrate EGS projects into a comprehensive rural development plan remains a major concern. A somewhat glaring reflection of this failure is the rapid decline in the relative importance of the SSGV in overall EGS outlays. Under certain conditions, a greater involvement of the community in the selection and implementation of EGS projects could reduce the risks of such failures.

3. As no separate provision for the maintenance of EGS assets is made, their potential benefits are not fully realized. Adequate provision must be combined with vesting of responsibility for maintenance in the local community, as that is likely to be cost-effective.

4. Given a fixed outlay, a lower wage would allow a wider coverage of the poor. However, if the wage is statutorily fixed, as in the case of EGS, it is not obvious how this constraint could be relaxed. More important, if welfare of the poorest is assigned highest priority, their deprivation must be mitigated to an acceptable level first.

5. If feasible, a mix of piece and time rates may be more appropriate as it would retain the flexibility to earn more and reduce delays in wage payment.

6. No economic justification is found for part payment of wages in the form of foodgrains when markets function efficiently.

7. Minimum labor-intensity requirements could be relaxed if the benefits of the assets created accrue to the workers.

8. Enhanced outlays under EGS are feasible provided other similar interventions (e.g., Jawahar Rozgar Yojana) are merged under it. As waste is smaller under EGS, it would be more cost effective than administering each intervention/program separately. If larger outlays are accompanied by a reallocation in favor of backward regions, the benefits to the poorest would be substantially greater without additional administrative costs.

Chapter 14

Public Works Employment and Rural Poverty Alleviation in the Philippines

Roehlano M. Briones

INTRODUCTION

In many developing countries, public works employment (PWE) in its various forms is a long-standing and well-recognized strategy for rural poverty alleviation. However, Philippine policy does not accord such status to PWE programs.[1] This study undertakes two tasks. First is a review of the country's experience to consolidate existing knowledge about PWE in the Philippines. Second, case studies are conducted to illustrate the "best practice" for PWE. Guidelines for PWE policy and implementation shall then be drawn from the review and casework.

Traditionally, the national government has been in charge of public works in the country. Executive line agencies (called departments) work independently or in tandem for public works provision. The agency most directly involved is the Department of Public Works and Highways (DPWH). The Department of Agrarian Reform (DAR) also initiates rural works, though the actual implementation of agrarian reform infrastructure projects is handed over to the DPWH. The Department of Agriculture (DA) finances the construction of farm-to-market roads and postharvest facilities. An agency attached to the DA is the National Irrigation Administration (NIA), which implements irrigation projects, including the establishment of communal irrigation facilities.

Meanwhile, the Department of Interior and Local Government (DILG), as well as the Department of Labor and Employment (DOLE), collaborate with the local government for rural works provision. Most line agencies have regional offices, which in turn are used to coordinate provincial and sub-provincial offices. The economic planning agency called the National Economic Development Authority (NEDA) has no direct involvement in project implementation, but is still influential in terms of broad policy thrusts.

Local government units (LGUs) are stratified into the provincial, municipal, and village *(barangay)* levels. Barangays may be further divided into *sitios* or *puroks*. Leaders at each level (purok leader, barangay captain, municipal mayor, and provincial governor) are elected by popular vote. Local government elections are held every three years.

The most innovative development in the public institutional framework has been the shift to decentralization. Initial moves are to form Regional Development Councils (RDCs), composed of regional heads of line agencies, provincial governors, city mayors, legislative representatives, and NGO representatives. The RDCs are involved in regional planning, interagency coordination, coordinating budgetary and project allocation from the central offices to regional offices and LGUs, and assisting LGUs in formulating plans and securing funds. Lower-level versions of the RDCs are the Provincial Development Councils (PDCs) and Municipal Development Councils (MDCs).

Infrastructure Policy

The official policy on employment generation in public works is to apply "labor-based equipment-supported" (LBES) methods. These methods pertain to a technology in which labor, supported by light equipment, is used as a cost-effective method of providing or maintaining infrastructure to specific standards. Public works agencies are to incorporate such methods within their regular programs on a nationwide scale, particularly for rural-based projects. Project implementers are given suggestive, elaborate guidelines (in the form of manuals) to effectively substitute labor for equipment. It is noteworthy that the definition of "labor-based equipment-supported" refers only to the value-added component of construction and is silent about the materials component. Thus, if labor accounts for the bulk of value added, then a project can be "labor-based" even though total project cost is accounted for mostly by materials.

Contracting System

For the DPWH, the type of small contractors accepted in LBES works are labor-only contractors, typically informally organized. These labor gangs are hired on the basis of *pakyaw* contracts, which are derived from traditional construction arrangements. The principal to the contract is the public works agency, while the agent is a pakyaw team represented by a leader. A pakyaw contract engages the team to provide labor for the completion of a specific task, within a specific time period and for a stipulated amount. The principal provides all equipment and tools. As the principal recognizes no employer-employee relationship with the pakyaw, legally the project is still undertaken under force account.

Workers should belong to the various barangay associations near the project site. Unskilled labor should be drawn from the host barangay, while semiskilled labor may be recruited from the municipality; skilled labor may be obtained from within the province. In any case qualified workers nearer the barangay are given priority. The teams are composed of groups of 20 workers. As many groups of workers as possible should be formed.

One obvious advantage of the pakyaw is that it lightens the recruitment and supervision load on government personnel. Another is that it transfers the burden of avoiding minimum wage legislation from a national government agency to the pakyaw leaders, in a rural setting where statutory wage enforcement is usually trivial.

PUBLIC WORKS EMPLOYMENT PROGRAMS IN THE PHILIPPINES

Until the mid-1980s, PWEs were localized programs. It was only in 1986, under the Aquino administration, that PWE schemes were implemented on a nationwide scale in the Community Employment and Development Program (CEDP). The Local Infrastructure Development Program (LIDP) under the Kabuhayan 2000 (1994 to 1996) included an employment component focused on reforestation, land development, and infrastructure. The most recent employment generation scheme is the Rural Works Program, initiated by the DOLE, in response to the Asian crisis and the El Niño phenomenon, which ravaged the agricultural sector in 1997.

The foregoing are distinct and completed programs with explicit employment objectives. Special mention should be made of the various Comprehensive Agrarian Reform Program (CARP) infrastructure projects, composed of numerous individual programs, which from the late 1980s became a major impetus toward PWE generation. The ongoing Agrarian Reform Communities Development Project (ARCDP) identifies as an objective the creation of wage employment through labor-based rural works, under the administration of LGUs.

The Community Employment and Development Program

The CEDP was aimed as a pump-priming measure to counter the employment stagnation from the 1983 to 1985 economic crisis. The CEDP involved all the major agencies. It emphasized small-scale infrastructure projects such as barangay roads, communal irrigation systems, and school buildings. Rural works projects under the CEDP were required to allocate at least 30 percent of project cost to labor. The specific design of the project (embodied in the program of work) is supposed to involve LGUs.

A distinctive feature of the CEDP was the involvement of NGOs in monitoring. A special monitoring committee consisting of NGO representatives submits a status report (distinct from the implementer's status report) on particular works. Labor intensity was supposed to have been enforced by the 30 percent labor content requirement. Water supply and school-building projects could not comply due to the nature of the work and the need for relatively expensive construction materials. Moreover, some private contractors did not follow the labor content rule. Monitoring and enforcement of the labor content rule was also impeded by confusion and unfamiliarity with implementation guidelines of the CEDP. After implementation, no provisions for maintenance were undertaken specifically for CEDP projects. Communities were not organized to provide maintenance and ensure sustainability.

Remuneration for labor was on a cash-for-work basis. This was higher than the prevailing agricultural wage.

The Local Infrastructure Development Program (Kabuhayan 2000)

The Kabuhayan 2000 was an umbrella interagency program involving various employment-generating thrusts, with LIDP being the PWE scheme. Projects under the Kabuhayan were targeted to priority areas, which were the poorest 19 provinces and the fifth- and sixth-class municipalities. Within provinces, identified projects would employ fishers, small farmers, landless workers, upland subsistence farmers, indigenous peoples, and underemployed urban dwellers. The LIDP sought to identify projects within the regular infrastructure programs of the DPWH and LGUs in which LBES methods were deemed feasible. Strategies to meet these targets included the introduction and transfer of LBES technology to all LGUs, as well as national government assistance in the form of training, supervision, technical, and financial support. The DPWH was selective in identifying the enrolled projects, making sure that the 35 percent labor content rule was followed.

The Second Rural Roads Improvement Project

The SRRIP was a project on rural roads, which began in 1986 and was completed in 1995. In constructing the roads, the DPWH implemented labor-based methods mostly with pakyaw contracts, based on the usual DPWH guidelines.

On average, over half of the construction cost was spent for labor. According to the SRRIP implementers, as the projects consisted of dirt roads, material obtaining them on-site brought down costs. There were no problems in generating a relatively high labor content for the road works.

The implementation of LBES methods realized cost savings; expenditures per kilometer were 12 percent lower than projected. Additional benefits included the training of workers in concreting, masonry, and even the repair and maintenance of vehicles and light equipment.

Local governments were involved to the extent of identifying projects, and providing inputs to the program of works. Actual construction work was handled entirely by the DPWH. Upon completion, the roads were turned over to barangay officials for maintenance.

The Food-for-Work Program

The FFW program was implemented in the province of Negros Occidental (Eastern Visayas Region), the major sugarcane province of the country, which had a major rural works component.

The project selection process is typically initiated at the barangay level. Members of farmer associations and cooperatives identified plans and programs and discussed these with barangay and LGU officials. Construction of the infrastructure was supposed to be labor intensive. Each project lasted around five to six months and was scheduled to avoid planting and harvesting seasons. Community works were undertaken by provincial LGUs, while irrigation projects were administered by the NIA.

Upon completion, maintenance was to be undertaken by the concerned LGUs; the irrigation facilities were to be maintained by the users associations. The provincial government continued to advocate continuation of the program approaches and procedures, despite its termination in 1996. The rural works component made no use of labor content requirements. The rural works component did have a high proportion of women participants relative to other PWE schemes in the country.

The Rural Works Program

In 1997, the Asian financial crisis as well as El Niño induced drought. The RWP consists of local government projects aimed at assisting displaced workers in the aftermath. The RWP was a cost-sharing scheme between national (DOLE) and local (LGU) governments. The project was deemed successful, mainly due to the close partnership between DOLE and LGU proponents. The projects selected were mostly brief, small-scale efforts.

Several problems were noted. Frequent heavy rains delayed implementation. Second, the works seemed to have not been scheduled to take into account agricultural seasonality, as the works reportedly interfered at times with activities such as planting, cultivation, and harvesting. The labor cost ratio of a typical RWP project averaged 45 percent. Aggregating over all projects, however, the total wage bill was only 34 percent of total cost. Wages were set at the legal minimum for the region, which was set far above farm and informal sector wages and was certain to attract many of the nonpoor to participate in RWP projects.

General Assessment

Rural-based PWE schemes in the Philippines were often scheduled so as to avoid competing with agricultural activities. The major PWE thrust was to incorporate labor-intensive methods in small-scale rural works. Current schemes (i.e., labor-based CARP projects) are clearly integrative, with employment generation strictly a secondary consideration in project design.

Targeting

Although poverty alleviation was a general intention for all PWE schemes, there had been no project with specific measures toward targeting of the poor. Rather, target groups were identified, and it was assumed that hiring from these groups was equivalent to assisting the poor. The problem was that these groups were usually too broadly defined.

Self-targeting (a consideration in PWE schemes) was missing in all the programs reviewed. The basis for remunerating labor was the statutory wage, which was always higher than the prevailing agricultural wage.

Sustainability and Replicability

Asset maintenance was essential for ensuring long-term productivity improvement from public works; the absence of a maintenance strategy was another serious gap in PWE schemes.

A final element in maintenance and sustainability that was typically ignored is the participation of communities. In most projects, they were involved in project identification; in the pakyaw system they were responsible for providing labor and other inputs to public works; however, there was little progress in incorporating communities in actual design, implementation, and maintenance. These two themes—decentralization and community participation—shall be the focus of our case studies.

THE CASE STUDIES: IN-DEPTH ANALYSIS

Introduction

The review of experience provides some broad generalizations that are useful in making policy and program recommendations. However, to obtain insights on best practices, more substantive primary data on the basis of the case studies are needed. Projects chosen here were the more successful ones that adopt approaches that are innovative in the Philippine context. Chosen here is one project administered by an LGU, and two administered by the national government. The projects all fall under the community works classification. The element of community participation is novel from the viewpoint of traditional rural works policy.

In addition to project documents, information was gathered directly through personal and group interviews of government officials, field staff, community leaders, and project beneficiaries. Households are classified as working if at least one member was employed in the project. Heads of 20 working and 20 nonworking households were interviewed. The respondents were selected randomly, based on a roster of workers (for the working households) or of villagers (for the nonworking households.)

Background of the Selected Projects

> *The Tumbaga Access Road Project (ARP).* The Tumbaga ARP is CARP-related, administered by the national government under the Agrarian Reform Infrastructure Support Project. It exemplifies the labor-only contract or pakyaw.
>
> *The Ciabu Water Supply Project (WSP).* The Ciabu WSP is a local government project exemplifying the implementation of PWE under a decentralized regime. The project made use of local labor, and upon completion was operated entirely by the beneficiaries organized in a water users association.
>
> *The Palanas Communal Irrigation Project (CIP).* The Palanas CIP involved the construction of a small (50 hectares) irrigation facility in the ricelands of the eastern Philippines. It exemplifies intensive participation of beneficiaries in project implementation, beginning

from the proposal and design stage, to actual construction and maintenance.

The Tumbaga Access Road Project

The Tumbaga ARP (located in Southern Luzon) consists of two gravel roads that connect a wide swathe of rice fields to a national road. Each road is 750 meters long, with two short bridges, accompanying culverts, and ripraps.

Overall, labor accounts for only 14.5 percent of the total cost. The project is nevertheless regarded as labor-based, as equipment rental contributes only 5 percent of the cost. Materials accounted for most of the project expenses (73 percent).

The evaluation reports attest to the high quality of the road and to its usefulness to the community and found no evidence or even suspicion of corruption.

Production and Employment Practices

There were four pakyaw contracts under different team leaders. The pakyaw groups were selected by the district office of the DPWH. Separate contracts were drawn up for construction materials and for equipment rental.

Supervision of the entire works was undertaken by the DPWH engineer and project-in-charge. Labor supervision was mostly undertaken by the pakyaw leaders. The pakyaw leaders brought with them their own gangs of skilled workers, none of whom were from the area. Meanwhile, for tasks done in conjunction with equipment operations, supervision of workers was also undertaken by the construction companies who leased out their equipment for the project. While pakyaw leaders received lump-sum trenches of their payment, workers from the community were paid cash wages on a weekly basis. Payments for pakyaw contracts were, however, made only monthly. The construction and material supply companies served as sources of liquidity for funding wage payments.

The financing role of the construction firms was not a formal arrangement. Rather, the pakyaw leaders, the construction firm, and the field staff reached this arrangement informally under the exigencies of project operations. Pakyaw leaders, who were just workers themselves, could not finance wage payments. The latter's interest in providing this service was to allow the project to proceed thus providing them profit from leasing equipment and providing materials.

Overall, the LGUs remained largely spectators to the proceedings. Occasionally the municipal LGUs sent inspectors to examine the projects in progress.

Community Associations

The main community association involved in the Tumbaga ARP was the IA. Consultation of the community was mainly undertaken by the DAR and was limited to general design and implementation features of the access road.

The village leaders (both in the barangay government and in the IA) helped identify the road layout, with access of IA members as well as easement of right-of-way as the primary considerations. In negotiating for right-of-way, village leaders had to exert moral suasion on the absentee landowners.

Selection of workers was directly the responsibility of the pakyaw leaders, usually upon recommendation of the IA members or the purok leader. All unskilled labor was hired from within the village; a few skilled carpenters and masons were taken in as well.

Sustainability

The barangay, being the recipient of the road, was immediately responsible for these maintenance activities. From the time the road was completed, monthly clearing was undertaken as part of the barangay's routine general cleaning drive. This work was usually done by women in the community, and recently was subsumed as a regular activity of the women's association.

Asset Benefits

The access road had been in operation for over a year at the time the interviews were conducted. No benefit-cost analysis of the project had been done by either the feasibility or the evaluation stage. A rough analysis was undertaken as follows. The first source of savings would be the savings in paddy rice transport cost due to the road. The charge for transporting from the field to the national road is now 5 pesos per sack, whereas before, the cost of hiring a laborer to haul a sack was 10 to 20 pesos. This amounts to around 2.5 percent savings in the cost of rice farming. (On the other hand, income to off-farm workers may have fallen.)

However, because of low yields and a small coverage area, the benefit from transport cost savings is quite modest; a reasonable estimate would be 225,000 pesos per year. This would be the lower bound benefit from the access road. From this source alone the rate of return of the project would only be around 1.1 percent. However, the access roads are beneficial to the purok residents, who previously had to cross the paddy fields to reach the national road. Moreover, there are dynamic gains in terms of new activities in the sitio made possible by road access.

The Ciabu Water Supply Project

The next case study is the Ciabu Water Supply Project (WSP), a case study in decentralized implementation of a PWE project. Barangay Ciabu is another agrarian reform community in the town of Baybay, Eastern Philippines.

The WSP is well known throughout the country for the success of the water distribution program of the municipal LGU. There remain remote sitios where only a few faucets serve numerous residents, who sometimes must resort to rudimentary wells. Among the underserved communities were three sitios (subdivisions of barangay) in Ciabu. The implementation of the project was completely undertaken by the municipal LGU. The first phase was the Spring Development Project, to identify a water source and construct water tanks, a reservoir, and main distribution pipes. The second phase involved the construction of a distribution network and is the subject of the case study.

The Ciabu WSP began in 1997 and was finished in just two months. It involved the construction of distribution pipes and communal faucets or hydrants. The local government was wholly responsible for implementation of the Ciabu WSP. The national government confined itself only to channeling project finance, and occasional monitoring of work in progress.

The ARCDP followed a matching grant policy, in which the LGU was required to shoulder a local counterpart amounting to 30 percent (Table 14.1).

TABLE 14.1. Breakdown of costs, Ciabu WSP, in Philippine pesos.

Item	ARCDP	LGU	Total	% of total cost
Indirect cost	–	49,372	49,372	16.8
Direct cost				
Materials	190,040	–	190,040	64.2
Labor	17,081	39,394	56,475	19.0
Total cost	207,121	88,766	295,887	100.0

Source: Municipality of Baybay.

Based on project estimates, the project was expected to generate only 437 person-days. Because of its small scale, no special equipment is necessary, as the work required only digging pipelines, burying pipes, and masonry work on the communal faucets. Only three skilled workers (foreman, plumber, and mason) were called for, to be hired for a total of 73 person-days. Materials accounted for 64 percent of direct project cost. The labor cost ratio was 19 percent. Incidentally, the labor cost ratio of LGU-implemented water supply projects in the RWP averaged 25 percent.

The completion report noted a major drawback in that the budget fell short of providing quality pipes appropriate for the pressure conditions of the reservoir. Project expenses hit the mandated ceiling then in place for this type of ARCDP project. Because of this experience of the Ciabu pilot project, the cost cap was removed for subsequent projects undertaken by the LGU for the other ARCs in the municipality.

Employment Policies and the Role of the Community

As early as 1995 (two years before commencement of the works), the municipal LGU was already consulting with barangay officials and community residents regarding the needs of the barangay. During project construction, the municipal engineer was assisted by the barangay captain in locating the water hydrants, who in turn consulted residents informally. However, the final project design (as codified in the program of work) was drawn up by the engineer with few (if any) specific inputs from the community. Decisions about the ratio of labor cost in total cost, material procurement, etc., were made by the municipal engineer.

The municipal LGU was subject to a cost-sharing scheme with the ARCDP. Likewise, it engaged the barangay government in a cost-sharing program, with the community counterpart consisting solely of labor. Labor provision was mobilized through a traditional community effort locally known as *pintakasi,* which involves community work by groups, typically on a rotating basis. On any given day, a group of 20 to 40 villagers was at work under the supervision of a municipal engineer.

In the earlier spring development phase, no payment for wages was made, which was acceptable to the community as the amount of work required was modest. Each worker contributed only a few days of work. Their only "incentive" was free lunch and snacks during the workday, prepared by the women of the community.

The water supply phase required greater labor outlay; the payment of wages was therefore warranted. The daily wage was disbursed on-site by the municipal project officials in the form of cash during Saturdays. The barangay and municipal governments set a one-day weekly contribution (usually taken on Saturdays). According to the workers, the importance of community ties and a sense of solidarity motivated them in part to donate the extra day.

For each week during the duration of construction, one sitio would be assigned to provide labor on a rotating basis. The sitio leaders recruited residents within their jurisdiction and set work assignments for individuals. The project engineer though had some leeway in keeping some productive laborers for a longer interval. No problems in availability of manpower were reported.

The farmer's cooperative in Ciabu, in addition to the barangay government, was active in preparation and social mobilization for the project. It should be understood that in this case the elected village leaders could hardly be distinguished from the community association leaders. The Ciabu farmer's cooperative was founded by the current barangay captain; while the WSP was being undertaken, he was also then serving as the cooperative chairman. Moreover, some of the barangay councilors and officials also served as leaders in the cooperative. Given the bias of the DAR toward extending assistance toward organized farmer groups in ARCs, it was natural that the barangay officials would also carry their identity as cooperative leaders in transacting with both national and municipal governments.

Maintenance

Upon completion, the water system was turned over to users, consisting of 220 households. The cooperative permitted water users to access the system using pipes directly extending into their homes. There were 60 piped-in users. The rest of the users gained access through communal faucets (20 in all).

Initially the system was managed by the same farmer's cooperative mentioned earlier. However, with a change in leadership in the cooperative, many water users raised complaints against the new management. Inasmuch as many users were not members of the co-op, the barangay officials convened a general assembly of water users, where they decided to constitute a new user association, the Ciabu Potable Water System. The association membership fee was 50 pesos, while the monthly fee was 10 pesos for households with piped-in water, and 3 pesos for households using communal faucets.

Some members of the farmers' cooperative objected to the new arrangement and refused to pay their user fees. The vast majority of villagers ceased their complaints about the water system management. The Ciabu

Potable Water System was performing its maintenance tasks well in terms of collecting water fees, repairing damages, and regulating the use of the communal faucets. These regulations included prohibitions against bathing and laundry washing. In the absence of a piped-in connection, households undertaking such activities must resort to a nearby river.

In the medium term, facility maintenance was enssured through the activity of the water association. However, there were threats to the project.[2] The reason given by field staff and barangay leaders is the fact that under the cost-sharing scheme the villagers were compelled to contribute toward project provision. The end users, having invested their time in construction, would therefore not waste their earlier efforts by allowing the system to deteriorate.

An alternative explanation would be the selection effect: under an equity scheme, only communities with a long-term interest in and capability for sustaining project benefits would accept a cost-sharing arrangement. From the viewpoint of economic rationality, this alternative explanation would be the more likely one, as earlier efforts were after all "sunk cost," which would therefore not be a factor in their decisions to maintain or not maintain the water system.

In the long term, a serious problem might confront the Ciabu water users. The rate of 10 pesos per month for piped-in water users was too low, given their consumption levels as well as depreciation of the facility. A major breakdown of the water system might require extensive rehabilitation requiring external assistance. To make the system completely self-reliant within the reasonable lifespan of the facility, higher rates might have to be collected, especially from the piped-in users. The most sustainable arrangement would be to meter the water usage and water fees the piped-in users should be charged metered rates.

Asset Benefits

The asset provided was of satisfactory quality, according to Ciabu residents. Currently the system shows signs only of ordinary wear on the pipes and faucets. At the source, however, upon ocular inspection one could observe that some pipes used had indeed been substandard PVC pipes. According to the project engineer, this was a result of the previously mentioned cost cap.

Among the households whose water source shifted from spring-fed wells to the communal faucet, the daily time saving was 31.5 minutes. This, however, concealed many other benefits. Previously, large household reserves of water were needed, which was very inconvenient. The presence of a water system conferred greater benefits in periods of drought, where water flow might weaken to a trickle, and queues could be long and time-consuming. For example, in a nearby village households suffered from endless queuing at a well during the El Niño drought.

The Palanas Communal Irrigation Project

The third case study highlighted an intensive participatory approach of farmer-beneficiaries. This approach was anchored on an innovative cost-sharing scheme.

The participatory approach had been a standard format followed by the NIA since the 1980s. Communities were organized into IAs, which were formal associations of cultivators who operated a communal irrigation facility. The principle behind the IA was the necessity of cooperation among water users. Irrigation facilities were common pool resources: whereas persons outside the coverage area might be easily barred from accessing water, abusers within the area could not be as easily excluded. For example, one type of problem was the "tail-end syndrome": in the absence of an enforced rotation system, downstream users typically got the least water supply.

The participatory approach consisted of the active involvement of an IA in all phases of a project (planning, construction, and maintenance) as well as in shouldering project cost. The NIA's approach was developed over years of experience in community-based irrigation works.

The NIA required the organization of the IA prior to the construction of facilities, with membership including at least 80 percent of farmers in the coverage area. The inclusion of the IA in the planning and construction phase might be seen as facilitating the flow of information between end users and implementers, as well as checking agency problems (i.e., incompetence or negligence of the NIA in construction).

Project construction supervision was still exercised by NIA staff, composed of a supervising engineer and an irrigation development officer, or IDO. While the supervising engineer oversaw the technical side of the irrigation works, the IDO focused on the community organizing, and played the key role in coordinating and even initiated the participation of communities in the irrigation system.

IA participation in irrigation works was anchored on cost sharing. The cost share of the user association is called *equity,* which (as in the Ciabu WSP) could take the form of labor contribution. There were two kinds of equity schemes. In the 10:90 scheme, the IA shouldered 10 percent of project cost by the time of completion, while the IA amortized the remaining 90 percent for a period not exceeding 50 years. Meanwhile, in the 30:70

scheme, the IA shouldered 30 percent of project cost by the time of completion, while 70 percent was shouldered by the NIA.[3]

The NIA offered these two schemes as a menu of options—itself an innovative feature in public works programs. In practice, the 30:70 option was seldom chosen. The reason was that this scheme would require large contributions from the IA members during the construction phase, as well as a high degree of cooperation between them. Few communities were optimistic enough to take this option.

The Palanas Communal Irrigation Project (CIP) consisted of an extensive upgrade of an existing facility. The previous system consisted of a small dam and earthen canals. The dam was washed out by massive flooding brought about by a massive typhoon in 1991. Much later (June 1999), the Palanas CIP began construction of a larger dam better suited to local weather conditions. The main canal was also widened and cemented. The works were completed in January 2000.

The computed direct project cost was 4.18 million pesos, whose breakdown is shown in Table 14.2. Of the direct project cost in Phase 2 (3.37 million pesos), one-third was the cost of labor. (The indirect cost was the amount charged by the NIA for administrative overhead. It excluded the salaries of the on-site staff for the project duration.)

Community Participation

The organizational work was limited to reconstituting the previous association and coordinating with the NIA. The members were highly motivated in the irrigation works, met often to make group decisions, and collec-

TABLE 14.2. Programmed and actual expenditures for the Palanas CIP, in Philippine pesos (as of December 31, 1999).

Item	Programmed	% of direct cost	Actual	% of direct cost
Direct cost	3,371,518		3,173,877	
Indirect cost	428,571		428,571	
Total cost	3,800,089		3,602,448	
Equity	1,011,455	30.5	1,011,519	32.0
Labor	589,663	17.5	424,745	13.4
Materials	437,010	13.0	586,773	18.4

Source: NIA Region 8 Office.

Note: The 1999 nominal exchange rate was Php 39.1: US$1.

tively enforced these decisions. No major organizational or managerial problems were encountered.

In the design stage, the Palanas IA insisted upon a strong and durable facility, even at some added cost (given their experience with typhoons). One of the major changes they insisted upon was a canal width that was wider than that specified in the original NIA plan. The project plan and schedule was made upon consultation with the IA. The IA was able within technical limits to design the largest possible labor and material contribution that could be mustered to meet the 30 percent requirement. This type of discretion was absent from the cost-sharing arrangements in the previous projects in Ciabu and Tumbaga. It was consistent with evidence that greater labor intensity accompanied community works characterized by intensive participation.

Table 14.3 breaks down the equity by type of work. Based on the equity programming, the IA contributed around 53 percent of labor cost; donated materials made up the remainder of the equity.

What happened, in fact, was a shortfall in labor contribution, but a significant overshooting in materials contribution. The latter was mostly composed of sand, gravel, and boulder. Materials were priced at their market value, i.e., 526.41 pesos/m^3 for sand, 507.14 pesos/m^3 for gravel, and 824.00 pesos/m^3 for boulder. During the data-gathering period, the IA was about to complete its equity requirement.

The involvement of the IA extended to details such as project finances, and material procurement, all according to NIA's guidelines on beneficiary participation. A process called "cost reconciliation" permitted the IA (through its financial committee) to audit project financial records.

The extensive participation of the IA was understandable given their interest in obtaining quality material, and properly accounting for the equity contribution. Moreover, these steps ensured that transactions were above-board, avoiding the rumors of corruption tainting many public works projects. No delays were reported because of the involvement of the IA in these procedures.

TABLE 14.3. Breakdown of equity contribution for the Palanas CIP, in Philippine pesos.

Item of work	Direct cost	Labor equity	Materials equity			
			Sand	Gravel	Boulder	Lumber
Diversion works	1,792,498	304,470	34,526	58,012	168,253	10,576
Canal system	1,258,689	285,196	111,696	45,993	0	7,954
Others	320,331	0	0	0	0	0
Total	3,371,518	589,666	146,222	104,005	168,253	18,530

Source: NIA Region 8 Office.

Some problems were associated with adoption of the intensive participation format. First, the remarkable equity contribution of the farmers relied greatly on accounting sleight of hand. In the case of materials, the farmer-beneficiaries did not actually finance any material purchases out-of-pocket; instead, these materials were dredged from the river bottom close to the facility's water source. Procurement and hauling of these materials relied mostly on the labor provided by the IA members, hence the crediting of these items to the beneficiaries' equity. However, the valuation of these materials was not based on dredging and hauling (actual) cost, but they were imputed a value based on the market price. This certainly overestimated the cost of materials.

Second, it turned out that the "community" that participated in this project was only a small portion of the total number of farmers covered by the Palanas communal irrigation system. Out of the 72 water users expected to benefit from the facility, only 23 were members of the IA (of whom 5 were recruited after the project started). This was contrary to the implementing guidelines on IA membership. (At the field level, such deviations were allowed as long as the IA voiced an intention to expand membership.)

This smallness was viewed by the NIA as a weakness in project implementation. Both the IA members and the NIA staff agreed that the narrow membership base is due to the selectivity of the leadership. According to the IA leaders, the IA was excluding farmers who might be a source of dissent within the IA. Most particularly, farmers who were "beholden" to their landlords, and might therefore introduce the influence of the elite within their group, were barred from membership. Moreover, the members were well screened so as to recruit only the reliable and cooperative individuals. The small size of the IA was itself conducive to cohesion, thus avoiding transaction costs of coordinating large groups, as well as the vulnerability of large groups to the free rider problem.

Responsibility for providing construction labor, both skilled and unskilled, was vested on the IA. Work slots were assigned to each individual IA member, who must fill it personally or recommend a substitute. Prioritization in recruiting non-IA members, as outlined in the standard procedures of the NIA, was as follows: the member's relatives residing in the coverage area, followed by other residents in the same area, and finally residents in the barangay. The NIA for its part hired only two skilled personnel (a machine operator and foreman) and no unskilled workers.

Wages are paid in cash twice a month on a 15- to 30-day schedule by the NIA. There was no evidence of underpayment. Despite the overall timeliness in project funding, there were occasional delays (up to one week) in the payment of the wages, which dissatisfied some of the workers.

Role of Government Units

According to NIA officials, its participatory approach was made possible because projects were implemented on force account (direct management of workers). Except for the IA, no other private entity was involved in the project.

The local government had no direct hand in the project. The municipal and barangay governments did make two important contributions, however: first, in providing equipment support, and second, in resolving right-of-way problems. LGU support was greatly facilitated by the fact that the IA president was also the barangay captain and headed the association of barangay captains in the municipality. The IA president was therefore able to coordinate easily with the town mayor and other local officials.

With regard to equipment support, the mayor lent municipal dump (rent-free) upon request of the barangay captain. She also assisted the IA in borrowing (rent-free) bulldozers and trucks from a nearby liquefied petroleum gas factory. This equipment support was not trivial: while NIA equipment rental ran to 66,370, private equipment costs in terms of fuel and parts alone amounted to 52,544. As for right-of-way, objections from landowners initially threatened to halt the project.[4] Serious opposition was forestalled through efforts of moral suasion, first by the barangay captain, and second by the town mayor.

Sustainability and Replicability

The IA was in charge of managing and maintaining the facility. Maintenance funds were generated by charging a user's fee, which were to be collected even from nonmembers.[5] Prospects for maintenance of the facility were favorable, barring severe calamities, based on the cooperativeness within the current IA, and the organization's track record in maintaining the previous facility.

Furthermore, the burden of paying water fees was small, as the cost of amortization has been eliminated. This raises the question of free riding by users who did not make equity contributions upon users who did make equity contributions. First, the wage payments were satisfactory enough to the members; second, the contributors argued that free riding was irrelevant. In the end, what mattered to them was that they personally stood to benefit from the facility even considering the cost of their contributions.

The characteristics of this group were quite idiosyncratic, raising some doubts about the sustainability and replicability of the intensive participation format. Their situation contrasted with that of other areas in which no

IA existed, and the IDO needed to undertake protracted community organizing efforts. In other areas in which IAs existed, farmers typically settled
for the easier 10:70 scheme; moreover, frequently members were apathetic,
or whether active, factious.

It was indeed unrealistic to expect this approach to be disseminated
widely in community irrigation works. The point, however, was not to repli-
cate this approach widely, but to adopt it systematically in locations where
cohesive and capable groups existed or could potentially be organized. The
merit of an equity scheme was clear, as it allowed such groups to self-iden-
tify in an incentive compatible manner.

Benefits from the Asset

The estimated total project cost was 5.24 million pesos (higher than actual because of additional overhead items and contingencies). Project benefit was expected to take the form of increased rice yield. The current rice
yield was 2.4 metric tons per hectare; with the project, the yield should increase by 92 percent to 4.6 metric tons per hectare, a 92 percent increase.
This corresponded to a 112 percent increase in net farm income per hectare,
reaching a total of 11,616. The IRR was estimated at 13.9 percent. It was
likely that variability of yield from season to season (and from year to year)
would decline as a consequence of the facility. For the poor farmer such
stabilization of income would be invaluable.

CASE STUDIES: WAGE BENEFITS AND TARGETING

The Tumbaga ARP

The average wage paid for the Tumbaga ARP was 143 pesos daily, as determined by the pakyaw leaders. This was approximately 5 percent below
the farm wage of 150 pesos prevailing in the area. The margin between actual and market wage was expected to lead to some degree of self-targeting.
Intertemporal targeting could have also been promoted by the timing of the
works during the months of September to December, which were mostly
lean months for rice farming (except September, which is harvest season).

Table 14.4 presents a profile of workers and nonworkers. The two groups
were demographically similar, except that the control group was on average
less educated than the worker group. Farm size was greater in the nonworker group. There were about the same number of landless workers
among in both groups, but duration of off-farm employment was greater
among the workers.

TABLE 14.4. Characteristics of workers and nonworkers in the Tumbaga ARP.

Characteristic	Average	
	Worker	Nonworker
Age of household head (years)	47.1	48.2
Schooling of household head (years)	6.6	4.9
Household size	4.8	4.6
Number of landless	12	13
Number of landowners	3	4
Farm size (ha)	1.1	2.0
Days off-farm work per year	98	68
Annual income (Philippine pesos)	40,604	43,025
Sources (%)		
Farming	14.9	26.9
Off-farm employment	41.1	35.4
Project employment	14.5	–
Other nonfarm employment	26.9	33.3
Other nonfarm sources	2.6	4.4

Source: Author's data.

Worker income was slightly lower than nonworker income. Farm income share was lower for workers (consistent with their greater reliance on off-farm income and employment). Even for the nonworker group, farm income accounted for only a quarter of total income; rather, nearly half of income arose from nonfarm sources. Most of the farmers in the sample did not report any nonfarm or off-farm occupations.

Of persons who were engaged in off-farm employment, nonfarm work accounted for 100 days of employment in both worker and nonworker categories over the previous year. The rest of off-farm employment was largely taken up in *hunus,* an informal agrarian contract in which the landless worker supplied labor to a farmer in exchange for a share of the harvest. The type of work provided for hunus was mostly light, except during harvest season. The table suggests that workers tended to earn relatively less of their income from farming and more from off-farm work compared to nonworkers.

The duration of project employment averaged only two weeks. The reason was the rapid turnover of workers, which led to an implicit rotation of work. One reason for the rapid turnover was that there were four different pakyaw leaders and teams; this implied discontinuity in worker selection. The rotation system reflected both the desire of the pakyaw leaders to appear accommodating to all the members of the community, as well as rationing of work.

Table 14.5 examines the distribution of household per capita incomes within worker and nonworker categories. We focused on per capita figures, as this was a better measure of the living standard of the household. For the worker group, both with- and without-project incomes of the worker group are shown.

To calculate the without-project income, it is first noted that project wage earnings accounted only for a small proportion of the workers' incomes. Seven of the workers worked for two weeks or less, while eight worked between three weeks and one month. For the eight, earnings were imputed for the few workers who insisted that some employment opportunities were indeed missed due to the project. The adjustment resulted in without-project per capita income of workers only 7.2 percent lower than the with-project income.

On a per capita basis, the without-project incomes of workers were surprisingly higher than the incomes of nonworkers. On first glance, this indicated imprecise targeting. However, without-project per capita incomes of workers were much lower than those of nonworkers if the top five per capita incomes were removed. This was confirmed in the poverty headcount: seven workers were classified as poor by an international poverty line (Balisacan, 1997), while nine of the workers would be classified as poor on a without-project basis. The disparity in the poverty headcount was largely

TABLE 14.5. Comparison of average per capita income and project employment in the Tumbaga ARP.

	Control group	Worker group		
		Without project	With project	Days of project work
Per capita income				
Bottom 10	4,200	2,991	3,574	15
Bottom 15	6,558	5,083	5,729	25
Whole group	9,126	9,258	9,926	22
Correlation coefficient with duration of project work	–	0.12	0.13	1
Number of the poor				
International line	7	9	8	
Official line	12	15	15	

Source: Author's data.

going by official poverty lines. The top five without-project per capita incomes of workers turned out to be mostly persons with high nonfarm incomes, who were hired because of their carpentry skills.

To a limited extent project employment to the unskilled seemed to have been well targeted. This could be explained by the self-targeting effect of the below-market wage. However, one should not discount the possibility that recruitment recommendations by IA members to pakyaw leaders were directed to underemployed persons in the community.

The project could be said to have reduced the poverty headcount by just one. This need not be held strongly against the project if the wage benefits were biased toward the poorer workers. However, the correlation coefficient between without-project income of workers and days of work in the project was positive though insignificant. The same characterized the correlation coefficient of with-project income of workers with days of work. Apparently, the rotation system adopted in the project essentially randomized the selection of the workers, removing any bias toward the poorest.

With rotation, the profile of workers would begin to reflect the profile of residents in the village, many of whom were not poor. The nonpoor still suffered from underemployment, and therefore might desire a stint in the project despite its lower-than-market wage as long as its schedule was flexible. The rotation was not entirely disadvantageous; while the "leakage" problem in targeting might have been worsened, the "undercoverage" problem might have been reduced. The poorest village residents, particularly the landless, did manage to join the project. Their work stints, however, were all too brief.

The Ciabu WSP

The modal wage among the sample workers of the Ciabu WSP was 75 pesos. This was lower than the planned daily wage of 83 pesos (adjusted for equity). However, there were seven skilled workers who were paid between 90 and 150 pesos daily. These persons pushed up the sample average wage to 103 pesos. These wage benefits arrived at a most auspicious time, when the El Niño drought was then wreaking havoc on the countryside.

The worker-nonworker profile is shown in Table 14.6. Workers and nonworkers were demographically similar, except for average age (which is much lower for the workers). Several young persons managed to enter the Ciabu WSP, including some nonhousehold heads. Tenurial status was similar across categories, except for smaller average farm size for the workers. The farm sizes in Ciabu were apparently much greater than those of other

TABLE 14.6. Characteristics of workers and nonworkers in the Ciabu WSP.

	Average	
Characteristics	**Worker**	**Nonworker**
Age of household head (years)	38.4	45.1
Schooling of household head (years)	6.5	6.3
Household size	6.3	5.0
Number of landless	4	4
Number of landowners	9	8
Farm size (ha)	2.1	3.5
Days off-farm work per year	81	73
Annual income (Philippine pesos)	39,307	19,850
Sources (%)		
Farming	45.1	53.1
Off-farm employment	26.4	17.1
Project employment	7.7	–
Other nonfarm employment	15.0	21.0
Other nonfarm sources	5.9	8.8

Source: Author's data.

case study sites, but the quality of these lands was much poorer, because of the hilly terrain.

The annual incomes for Ciabu were low. Several facts should be noted: the year was 1997, hence the purchasing power was somewhat lower than in the other projects, which came later. Also, the phenomenon of El Niño reduced coconut and rice yields.

Nevertheless, it was safe to conclude that the Ciabu residents were poorer than those of the other project sites, consistent with the fact that Ciabu was a marginalized upland village devoted to coconut growing.

Farming was the biggest income source for either category. For workers, off-farm employment was the next biggest source of income, while for nonworkers it was nonfarm employment. Project employment had only a small contribution to worker income. The very low nonworker incomes suggested that the poorest members of the community were covered by the project.

The tabulation of per capita incomes is shown in Table 14.7. Using a method similar to that in the Tumbaga case study, wages and employment for the counterfactual without-project situation were estimated. The incremental contribution of project wages to income was only 3.2 percent, implying that nearly 60 percent of project wage earnings simply replaced alternative earnings. This is mainly because the skilled workers earned a

TABLE 14.7. Comparison of per capita incomes, Ciabu WSP.

	Control group	Worker group		
		Without project	With project	Days of work on the project
Per capita income				
Bottom 10	2,285	2,196	2,367	13
Bottom 15	2,772	3,262	3,420	12
Whole group	4,310	6,768	6,986	14
Correlation coefficient with duration of work	–	0.49	0.50	
Number of poor				
International line	14	9	9	
Official line	18	17	17	
Coefficient of variation	0.87	1.19	1.16	

Source: Author's data.

greater part of the project's wage benefits, and claimed that they could have found employment elsewhere.

Average per capita incomes of the workers (with or without the project) were much higher than those of nonworkers. Most of the nonworkers were poor on the basis of the international line; this same line classified only a minority of workers as poor. (The official line fails to distinguish significantly between the poverty headcount of the two categories.) The superiority of workers' incomes originated from the top half of sample workers. The bottom half of households were quite similar (in fact, for these households the workers are slightly worse off). The difference became increasingly pronounced as the higher income households were added. The coefficient of variation of worker income was large relative to that of the control group, confirming the relatively larger dispersion of per capita incomes in the former category.

Furthermore, within the worker category, the distribution of wage benefits did not favor the poorer individuals; in fact, a mild tendency existed for project employment to rise with rising household income. This was seen in the fewer days worked by the bottom half of the workers, and more clearly, the positive value of the correlation coefficient for with-project or without-project incomes.

As in the Tumbaga ARP case, the mistargeting arose from accommodating skilled workers and rotating work assignments in compliance with the pintakasi tradition. Of the three workers with the highest per capita incomes, one was a recordkeeper and a high school graduate, coming from a family with one overseas worker remitting foreign earnings, while two were skilled carpenters. (When these workers were excluded, the correlation coefficient of without-project per capita income is –0.15.) Furthermore, excluding the poorest worker (who worked continuously on the project for two months), the poor workers contributed only seven days of labor on average. While work rotation did allow many of the poor to be included in the project, it also ensured that wage benefits could not be adjusted to provide greater assistance to the most needy.

The Palanas CIP

For the Palanas CIP, the wage was set at 200 pesos, which was the lowest salary in the government pay scale but way above the minimum wage. Effectively though, the wage was 100, as 50 percent was collected as equity. This is obviously another case of overvaluation (just as with the materials component of equity). The daily farm wage in the locality is estimated to range from 60 (slack) to 75 (peak). Thus, the effective wage (equity excluded) remained significantly higher than the alternative wage. Targeting therefore might be a problem unless effectual means existed to ration out the nonpoor.

Based on the standard NIA procedures, the work program was scheduled to avoid conflict with the peak agricultural season around November to December, and June to July. In reality, delays in the project led to conflict with farm occupations. The farmers who worked actively on the project claimed that their role in farming over the project year was confined to overseeing the farm work of their wives and children, and occasional fieldwork on Sundays.[6]

A comparison of workers and nonworkers (Table 14.8) shows noticeably different characteristics between the two categories. Workers were younger than nonworkers and were somewhat better schooled. Only a few workers were landless, while none in the control group were landless.

Income of workers was much higher than that of nonworkers. The biggest chunk of workers' earnings came from the project, in contrast to the earlier case studies in which project work hardly made a dent in worker incomes. Meanwhile the bulk of earnings of the control group came from farming. Off-farm work (agricultural work excluding farming) frequently took the form of *angkon,* a local version of the hunus.

As for nonfarm employment, the project had only two workers with nonfarm occupations. The first had a nonregular job that was not affected by

TABLE 14.8. Characteristics of workers and nonworkers in the Palanas CIP projects.

Characteristic	Average	
	Worker	Nonworker
Age of household head (years)	47.1	56.5
Schooling of household head (years)	6.6	5.9
Household size	4.8	4.2
Number of landless	3	0
Number of landowners	6	6
Farm size (ha)	0.8	1.1
Days off-farm work per year	24	59
Annual income (Philippine pesos)	52,258	37,622
Sources (%)		
Farming	26.6	55.2
Off-farm employment	6.3	3.2
Project employment	44.8	–
Other nonfarm employment	21.0	30.5
Other nonfarm sources	0.9	14.2

Source: Author's data.

the irrigation works, and the second suffered some income loss due to his participation in the works (owing to the moral suasion of his brother, who happened to be the IA president). Other worker households with large nonfarm incomes actually earned them through household members not working in the irrigation project.

Per capita income figures are summarized in Table 14.9. Calculation of without-project incomes requires some care, given the substantial time that individual workers devoted to the project. The number of days employed and wage were imputed as follows:

> For the landless worker: Average off-farm wage × 150 days
> For the farmer: Average off-farm wage × 20 days

Angkon work accounted for the remainder of the man-year of the landless worker. The imputation of 150 days for landless workers was based on their own estimates, while for the farmers the off-farm work estimate was based on the off-farm employment of the control group.

On the aggregate, the increase in per capita income due to the project was 47 percent. The ratio of incremental contribution to project income was 84

TABLE 14.9. Comparison of per capita incomes for the Palanas CIP.

		Worker group		
	Control group	Without project	With project	Days of employment
Per capita income				
Bottom 10	4,018	2,211	5,688	177
Bottom 15	5,757	3,905	7,893	167
Whole group	9,615	9,340	13,690	155
Correlation coefficient with duration of work	–	–0.44	–0.04	1.00
Number of poor				
International line	7	10	5	
Official line	13	14	13	

Source: Author's data.

percent, meaning only 16 percent of project wage earnings replaced incomes that would have been earned in the absence of the project. Averaging incremental incomes over workers, the proportion shot up to 142 percent. That is, per capita incomes of workers increased by nearly two-and-a-half times on average. This followed from the prolonged employment duration, the relatively high wage offered by the project, and the limited off-farm and nonfarm opportunities in the area.

Without-project incomes of workers were much lower than those of the control group. The disparity was larger for the bottom half of workers. There were only six poor persons in the worker category compared to 13 without the project, representing a poverty incidence reduction from 65 percent to 30 percent (at least during the duration of the work). Meanwhile only 35 percent of the control groups were poor—higher than the without-project poverty incidence of the worker group, but lower than the with-project poverty incidence. The project appeared to have been well targeted to the poor as well as to the poorer workers.

It was noteworthy that targeting performed well despite the high wage and the absence of a deliberate policy to hire the poor. Hence, the most important factor in the targeting performance of the project must have been the characteristics and hiring decisions of the IA members. If the IA member or his son worked, he tended to be poor; if another, then the designate tended to be poor, e.g., a landless relative. Targeting happened to coincide with the fact that the project had to be located in a very poor area with highly fragmented landholdings and limited nonfarm opportunities.

Within the worker group, there seemed to have been some exercise of targeting, i.e., the poorer workers got to work more days in the project. The correlation coefficients confirm this pattern: the without-project incomes of the workers were negatively correlated with days of project employment; the with-project incomes, however, were very weakly correlated with the employment duration, suggesting that the project contributed to equalizing worker incomes. The correlation coefficient between days of project work and share of nonfarm income in total income was –0.75. Clearly, the better-off workers were less willing to devote more man-days to the project. The targeting of wage benefits toward the poorer workers was not seen in the other two cases, as these implemented a rotation system.

Overall Evaluation of Targeting

Earlier we had stated that no explicit antipoverty targeting methods had been adopted in the projects. However, specific practices in each project in-advertently determined targeting performance. In the Palanas CIP, the fact that the IA members and their relatives were among the poorest in the com-munity led to a superior targeting of wage benefits. Targeting was weaker in the Ciabu WSP, and more so in the Tumbaga ARP, first due to adoption of rotating work assignments, and second due to the premium placed on hiring skilled labor.

However, the recruitment practice was not the only location-specific variable; other variables such as income, demographic characteristics, ten-ure, etc., varied with location. To bolster this contention it was necessary to control for these other variables. This was done through ordinary least squares analysis. The lefthand side variable is the days of project employ-ment. Correlates on the righthand side were worker characteristics, house-hold per capita income, and the location dummies. (The Palanas dummy ap-peared only as an interaction term with household per capita income, making it a "slope shifter"; the Ciabu and Tumbaga dummies were repre-sented as "intercept shifters.") The log transformation of the variables (ex-cept for the dummy variables) was made to capture nonlinearities in the cor-relations.

The terms "correlation" and "correlate" were chosen carefully; this was not a causal model, particularly because worker characteristics and incomes were related, and the sample collection was more purposive than random. Hence, the coefficients should be interpreted as indicators of correlation; the t-values likewise should be interpreted in terms of correlations, rather than statistical values for testing hypotheses. The results of the OLS correlation are reported in Table 14.10.

TABLE 14.10. Results of correlation analysis.

Worker characteristic	Pooled workers		Pooled workers (Palanas dropped)	
	Coefficient	t-value	Coefficient	t-value
Distance from project (km)	−0.512	−1.427	0.029	0.051
Age (years)	−0.281	−0.623	−0.348	−0.677
Schooling (years)	0.129	0.451	0.041	0.124
Household size	−0.369	−1.283	0.112	0.318
Size of landholding (ha)	0.296	0.940	0.062	0.175
Is a landowner (landowner = 1)	−0.416	−1.318	−0.511	−1.329
Income (w/o project, in Philippine pesos)	−0.020	−0.150	0.056	0.405
Palanas income (interaction term)	−0.455	−2.173	–	–
Tumbaga sample	−5.928	−3.090	0.286	0.634
Ciabu sample	−6.526	−3.632	–	
Constant	10.550	4.379	3.122	0.212
adj-R^2	0.62		−0.06	

Source: Author's calculations.

FINAL EVALUATION: LESSONS LEARNED AND BEST PRACTICES

In this concluding chapter, we synthesize and consolidate lessons learned based on the review of PWE experience and specific observations from the case studies. Some recommendations for project design have been incorporated.

On Employment Generation

The Philippines has pursued an integrative approach to PWE schemes: instead of providing make-work schemes that happen to generate productive assets, the country has tried to incorporate labor-intensive methods in its general infrastructure program. The country does not adopt as a matter of policy explicit quantitative indicators of labor intensity to set apart PWE schemes from other infrastructure projects. Such an approach has blurred the distinctiveness of PWE schemes.

The CEDP experience reveals instances of tradeoffs with technical adequacy. However, other experiences (i.e., the LIDP) have shown that labor content rules are not inconsistent with technical adequacy as long as the implementer is selective about the types and locations of the projects to be enrolled under PWE objectives. Despite this, hard-and-fast labor content rules have not been adopted since the LIDP. Instead, policy embraces LBES technology; in practice, the vagueness of the LBES rubric has permitted numerous departures from optimum labor content in public works.

The imposition of strict guidelines on labor content for priority PWE schemes should be considered. A benchmark figure of 50 percent project cost paid to labor is reasonable. To preserve asset quality, the enrolled priority projects must be chosen carefully for technical soundness and suitability for labor-based construction. Table 14.11 presents a checklist of the types of works that may be suitable for LBES technology. Meanwhile, other PWE schemes can follow the usual integrative approach advocated in government policy.

Remuneration Systems

PWE projects currently hire on a cash-for-work basis. Earlier projects under a food-for-work format were bogged down by the high cost of food distribution. As for basis of remuneration, the experience with piece-rate payment and task-based payment has generally been negative from the implementers' viewpoint.

The payment should typically take the form of cash, except for special food-for-work schemes to support specific emergency or relief operations within a particular time frame. Payment should also take the form of daily

TABLE 14.11. Projects suitable for labor-based methods.

Item	Typical labor content	Potential labor content
Low-cost housing	25-35	30-40
Social buildings	20-30	25-35
Water reticulation	5-15	25-35
Stormwater	5-15	40-50
Sanitation	5-15	25-35
Secondary roads	5-15	30-70
Dams	10-20	50-70
Railways	5-15	20-30
Electrication	10-15	35-45
Irrigation	15-25	30-70
Forestry	25-35	35-45

Source: DPWH.

or weekly wages. In terms of Philippine experience, such remuneration schemes are too difficult to administer.

The Pakyaw System

From the viewpoint of the public provider, the pakyaw contract is intermediate between daily wage payment and task-based payment system. The pakyaw system works because the provider saves on supervision effort, while the agent faces incentives to produce specific outputs within a reasonable time frame.

The pakyaw system should be widely disseminated through collaborative PWE schemes implemented by the DPWH and LGUs. The national government should establish a uniform policy exempting pakyaw contracts from tax and pakyaw workers from the usual labor market regulations.

Further advantages from the pakyaw system can be realized if greater emphasis is placed on preparatory organization of communities. Organizational work could also solidify the pakyaw groups from loose associations to active coalitions of workers.

Special guidelines should be drawn up for additional personnel and generous time allocation (say, one to three months) for community organizing prior to pakyaw contracts. To ease the administrative burden, the special guidelines could be applied only to priority PWE schemes.

Timeliness of Wages

Adequate preparation should be made to ensure frequent and timely payment of wages. The pakyaw contractor has no capacity to fill in the liquidity gap created by the irregularity of payments by the government.

A special revolving fund large enough should be constituted to finance three months' worth of pakyaw contracts. The possibility of creating a new contracting system for small and medium companies owning light equipment should be carefully studied, so as to provide an alternative to cash-strapped pakyaw contractors.

Local Governments

The local government project (WSP) had the ingredients that are supposed to make decentralized implementation work: an LGU sensitive to the basic needs of its local communities, with the means to obtain funds, and the know-how to conduct community works. The WSP project, however, pertains to a small-scale project that placed few administrative demands on the municipal LGU. For larger and numerous projects, financial resources and technical capacity of local government must to be strengthened.

At the village level, elected barangay officials play a crucial role in implementing development projects. Often these officials are simultaneously leaders of the major village associations. They do not participate in project preparation and design, essential in negotiating access rights, as well as part-organizers of labor recruitment efforts. Doubts may be raised as to how reliable these leaders are in representing the wishes of the villagers. However, the question can be raised at any level of leadership or of government; the question is at which level leaders are most sensitive to the wishes of their constituency with regard to local rural works.

To overcome LGUs' lack of capacity experience and resources, it may be necessary to engage national agencies in collaborative ventures with LGUs. At the village level, barangay officials should take a lead role in mobilizing and coordinating with villagers.

Cost Sharing or Equity

Participation of communities from inception, planning, design, to construction and maintenance of rural works, is integral to project success. The concrete mechanism for generating and shaping this participation has to be carefully designed.

The self-help format, involving equity contribution from communities, is one such mechanism. Cost sharing should be motivated by more than just stretching a tight government budget. It should serve as an *incentive mechanism* for galvanizing communities to supply inputs in which they have comparative advantage, namely information, labor, and monitoring effort. Collective effort should be seen as more than a by-product of communal ethos, or of traditions of mutual help (e.g., pintakasi), but rather of incentive design. Of course, the prerequisite to all these is that the community must clearly understand that equity contributions are compulsory for recipients of public assistance.

It is often argued that cost sharing encourages communities to take responsibility for infrastructure maintenance. A better explanation would be that cost sharing is also a *screening device*: only those communities ready to ensure the enjoyment of long-term benefits from an asset would be willing to accept an equity scheme.

The foregoing arguments about cost sharing apply as well to local governments. This justifies the requirement of matching grants for specific development projects. In turn, LGUs can extend equity schemes to beneficiary villages. Further study is needed regarding the optimum structure of these cost-sharing arrangements.

Labor Contribution and Self-Targeting

In providing equity contributions, communities should be given the option of contributing labor (rather than cash), a factor they possess in relative abundance. The cash equivalent of labor contribution as equity can be computed at prevailing market wages. This is an important way of achieving self-targeting wage rates, particularly if there are political and regulatory problems in evading statutory wages.

Location and Targeting

Geographic targeting is widely practiced in Philippine PWE schemes. Despite the broad intention of alleviating poverty, selection of workers is merely "targeted" to the unskilled members of the community. Thus, targeting of the poor depends in a significant way on prior selection of a poor village.

A rigorous procedure of geographic targeting is adopted in identifying the beneficiary communities for rural PWE. That is, the poorer areas may be selected in succeeding order: poor provinces, then poor municipalities of the poor provinces, and poor villages of the poor municipalities.

Rotating Work and Targeting

In two of the case studies, work assignments were rotated among numerous villagers explicitly to encourage community participation. Thus, many of the poor workers' stints were cut short, whereas some of nonpoor were able to join the works during their idle days. Although coverage of the poor may have increased commensurate with an increased number of workers, it remains likely that wage payments were less efficiently allocated toward alleviating poverty.

The communities in self-help schemes should refrain from adopting rotating work assignments. Instead, persons paid below-market wages should be allowed to work for the duration of a project.

Maintenance

That assets characterized as common pool resources be managed and maintained by users' associations is hardly debatable. However, other assets benefit more than a narrow group of users, e.g., feeder roads, flood control structures, etc. Thus, a user association may not be held solely responsible for upkeep; some joint maintenance between "frequent users" and the government, particularly the LGU, may be required. However, the absence of maintenance provision is endemic to public works projects. A clear procedure that appropriates funds, sets up a monitoring system, and identifies lines of responsibility, with respect to maintenance, should be laid down. The procedure should be adopted simultaneously with the program of work for constructing the asset.

Final Remarks

In the Philippines, rural works is only weakly linked with employment generation directed at poverty alleviation. This link can be strengthened considerably. The production of rural infrastructure and other community works can be feasibly undertaken with a high employment content oriented toward the poor. Funding agencies can aid in strengthening these links, by devoting funds for rural works programs that have high employment content, and are directed toward hiring the poor without compromising asset quality.

Notes

Chapter 1

1. We consider as rural any locality that exists primarily to serve an agricultural hinterland. In contrast, urban economies are driven by manufacturing, government, or some other economic base independent of agriculture. Given this view, "rural" areas include all the rural settlements, central marketplaces, and rural towns that are heavily integrated with the agricultural economy.

"Nonfarm" activities are defined by most countries to include all economic activity other than crop and livestock production. Therefore, they include agricultural processing and trade (conventionally classified as part of the manufacturing and commerce sectors, respectively), as well as construction, mining, transport, and financial and personal services.

The rural nonfarm sector in Bangladesh accounted for 28 percent of all rural employment. During the 1990s, employment in the rural nonfarm sector grew at an annual average rate of 2.8 percent compared with 1 percent growth in agricultural employment, and 1.5 percent growth in rural employment (Shilpi, 2003).

Chapter 2

1. The Grameen Bank, BRAC, PROSHIKA, Bangladesh Rural Development Board, and ASA in Bangladesh have been subject to considerable in-depth study and analysis while many others have been studied less intensively. Similarly, in India, the long-established projects such as the IRDP as well as a large number of self-help groups providing microcredit have been subjected to in-depth study.

2. However, the unwillingness on the part of the poor cannot be easily verified or proven because the operational procedure followed by most microcredit institutions in Bangladesh is such that the groups or individuals who become members are first chosen informally by the field officers and then they are formally enrolled. Consequently, usually no exclusion occurs after applications are made. Thus, no easy method exists for identifying whether a household has self-selected to be excluded or if it has been imposed by the organization (Rahman, 2000).

3. Some of the MFIs, for example, ASA (Association for Social Advancement), in Bangladesh, follow the practice of uniform minimum loan size which implies that the installments for repayment by each borrower are the same, thus making it easier for the borrowers to understand the calculations and for the field workers to implement the financial discipline. The practice of uniform loan size of the ASA helps the unit offices to expedite the loan approval process.

4. The risk exists that the current high loan recovery rates may not be sustained when larger amounts are lent to better-off clients. The poor recovery rates obtained by the agricultural banks in South Asia and elsewhere when lending to medium/ large farmers pose a cautionary note in this context.

5. The assumption is not that if "x" was borrowed, it was invested entirely in project "y" and that this led to a "z" change in consumption. The line of argument is that "x" was borrowed, it was spent by a utility-maximizing household, and that controlling for other factors one could roughly attribute "z" change in consumption or welfare to this loan.

6. This study also highlighted other factors associated with poverty, lending weight to the agencies' adoption of a multidimensional antipoverty strategy. Education, family-planning services reducing dependency ratios over time, and raising earners ratios, as well as health intervention resulting in improvements in health status, all contributed to the reduction of poverty. Microcredit enabled the poor nonfarm entrepreneurs to become independent from exploitation by the financial market intermediaries. Among the target group in the control village for the study of the Grameen Bank, only 9 percent of the households received institutional credit. With Grameen Bank intervention, the share of institutional loans to total loans was 78 percent, and the proportion of households receiving institutional credit was 30 percent (Zaman, 1999; Khandkar, 1998).

7. Although there are several channels by which microcredit can reduce vulnerability, there are fewer ways by which it can single-handedly reduce poverty. This is partly because the concept of vulnerability is a somewhat broader one than that of income poverty, and as such, more channels are available through which "impact" can be achieved.

The savings collected by microcredit organizations could have a greater impact on reducing vulnerability than they currently do. Usually, they collect compulsory regular savings from their clients with a view that the money would act as a de-facto lump sum "pension" when a client leaves the organization. Access to these deposits is otherwise limited, thus curtailing a potentially important source of consumption smoothing. Recently, an increasing number of microcredit agencies have begun to provide more flexible savings methods.

In 1998, Bangladesh Rural Advancement Committee (BRAC) initiated a current account plan, which is independent from its existing long-term savings system. In 1997, the Association for Social Advancement (ASA) in Bangladesh decided to allow its clients to freely withdraw their savings and the following year opened a savings service to everyone in the village. Grameen Bank also recently introduced an open access current account plan that can be used even by nonmembers. Depositors earn a competitive market interest rate and are allowed to withdraw money irrespective of whether they have an outstanding loan. The experiences of savings and deposit services by the microcredit agencies raise in turn issues of prudent regulation and deposit insurance which need to be carefully considered (Hashemi et al., 1996; Kabeer, 1998, 2001; Khandker, 1998; Goetz et al., 1996).

8. The indicators used to study the impact on women are many and varied. Many of the studies take into account women's income and savings, but do not consider their employment and control over loan use. Many studies ignored women's

access to food and nutrition and how the shortages are shared. Inclusion of such indicators is important because a general concern exists among nutritionists about widespread intrahousehold disparity in food intake. Some studies provide in-depth analysis of decision making and awareness, but they ignore many of the material aspects that may be no less important. The conclusions based on only one dependent variable or indicator should not be used as a basis for drawing conclusions about the overall impact. When a particular part of the outcome is chosen, more than one indicator should be examined. This provides a more comprehensive understanding of the type of impact but also provides scope of cross-checking of the impacts.

9. A recent study analyzes empowerment of women on the basis of three composite indicators: (1) decision making, (2) mobility, and (3) control of loan. The composite indicator of decision making includes the role of decision making about the purchase of family's food and clothing, family's investment activity, spending one's own income, children's education, and children's marriage. The indicator relating to mobility includes the polling center, the marketplace, and the health center. The MFIs inputs included number of loans, cumulative amount of loan, and length of membership. The other explanatory variables included were the female characteristics (age, marital status, education, whether she is the head among the women), the head of the household's age and education, and household's resource position (ownership of land and nonagricultural capital). The cumulative total loans and membership length seem to have positive impact on each of the empowerment indicators. The women's decision making role increases with the amount of loan but at a declining rate and after a point further increase in the amount of loan has a negative impact. When the total loan is at a high level, new loans may be used in such projects which may not require women's participation and thus may not enable them to achieve a greater empowerment. However, unlike the amount of loan, length of membership does not have a gradually declining incremental impact on women's empowerment. It has a continuous positive impact on decision making and control over loan. Women's age is not an important factor in determining their decision making role, mobility, or control of the loan. Women's education has a positive and significant impact on mobility. Woman who are the heads of households (vis-à-vis those women who are not) have positive impact. Women's own income has a positive and significant impact on empowerment (Rahman, 2000).

10. The issue of complementarity also arises when considering the effect of microcredit on the empowerment of women. Although microcredit can enhance a woman's status in the eyes of other household members, as she is the origin of an important source, social mobilization and legal education interventions in conjunction with credit are likely to have a more significant effect than credit alone.

11. The appropriate interest rates in Asia for MFIs have been calculated at 35 to 51 percent per income on the assumption of (a) at 15 to 25 percent, (b) at 20 percent, (c) at 12 to 15 percent, (d) at 8 to 10 percent, and (e) at 2 percent (Gibbons and Meeham, 2000, 9-10, 22).

12. "Social intermediation" (raising awareness and motivating the poor to borrow and demonstrating through group discussions how to use a loan) is usually regarded as a financial service, because it is inextricably tied up with credit delivery

for the poorest of the poor, making possible "credit reception," and constituting the other side of the coin of credit delivery.

Chapter 3

1. Grameen Bank is the pioneering and largest microcredit agency in Bangladesh. Its innovative approaches toward financing the assetless, rural women borrowers, without any collateral, have been widely acclaimed and followed worldwide.

2. Organizationally, the Microenterprise Customer Relations Officer (MECRO) who is supervised by a branch manager (and the latter in turn by the zonal officer) is responsible for selecting borrowers on an assessment of their financial viability.

3. As long as a branch maintains a floor recovery rate of 90 percent and a minimum disbursement of 75 percent of the target, it is rewarded percentages of the total loan recovered according to a prefixed scale. There is no precise module to distribute the total incentive money among the branch officials, but the branch manager and the MECRO receive the larger share of the amount. The zonal officials are also included in the incentive plan.

4. On-the-job training and workshops are believed to strengthen the officers' eligibility for future promotion. Also, every branch receives a motorcycle.

5. For example, thana-level (i.e., village or local level) government officials, especially those in the agricultural extension, livestock, and fishery ministries and the Bangladesh Krishi Bank (BKB).

6. The Bangladesh Krishi Bank (BKB), an agricultural credit agency, has been the major partner so far.

7. The SOs are responsible for organizing groups (VOs); they oversee the election of the manager and president, motivate VO members for social mobilization, supervise cultural programs to propagate the idea of savings mobilization, regularly attend group meetings, monitor records on financial transactions maintained by the VO managers, approve loan applications and forward them to the project coordinator, assist subject specialists in organizing training and select participants for the trainings, liaison with bank officials and other local-level government officials. The SOs are accountable to the project coordinator. During the initial years, the SOs took the initiative to organize people in a neighborhood by seeking assistance from elderly/influential individuals, who called meetings. Even though the participants in the meeting were asked to raise their hands if they supported the candidacy of particular individuals for the positions of managers or presidents, very often such candidacy was sorted out prior to the meetings. The president of a VO is normally an elderly person who is respected by all groups in the community while the manager is the person who effectively runs the VO. Later on, when the KST credit operation was widely known in the area, often the local youth have contacted the project office and assisted in doing the groundwork for formation of a VO in their area.

8. Some of the conventional MFIs have plans on term deposit whereby they mobilize funds from the rural rich (nonbeneficiaries). The KST approach internalizes such activities.

9. The effective interest rate stands around 34 percent per year; both principal and interest are paid together from the second week.

10. For example, for non-BKB loans, one installment is for beef fattening; two for banana and papaya plantation; four for poultry, fishery and nursery; and 50 for dairy cow, rickshaw, mortgage, and hand tubewell.

11. Because of the minimum education requirement for VO managers, they often belong to a relatively rich family and thus have extra leverage to put pressure on a defaulter. In order to put social pressure on the defaulter to repay, VO managers often visit his or her house with some other members.

12. The dollar equivalent of the amount of loans is estimated at the prevailing rate of exchange. The (effective) interest rate of MFIs is 20 to 25 percent. The effective rate of interest is different from the nominal rate when the latter is charged at a flat rate and not in the declining balance even if repayment is being done bimonthly or weekly.

13. The average cost for borrowers of processing and obtaining a loan is 2.16 percent of the average amount of loan. There is no need for repeated visits to the branch because the FO performs the role of the link person between the bank and the borrower, thus reducing the borrowers' transaction costs.

14. Ninety-five percent of GU products are exported either directly as fabric or as garments which use GU fabrics and 5 percent are sold through its retail outlets, which largely replace imported fabrics.

15. The figures are 6.4 months versus less than 3.7 months per year.

16. Its powerloom unit produced 20 percent of GU's gray cloth production during 1999-2000.

17. A high proportion of the borrowers in the indirect lending program are not graduates of NGO programs and run businesses involving large sums of money. The reasons for this are not clear. Several hypotheses may be proposed. The defaulters on loans taken from the commercial banks who are no longer able to borrow directly from them resort to NGOs. Also, one may decide to borrow from NGOs, even if he or she fulfills all the requirements for direct lending, if NGOs share with the borrower a part of their 3 percent commission.

18. Other than delays in processing loans, which are often instruments to raise the premium, the transaction cost mostly includes the unauthorized charges charged by bank officials.

19. The project has three main sources of income: interest on loans disbursed to enterprises, interest on credit funds transferred to branches, and interest earned on the credit fund account. The loans disbursed to enterprises are the loans disbursed to the targeted population in the rural areas. The credit fund transferred to the branches is the amount transferred from the headquarters to the branches, but is not used yet for disbursing loans to the target population. Finally, the credit fund account is the amount that stays in headquarters until it is disbursed to the branches.

20. Alternative employment opportunities for VO managers in the countryside are slim. The VO manager's job is regarded as a good opportunity, not only because of the income but also because of the additional benefits (training, networking, and easy access to credit).

21. Monitoring is the responsibility of the auditors who have considerable work-loads. Moreover, auditors are also a subset of the VO managers, and thus, financial irregularities of the VO managers are unlikely to be properly reported.

22. As in the case of KST, generally many loans in SEDP go to postharvest pur-chases of paddies. Rice is sold in Bangladesh in a large, countrywide integrated market, and such marketing is a year-round activity. Consequently, incomes in rice milling and paddy trading do not have the seasonal fluctuations characteristic of many farm commodities. As well, petty trade in paddy and rice is a flexible vocation that lends itself to a wide variety of contexts of scale, technology, and division be-tween hired and family labor. This is why a large proportion of businesses in any given rural market town at any given time happen to be in this particular trade. Also, rice trade is among the most labor-intensive businesses in Bangladesh. It is not sur-prising therefore that the largest proportion of existing businesses that have received loans from SEDP are rice mills and paddy processors. Working capital in an agroprocessing industry such as rice milling is possibly no different from an invest-ment in trade. Several studies point to the increasing importance of tailoring and in-formal garment making as a small manufacturing/service activity in the semiurban and rural areas. It owes its existence to the availability of local demand (a compara-tively youthful population with a fairly strong demand for ready-to-wear garments), ready availability of raw material, especially fabrics often at slightly discounted prices from the ready-to-wear garments industry.

23. Trading enterprises in the sample on which this analysis is based still have re-ported quite decent profits which may have been due to the SEDP clients wresting market shares from other merchants (who may be less advantageously situated in terms of credit supply). When SEDP changed policy in 1998, as much as 35 percent of its borrowers were engaged in trading of one kind or the other. It was probably felt that at that stage this subsector was saturated with capacity.

24. About 70 percent of the borrowers have more than 0.5 acres of land and 30 percent of the borrowers are medium and large landowners. The average years of schooling are seven years—that is substantially higher than that of MFI borrowers. Eighty percent of borrowers are literate and numerate—that is higher than the na-tional average.

25. The average loan size for women was Tk 28,200. For men it was Tk 42,184.

26. That is why in many instances women borrowers, who constitute 80 to 90 percent of total borrowers of microcredit agencies, hand over money borrowed by them to the male family members. Frequently, when men do not make timely repay-ments, they are not subject to the credit discipline of the bank because they are not legally considered borrowers. In such a situation, in the case of NGO/MFI borrow-ers, women repay by drawing from other sources of income. In the case of SEDP this is not always a possible alternative for women because the average loan size is much larger than the average NGO/MFI loans.

27. A small loan in relation to the combined fixed and working capital of an en-terprise (i.e., average Tk 200,000) involves a small risk.

28. Most of these institutions had access to donor grants. Along with it, they have an accumulation of assets in both tangible and financial forms. In the case of micro-

credit programs, they provide two additional assets: access to additional financial resources generated through members' savings and own accumulated capital.

29. That is, reduces cost per program or activity undertaken.

30. For example, the fruit juice processing industry in Bangladesh, promoted by this project, has successfully entered the overseas export market for fruit juices in the United States, Europe, and South Asia, and also exports canned and processed fruits in collaboration with companies in Europe.

31. While both MEDU and SEDP are intended to reach the "missing middle microentrepreneur," there are several differences, e.g., interest rates, sectoral targeting, and performance incentives. First, MEDU's interest rates vary from 12.5 percent for fixed capital to 14.5 percent for working capital, while in SEDP, interest rates vary from 10 to 14 percent, increasing with the size of the loan. In the case of SEDP, investment in fixed capital is likely to be discouraged, because larger loans carry higher interest rates. Second, while MEDU provides loans only for nonfarm activities, SEDP provides loans for both noncrop agriculture and nonfarm activities. Both concentrate on existing enterprises, with a comparative neglect of new entrepreneurs. Although MEDU is located in urban and semiurban areas but is intended to reach borrowers in the neighboring rural areas, SEDP in both location and clientele focuses on semiurban areas close to the markets precisely because they want to target semiurban enterprises. Third, regarding performance incentives, MEDU relies on training and financing incentives for the bank officials. In the case of SEDP, the field officers of the project, rather than the administering bank, monitor loans and ensure their recovery. Their incentive is the success of the pilot project for which they work. While MEDU emphasizes training for bank officials, SEDP provided two types of training for borrowers (general banking practices and specific technical skills) and the costs are borne by the project and not the bank.

Regarding their comparative performance, the average sizes of loans for both projects are broadly similar—an average of Tk 30,000—and are collateral free, although there are no gender-specific targets. Lending to women is significant in the case of SEPD, unlike MEDU. This is partly because SEDP actively encourages women to borrow. The availability of small loans at low rates of interest is more attractive to women. Finally, both programs fail to reach MFI's graduates and to promote new entrepreneurs.

Chapter 4

1. The participating commercial banks in the project disbursed the credit under the general administration of the National Bank for Agriculture and Rural Development (NABARD). To achieve these objectives the project provided technical, financial, and infrastructure support for microenterprises; provision of a line of credit to enable commercial banks to support all viable income-generating activities; institutional support to strengthen credit delivery and reception; promotion of savings and credit operations by building the capacity of SHGs; and management support for monitoring and evaluation.

2. The poverty line is defined as annual household incomes of up to Rs 11, 000 at 1991-1992 prices.

3. The poverty prone groups included mainly schedule caste (SC), schedule tribe (ST), and other backward classes (OBC), as well as agricultural laborers and small holders (i.e., households owning and/or cultivating less than 2.50 acres).

4. SHGs were designed to inculcate financial discipline among the rural poor (especially women), mobilize savings (as a prerequisite to borrowing from a bank), and to distribute loans among the members in accordance with collectively agreed-upon priorities and rules. Joint liability of the members ensured prompt repayment of loans. Pooling of credit requirements helped lower transaction costs of the poor borrowers as well as of the lenders (e.g., mostly banks).

5. VDC had the custody of the below poverty line (BPL) list. A household must be included in this list to be eligible for a loan.

6. The monthly savings contribution is deposited in a bank. The credit-savings ratios range from 1:1 to 4:1, depending on the SHG financial discipline.

7. Periodically, the group members and the NGOs verify the records.

8. Equivalent to 24 to 36 percent per annum. The difference was credited to the members' accounts.

9. Sahyoginis, or field workers, are the link between NGOs and SHGs. They perform a wide range of functions, e.g., awareness building, attending group and block meetings, maintaining records, training women in nonfarm activities, and arranging trips to the banks. Sahyoginis are trained both by MAVIM and NGOs.

10. A credit rating agency carries out the capacity assessment rating of MFIs.

11. Whereas in DHRUVA there is no legal exclusion for women, in IIED the group was meant to be composed of women only.

12. Beneficiaries of individual loans were emphatic about the inappropriateness of group loans for investing in income-generating activities. As an estimate, in the sample the average individual loan size was Rs 15,600, while group loan sizes ranged from Rs 500 to Rs 4,000.

13. The exceptions were as follows: One poor borrower sold off the buffaloes after three months to repay a loan from a moneylender and another relatively affluent borrower abandoned his textile trade to concentrate on other business.

14. The poor individual beneficiaries used the loans in activities such as selling snacks, vegetables, dried fish, cutlery, buying buffaloes, and making picture frames. The nonpoor individual borrowers used the loans in activities such as buying goats, sewing machines, welding equipment and grinders, and trading in textiles. Finally, SHGs borrowers used the loans for ceremonial purposes (e.g., Diwali), tailoring, and trading in vegetables and legumes such as lentils and beans.

15. Some notional estimates of gross returns in activities pursued by the poor included: cutlery (about 43 percent), picture frames (120 percent), snacks/tea stall (109 percent), and vegetables and dried fish (about 128 percent). A larger range was observed among the nonpoor: textiles (30 percent), oil (90 percent), and tailoring (200 percent).

16. Among the individual beneficiaries, 47 percent were poor but the majority was nonpoor. All poor beneficiaries except one belonged to backward classes. Among the nonpoor, 50 percent belonged to upper castes. The share of female poor

beneficiaries was slightly higher than the nonpoor beneficiaries (57 percent versus 50 percent).

17. Among the poor, 43 percent owned land and among the nonpoor, 62 percent. The average land owned was greater among the nonpoor (5.83 acres for the nonpoor versus 2.66 acres for the poor). None of the poor owned additional assets while 75 percent of the nonpoor did. The poor were less educated than the nonpoor.

18. Also, individual loan sizes were larger in cases of individual lending than in cases of SHG: The average amount borrowed by poor individual beneficiaries was smaller than the average amount borrowed by the nonpoor (Rs 10,000 versus Rs 15,600). Among the SHG respondents, however, the loan amounts were considerably smaller, ranging from Rs 500 to Rs 4,000. This difference is partly a reflection of the absorptive capacities of the poor and nonpoor but could also be attributable to the influence of the nonpoor in the implementing agencies (VDCs and banks).

19. There was often inadequate knowledge about both the existence of the project and about its functioning characteristics. It is also plausible that the dissemination of information among the poor was weak and slow, given their low rates of literacy and limited social interaction.

20. In violation of the MRCP guidelines, the domination of VDCs by influential people and the participation of more than a few of the affluent suggest that violation, in complicity with VDC and bank officials, was very common.

21. The functioning and accountability of VDCs was less than satisfactory. The relationship between the VDA and the VDC and the role of VDA in vetting loan applicants was unclear. Decisions were seldom participatory and the selection of beneficiaries was usually a ritual with the gram sevak (village-level worker) playing a decisive role.

22. Generally, in the case of SHGs, a higher percentage of the members were poor (as of March 1997, 32 percent of the 10,000 members of SHGs were above the poverty line).

23. Greater participation of women in local institutions such as the panchayats was one manifestation of the growing self-confidence and assertiveness among them.

24. The survey indicated that out of the four poor female beneficiaries, one (a widow) had complete autonomy in the selection and use of assets; the rest (three) declared that it was a joint decision. The income was shared among the family members but in most of the cases loan repayment obligations were women's responsibility.

25. The consultants were expected to assist at least 100 entrepreneurs (microenterprises and not household livelihood enterprises) in setting up new or expanded enterprises per year. They were to obtain part of their remuneration by charging the entrepreneurs 5 percent of their profits. However, several difficulties were apparent as implementation proceeded. Among them, it was difficult to assess profits for microentrepreneurs, and willingness to pay was low unless the consultant had made an identifiable contribution to profitability (such as solving a specific technical problem or removing a marketing or supply bottleneck). While the role played by district-level consultants was appreciated by both the microentrepreneurs and banks

and facilitated bank lending, it was not sufficiently appreciated to result in a willingness to pay for the services of the consultants (ADB, 1997).

26. They would probably still be enough to more than cover the loan cost.

27. Two-thirds of the enterprises (8 out of a total of 12) were originally financed either from personal savings and/or loans from relatives or banks. None borrowed from a local moneylender.

28. Half have been engaged in microenterprises for one to three years, one-fourth for three to six years, and one-fourth for 17 years. One-fourth of the enterprises was entirely family run without any hired workers. The ones with hired workers also relied on family for support (usually the wives looked after the business if the men were absent).

29. The remaining activities were concentrated in repairing, flour milling, and other services.

30. It was reported that the surplus (over and above all expenses, including loan repayment obligations and "compensation" to family members) ranged from 10 to 15 percent.

31. The beneficiaries were essentially piece-rate workers and although they were protected from market uncertainties (e.g., unavailability of inputs, demand fluctuations, price variability), they had no inputs in the selection of the microenterprise they would like to participate in nor the skills they would like to acquire.

32. Denial of information about the benefits and procedures of SFMC to those not considered suitable was not unlikely. Although it could not be confirmed, it was plausible that the field staff of IIED concentrated on a subset of the trainees who were more motivated and displayed greater potential for the activities that were promoted by the NGO.

33. NGOs charged a relatively low rate of interest (12 percent per annum) when they borrowed at 11 percent per annum. The repayment period for the NGO was four years while the repayment period for the individual beneficiaries was two years. Another plausible explanation of why NGOs could afford to charge low interest rates was that SFMC provided both financial and training assistance. General financial assistance included grants to MFIs to ensure adequate and intensive capacity building in operational, organizational, and managerial areas. Specific financial assistance comprised need-based assistance to client MFIs to cover the program expansion costs in order to make them sustainable in the long term. In addition, SFMC provided financial support to technical and management training institutes for conducting orientation and training programs for the staff of MFIs.

34. In order to appropriately assess the impact of these programs on the poor, it would be necessary to supplement the present analysis of (notional) returns and their fluctuations with an investigation of types of assets financed by the schemes, their retention rates, and variability of yields.

35. Village panchayats have informational advantages (i.e., they are fully aware of the economic status of a household), they are more representative than ad hoc community organizations because they have quotas of participation for backward sections (which the Ayojan Samatis constituted by DHRUVA do not have), and also they are accountable to the village communities (because the decisions made must be ratified in a village assembly).

36. See appendix to Chapter 5 on PKSF—an apex microcredit organization in Bangladesh.

Chapter 5

1. In the case of SFMC-IIED in India, SHGs were assisted by the SFMC project funds in the following ways: (1) General grants for capacity building in operations, organization, and management; (2) grants, whenever needed, for the expansion of the program for meeting its corresponding increase in the costs of administration; and (3) grants to technical and management training institutions for conducting training programs for SHGs.

2. Organizationally, Palli Karma Sahayak Foundation (PKSF) has a general body that usually meets once a year for overall policy guidance, a governing body that guides the management and creates and reviews policy when necessary, a chairman, a managing director (who is the chief executive officer of the foundation), and two general managers heading the divisions of loan operations and administration and finance. The maximum number of members in the general body is 25, out of which government may nominate not more than 15 members from among persons associated with government agencies, voluntary organizations, or private individuals. The remaining ten members may be from among persons representing the partner organizations (NGOs and other agencies that received credit from PKSF) and/or private individuals. Presently, PKSF has a general body of 15 members consisting of distinguished personalities in the country.

The governing body is composed of seven members: chairman of the foundation, the managing director, two members nominated by the government, and three members elected by the general body. Presently, the governing body comprises persons of repute such as Mohammad Yunus, Managing Director of the Grameen Bank.

The chairman is nominated by the government from persons not in service of the republic, usually for a term of three years. The present chairman is a well-known economist and a professor of Dhaka University.

The governing body appoints the managing director. He is the ex-officio member of the governing body.

3. For details on PKSF, see S. Ahmed, *Creating Autonomous National and Sub-Regional Micro-Credit Funds*, PKSF, Dhaka, 2000; and *PKSF: Impact of Micro Credit*, Dhaka, 2000. As of December 2000, PKSF had received funds for a total of US$167.9 million. Of those a total of 39.3 million are grants (US$21.6 million from the government, US$12.7 million from USAID, and US$5 million from the World Bank). The remaining US$128.6 million are loans (US$100 million from the World Bank, US$18 million from ADB, US$10 million from the government, and US$0.6 million from others).

4. Landless and assetless are defined as those residing in rural areas owning less than 0.50 acres of cultivable land or having total assets of the value less than one acre of land in the locality.

5. The organization should have at least 400 organized members, Tk 0.2 million operating loan outstanding at field level and should have experience of at least six

months successful microcredit operation. Members are organized into groups following horizontal links, focusing on group discipline, regular attendance, and savings deposits. Members should have a minimum of six months' practice with a regular savings deposit. The organization must maintain a minimum loan recovery rate of 98 percent on a continuous basis. For a program operating for more than three years the minimum loan recovery rate is 95 percent. Finally, overlapping with the activities of other organizations in the same area must be avoided. The organization should have regular and full-time staff as well as trained and skilled manpower; the chief executive should work full time and be acceptable to the staff, group members, and to the community in general.

6. The groups should be organized within a 10 km radius of the project office and in the case of a local organization the head office should be situated in the working/operational area.

7. Usually, the main reasons for rejection are financial mismanagement and gross inconsistency between information in the application and information gathered from field verification.

8. Five percent for sanctioned loans of up to Tk 5 million, 4 percent for loans between Tk 5-75 million, and 4.5 percent for loans above Tk 75 million.

9. PKSF is considering imposing 2 percent additional service charge on POs as penalty for late payment.

10. PKSF collects information on changes in borrowers, savings, loan disbursement, and recovery every month. POs submit cumulative and monthly income, expenditure, and cash flow statements to monitor the financial health of the PO. POs regularly send the list of borrowers to PKSF. PKSF places utmost emphasis on field visits. Usually, the PKSF officer visits each PO every three months. However, if the PO is big and has multiple branches, a team of PKSF officials visits the program. The field visit is used for verification of the program as well as an effort for institutional development of the PO, and thus during visits, information submitted by POs is verified and suggestions are made for improvement. PKSF conducts a 100 percent audit of borrowers, usually annually, before embarking on major loan expansion. The audit reports are submitted to the managing director directly. As a part of annual financial auditing of PKSF, an external audit firm is engaged to verify the financial position of sample POs.

11. Main theoretical training on poverty alleviation programs, credit programs, and auditing techniques are provided within PKSF. Officers are sent to Bangladesh Rural Development Academy and Public Administration Training Centre to have an overview of rural development and poverty alleviation programs. On the other hand, practical training is acquired from Grameen Bank and other NGOs. Officers stay at least two weeks at the branch level of Grameen Bank, which gives them exposure to credit programs for the poor. PKSF also sends its officers to stay in the POs and learn about the program. After each visit either to Grameen Bank or to a PO, trainee officers submit elaborate reports on the organizations. Additionally, trainee officers visit the POs along with senior officers; thus, they learn PO appraisal process, auditing techniques, minor details of field activities, work ethics, and culture of PKSF.

12. By early 2004, PKSF has lent about $200 million to 210 microfinance institutions covering 3.8 million borrowers. More than 60 percent of borrowers are women. As the biggest microcredit funding apex organization in the world, PKSF's standards, guidelines, and modalities are being studied by various countries in relation to their policies for the development of the microcredit sector (S. Ahmed, 2000).

13. Other sectors where borrowers invested included 10 percent poultry rearing; 10 percent milk cow, cattle fattening; 7 percent land lease/purchase of house; 6 percent rickshaw/van (see World Bank Impact Study, 1999).

14. As an estimation of the percentage of poor borrowers: In the sample of the study previously mentioned, about 68 percent of the borrowers lived in poor housing conditions before joining the PO (either thatched, or wall and roof with substandard material houses) and only 7 percent had houses with tin walls and roofs.

15. Various indicators have been suggested to study their impact for current gains including the following: income (percent of beneficiary household members living below the poverty line income); food and nutrition intake (percent of beneficiary household member below poverty line intake); housing (percent of beneficiary households having below poverty level housing). Longer-term material gains indicators include land (percent of beneficiary households owning below poverty level arable land) and for crisis proneness (percent of beneficiary households facing deficit months in a year). Social gains indicators include education (education of beneficiary household heads and primary school enrollment rate of the children of beneficiary households); sanitary condition (percent of beneficiary household members using sanitary toilets); and for drinking water (percent beneficiary household members having access to safe drinking water). Specific standard/norms levels for each of the indicators were also defined.

16. Some of the financial data of PKSF as on June 30, 1999, are as follows: debt-equity ratio: 1.63:1; cumulative loan recovery rate: 98.16 percent; reserve rate (loan loss reserve/loan outstanding): 2.67 percent; operational expenses to loan disbursed (operational expenses/disbursement of the same year): 5.28 percent; operational self-sufficiency (total income/operation expenses including financial expenses + DMR): 161 percent; Percentage of total expenditure to total income is 36.43 percent.

17. To compare with the Indian experience as recorded earlier, PKSF shares the same objective with SFMC in India, i.e., fostering sustainable NGOs/MFIs by wholesaling credit and by providing capacity building to them (the terms and conditions PKSF seem to be more advantageous). Also, PKSF has some resemblance to MEDU indirect lending. The interest rate for MFIs is 3 to 5 percent in the case of PKSF and 11 percent in the case of SFMC. In SFMC, the repayment period is four years, including a six-month grace period. In the case of PKSF, regarding loans to small MFIs, the repayment period is three years with a grace period of six months, and in the case of loans to bigger NGOs/MFIs, the repayment period is ten years with a grace period of four years.

18. At the time PKSF started operations (1990), 90 percent of today's NGOs/MFIs did not have microcredit programs and some of them did not even exist. The existing ones were not even reasonably close to operating sustainability. In 1990 the

total portfolio of NGO/MFIs was Tk 0.4 billion (US$7.4 million). The size of this portfolio increased by 4,500 percent in a decade but it only grew 900 percent within the first half of the 1990s. Thus, the boom of the microfinance sector in Bangladesh took place in the 1990s, and especially in the second half of the decade.

Chapter 6

1. In India, self-help groups of ten members rather than five members, as in Grameen Bank in Bangladesh, have been found to be more appropriate by enabling a larger coverage of borrowers. Moreover, in India SHGs keep their savings with the group rather than pooling them together with the savings of other groups. Therefore, both the savings and their subsequent microcredit ventures are kept within each group in India (*The Daily Star,* 2004).

2. An illustrative exercise in Bangladesh shows that three criteria used in combination (landownership, housing, and occupation) succeed well in identifying the nonpoor, the moderate poor, and the poorest. In Bangladesh, the poorest were identified as agricultural laborers who reside in a single structure (thatched) and own no more than 0.5 acres of land (Sen and Begum, 1998).

3. Statement by IFAD: "In India, grants have been provided to the poorest members of the SHGs on the condition that the members work in infrastructural development projects" (*The Daily Star,* 2004).

4. A number of other factors did influence the inclusion of poor households in the program, such as help of patrons in the context of patron-client nexus in the rural area with landowners or moneylenders or village leaders patronizing their clients; efforts of the enterprising among the poor to canvass and get the support of influential persons in getting access to food rations as well as preference of the BRAC field workers for potential microcredit clients, i.e., those who had the potential to develop income-generating activities.

5. Recently, PKSF in Bangladesh has introduced a pilot scheme titled "Financial Services for the Poorest" with a flexible mechanism: weekly repayment is not compulsory; borrowers can borrow for a shorter period (not for the standard 52 weeks); loan sizes vary according to the needs of a particular borrower, coupled with variable grace period and interest rate according to the nature of the client and the geographic area where the borrower resides. A disaster management fund exists to protect the poor during unforeseen disasters (Ahmed, 2004).

6. Borrowers go to banks for deposit, borrowing, and withdrawing money but MFIs have to go to their clients for such purposes, which is costly (Matin et al., 2003).

7. An NGO in Bangladesh called the Association for Social Advancement (ASA) has sought to introduce a few of these innovations or procedures with a view to reducing costs. They include simplification of bookkeeping and documentation procedures, standardization, and common use of facilities and of reporting; decentralization of decision making including loan approvals to local branches; same amount lent to every borrower, and same number of clients in each branch; standardized guidelines for group formation, loan size, repayment amounts, and eligi-

bility for credit are documented in detail for branch staff and the procedures for recruitment and training of staff are simplified. Above all, ASA does not undertake any social intermediation and training functions (Chowdhury, 2004).

8. Leasing can be a source of medium-term financing for microenterprises to expand their operations. Leasing provides an alternative way for microenterprises to acquire the use of an asset instead of an outright purchase. Leasing is focused on the lessee's ability to generate cash flow from business operations to make lease payments because the lessor retains the ownership of the asset during the term of the lease. Capital base or credit history as well as knowledge about the results of business operations generates indicators of the adequacy of prospective cash flows. Typically, the borrower/lessee provides about 10 percent of the asset cost as an upfront security deposit. It is usually less than the equity component required in conventional bank financing. NGOs and other microfinance institutions can serve as leasing agencies. They often obtain funds from international aid agencies such as World Bank. Aggregate periodic lease rental payments (which are a combination of interest and repayment costs) are considered as a business expense against tax liability on income. Moreover, the asset is depreciated over the life of the lease, which is shorter than its economic life (Gallardo, 1997).

9. Should job or skill-specific training be paid for? The private gains to the trainees can at least partly pay for the costs of training. There is one reason for this. Payment will produce a feedback from the trainees, which in turn improves the quality of training. In a Latin American country, the entrepreneurs receive vouchers from the government for payment for training. They combine free vouchers with their own funds to pay for the training of their choice. Apart from professional training institutions and specialists in specific skills and jobs, entrepreneurs already engaged in existing enterprises transfer "hands on" specific skills in various jobs to the new entrepreneurs. Specialized institutions need to provide training to a wide variety of microenterprises that are publicly funded so that learning can be acquired for wider applications (Schreiner et al., 2003).

Chapter 7

1. Fifteen women in each union (the smallest administrative unit in Bangladesh) constitute a crew. A crew consists of three groups of five women who work in three segments of each union. RMP undertakes about 95 percent of all the rural road maintenance in the country. Each woman is required to save about 18 percent of her daily wages in the bank account where daily wages are transferred. Through the operations of the savings account in banks, the participants gain credibility with local banks for future transactions such as gaining access to credit from banks. CARE has developed various methods to train the participants in skills so that they do not fall back into destitution after employment is over. Equipped with savings and skills, the participants are able to launch income-generating self-employment activities of their own after the RMP employment is over. As the existing participants depart (survey of RMP workers in mid-1990s showed that each group works for an average of 6.3 years), a new group of workers is selected for employment. The same survey

evaluated that 52 percent of RMP workers were widowed; 10 percent were divorced; 12 percent were separated from their husbands; and 26 percent were married. Among the married women, 80 percent of their husbands were either disabled or had no income or employment. The additional income of workers over alternative earnings was more than 68 percent (Ahmed and Shams, 1994). There was, however, one negative side effect in that the 14 percent of the women workers could not take proper care of their children because of their involvement in RMP and 5 percent brought their children to their work site.

2. Among the nonparticipating households, 43 percent were unable to participate in work due to its arduous nature, 13 percent did not like the work, and 28 percent preferred alternative job opportunities.

3. Another survey in mid-1990s showed that income transfer constituted an increase of 50 percent over prework income (Ahmed and Shams, 1994).

4. Since the volume of construction-based employment does not coincide neatly with periods of the greatest demand, a complementary effort to create self-employment opportunities would bridge periods of hardship.

5. The scheme was restructured in 1999 and named Jawahar Gram Samriddhi Yojana (JGSY). The Employment Assurance Scheme (EAS) introduced in 1993 was also restructured in 1999. EAS focused on employment provision during the agricultural lean season. Any two members or a family between the ages of 18 and 60 who are in need and desire work are eligible. The difference between JRY and EAS was that the former was designed to provide employment to these rural poor who were below the poverty line, whereas EAS was to ensure employment of 100 days during the agricultural lean seasons to whoever among the rural poor sought employment at that time. The ensurance was extended to men and women between the ages of 18 and 60 in the selected villages covered by EAS. While JRY was a countrywide program, EAS covered selected disadvantaged villages such as flood-prone areas or areas with heavy concentration of disadvantaged classes such as tribal people and low-caste populations. Third, no more than two adults per family were to be provided 100 days of employment under the program. Those who needed and sought employment under the EAS were required to register with their village panchayats and a family card was to be issued to the family containing the two eligible members. With the restructuring of JRY into JGSY, priority was given to developing infrastructure for SC/ST habitations, education, and public health. Since 1999, a total of 1.57 million works have been completed up to March 2001. In addition to rural infrastructure, it has provided individual assets to the poorest of the poor SCs'/STs' families as 22.5 percent of funds are earmarked for them, to the amount of 598,000 works. Around Rs 43 billion was spent between 1999-2001 which resulted in some panchayats getting an amount as Rs 5,000 per annum—an amount with which no meaningful infrastructure can be built. More reasons are needed for any significant impact on poverty.

6. The households were classified as poor with an annual family income of Rs 6,400 or less (up to Rs 2,265 for destitute; Rs 2,266 to 3,500 for very, very poor; Rs 3,501 to 4,800 for very poor; Rs 4,801 to 6,400 for poor; and Rs 6,401 and above for above the poverty line). Women were to be provided 30 percent of the generated employment.

7. Scheduled castes/scheduled tribes were considered especially disadvantaged classes in society.

8. An illustrative list of types of projects was provided by the government for the guidance of the implementing agencies, ranging from social forestry or water conservation, minor irrigation, flood protection, and drainage and roads, to construction of sanitary latrines, land development, reclamation of wasteland and degraded land, and houses for SC/ST (Neelakantan, 1994).

9. The contractors were also involved in the identification of projects to be executed, and in helping the beneficiaries to register in addition to providing labor for employment on projects.

10. Surveys were conducted in 1993 to 1994 for the selection of IRDP (Integrated Rural Development Program) beneficiaries. Later on, the IRDP lists were intended to be available to the executing agencies.

11. The use of minimum wage raises several issues. If minimum wage is above market wage, persons with other market-based opportunities will be attracted to JRY. This inflates demand for JRY, probably requiring rationing when market wage exceeds minimum wage, and demand for employment will be very limited. If there is no demand for employment, resources may well be reallocated from these areas to poorer areas, in view of the overall limitations of resources. One way of expanding resources for the JRY program is to link the program with the large accumulation of public food stocks, resulting from the government's food procurement and price support policy. The amount of food stocks amounting in some years as much as 40 to 50 million tons inflicts heavy losses on the public exchequer. The use of food in payment of wages in JRY employment schemes is the most productive way of reducing unwanted surplus food stocks (Thamarajakshi, 1997).

12. The total employment generated by JRY/JGSY and EAS provided 1.24 billion workdays in 1995/1996. Since 1996, there was a decline in the provision of employment; by 1998/1999 there was a 36 percent drop in total employment. This was the joint result of a reduction in program expenditure and a rise in wage rate (World Bank, 2001). The average man-days of employment per person under EAS during 1993-1994 and 1998-1999 ranged from 1 to 73 in different districts covered by the sample checks undertaken by the auditor general of India in December 2000; there was a very wide variation among the states. The average man-days of employment created per person under JRY/JGSY ranged between 2 and 27. The ministry figures indicated 7 to 21 man-days of employment per person under JRY and 9 to 18 man-days per person under EAS (Ministry of Rural Development, 2000).

13. Productive works are, however, somewhat loosely defined as those that directly or indirectly lead to an increase in production or which, if not undertaken, would cause production to decline.

14. Over the years, in certain instances and at certain times, policies were adopted to focus certain infrastructure projects on the generation of employment for the poor. Examples were the Community Employment and Development Program (1986-1988) following the 1983-1985 economic crisis, the Second Rural Roads Improvement Project 1986, and the Kabuhayan (the local Infrastructure Development Program) 2000.

15. In practice, the membership of the IA was not as extensive or comprehensive as to include 80 percent of the users; in fact only 35 percent of the prospective water users were included as members. The selection of the members was designed to bring together a homogenous group in terms of economic status and other social characteristics to promote cohesiveness. Moreover, small size was initially conducive to cooperation and minimized the transaction costs of organizing a big group.

16. The first phase involved the construction of water tanks, a reservoir, and main distribution pipes. The second phase included the construction of the retail distribution pipes as well as communal faucets or hydrants.

Chapter 8

1. Even those who were marginally but not substantially above the poverty line went to work in employment projects in bad years, when they suffered a temporary decline in income.

2. In South Asian countries, it was found that with the expansion of the public works program, the availability of projects requiring a large amount of unskilled labor declined and the proportion of projects requiring skilled labor and equipment increased. Also, with a change in the seasonal pattern of unemployment occurring more in rainy than in dry seasons, construction projects involving earthwork become inappropriate. Projects that can be implemented year-round are required. For example, repair and maintenance projects that can be carried out in several seasons become more suitable. Moreover, if the objective was to benefit the poor, restriction on labor content could be relaxed if the projects created assets predominantly that benefited the poor, either being located near where the poor live or rendering services (such as water/tank/wells) mainly used by the poor.

3. It has also been suggested that if public employment schemes are financed predominantly by an additional tax on the rich (as in EGS in India, on the urban professional and big rural landowners, etc.), the chances are that, over time, those who largely bear the financal burden may press to increase the proportion of projects that create private assets (like the Jawar well scheme in EGS in India). The pressure to deemphasize public employment projects that are in the nature of public goods may be less strong if general revenues finance them.

Chapter 9

1. The small-scale industries sector in India covers a wide spectrum of industries categorized under small, ancillary, tiny, small-scale service and business enterprises, women enterprises, and cottage segments ranging from small artisans/handicraft units, on one end, to modern production units with significant investments, on the other. Their definitions have undergone changes in recent years. Note that there is no specific definition of microenterprises. In most Asian countries, however, a microunit is one that employs up to 19 workers (Rutherford, 2000).

2. These requirements include
   1. societies registered under Societies Act, 1860, or similar State Acts;
   2. trusts registered under Public Trusts Act, 1920;
   3. federations of self-help groups (SHGs);
   4. non-banking finance companies focusing on banking with the poor;
   5. specialized cooperatives such as Mutually Aided Cooperatives Societies;
   6. other cooperatives; and
   7. new institutions focusing on banking with the poor, e.g., local area banks.

3. M-CRIL—a credit rating agency—carries out capacity assessment rating (CAR) of MFIs.

4. The poverty threshold is determined to be Rs 12,000 for a household of five members. But since income estimates are essentially notional, account must also be taken of the assets/consumer durables owned. If a household earned slightly less than Rs 2,500 per capita annually but possessed a few assets/consumer durables (e.g., a TV), it would be appropriate to classify it as nonpoor. However, in the present sample the classification was easy as *all* beneficiaries belonged to households with incomes well above the poverty threshold.

5. This is somewhat inflated because of unusually high returns in trading activities (e.g., consumer goods). Note also that all of the revenue is not attributable to investment financed by SFMC.

6. Even those who had a stake in agriculture derived most of their incomes from nonfarm activities.

7. Members of ayojan samitis are elected representatives of beneficiaries of Abhyuttan Yojana. Under this Yojana, participants are given credit in the form of fertilizer, seeds, and pesticides. (More details are given later.)

8. It is somewhat intriguing why the NGOs charged a relatively low rate of interest (12 percent per annum) when they borrowed at 11 percent per annum. The margin was much too small to cover the lending costs. Moreover, although the repayment period for the NGO was four years, the period fixed for the (individual) beneficiaries was only two years.

9. In general, however, the financial estimates reported here must be treated as illustrative.

10. Abhyutthan Yojana was launched by DHRUVA in 1996. Disbursal of credit in the form of agricultural material such as fertilizer was meant to enhance productivity. A service charge on credit is fixed at the current bank rate and is calculated on the participant's daily balance. It is repayable over a period of ten years. An issue is whether distribution of these inputs through DHRUVA is more efficient than going through normal market channels. It is not obvious, for example, that economies of scale in buying these materials would be passed on to the beneficiaries. Moreover, given the selection procedure, the incentive to nominate those with repayment capacity for SFMC loans is likely to be strong. Unfortunately, although the annual report of DHRUVA for 1998-1999 cites impressive figures of loan disbursals and number of beneficiaries under Abhyautthan Yojana, there is no reference to the target group and its participation in this scheme.

11. IIED was set up in 1994 to promote self-employment for women from the lower strata.

12. No payment was made during the period of training. This is justified by Meena Jairam, Director of IIED, on the ground that genuinely interested women would show up while others would be discouraged. But it was arguable that it could deter genuinely interested but acutely poor women if the income foregone during the training period was nonnegligible.

13. The loan had to be repaid in 15 equal monthly installments at 12 percent annual interest rate.

14. Whether different activities in the nonfarm sector are rankable in the context of a village or a cluster of villages so that promotional measures are undertaken is an important policy concern. Unfortunately, while the sample respondents were able to identify several profitable activities, their ranking was constrained by the small sample size and the absence of consistent criteria.

15. For example, a Project Advisory Committee comprised the representatives of SIDBI and participating NGOs. The meetings were held when necessary and the recommendations were forwarded to the manager (SFMC), SIDBI. The manager then decided whether it was necessary to implement any of the recommendations. In addition, an independent agency monitored SFMC. Five to ten percent of the beneficiaries were surveyed periodically and the feedback was given to the head office of SIDBI.

16. It must of course be admitted that, if the objectives were fuzzy, it was not easy to devise performance indicators. For an explanation, see Tirole (1994).

17. This, of course, did not imply that all was well with the panchayats. In fact, there was some evidence of capture of village panchayats by a few locally influential persons, resulting in diversion of Jawahar Rozgar Yojana and Integrated Rural Development Program funds. In spite of such aberrations, however, panchayats were likely to perform better than ad hoc organizations created by NGOs, as they exhibited greater honesty in their functioning and were more accountable to the village communities.

18. This follows from the fact that savings are obligatory.

19. Hina Shah, Director, International Center for Entrepreneurship and Development (ICED), made this observation.

20. Among the group beneficiaries, seven involved in agarbatti making received practical training for two months and on-the-job training for two months. Out of the four in tailoring, three were trained for six months. It may be recalled that most of the group beneficiaries were nonpoor and some in fact were relatively affluent.

21. The report of this task force was released in December 1999. For details, see NABARD (1999).

Chapter 10

1. The Maharashtra Rural Credit Project (MRCP) scheme was designed as an alternative to the Integrated Rural Development Program (IRDP)—a major credit subsidy program. Specifically, the purpose was to develop and test through field im-

plementation, an alternative approach to the IRDP that could efficiently provide improved financial services to the rural poor.

2. In most of the Asian countries, a microenterprise is a very small unit that employs up to 19 workers, a small unit is one that employs between 20 and 100 workers, and medium-sized units employ between 101 and 500 workers (SIDBI, 2000). The project with an outlay of US$48.35 million was supported by an International Fund for Agricultural Development (IFAD, 1997) loan of US$29.20 million. The contributions of the Government of India (GOI, 1999)/Government of Maharashtra (GOM) and of the participating banks were US$14.97 million and US$1.65 million, respectively.

3. In principle, the selection is by consensus.

4. MAVIM is a government agency with the responsibility of promoting income-generating activities of women.

5. The credit-savings ratio could range from 1:1 to 4:1, depending on the financial discipline displayed by the SHG.

6. Apart from lowering the transaction costs for both borrowers and lenders, it is expected that bank staff will overcome their resistance to lending to the poor in time.

7. As of September 1998, the total number of SHGs formed under the project is 2,006, out of which 815 have been linked to banks. The loans sanctioned to these SHGs were Rs 12.64 million with an average of Rs 15,460 per SHG. Over 2,500 SHGs have opened savings accounts with banks. The repayment of loans is 100 percent (UNOPS, 1998).

8. See, for example, IFAD (1997). During the fieldwork for the present study, the author learned that one of the cabinet ministers in the present government in Maharashtra is listed as belonging to a BPL household.

9. About 10 percent of VDCs were inactive and nearly 10 percent did not provide any representation to women (UNOPS, 1997). Exceptions include a village in Pune, in which an assertive SHG threw out the VDC and established an all-women VDC. In a few other cases, SHG members appealed to sympathetic bank managers and succeeded in removing undesirable members. For details, see Shankar (2000).

10. A list of individual beneficiaries engaged in nonfarm activities in the sample villages was compiled. From this list, two-thirds of the sample households were chosen on the basis of whether they belonged to any of the poverty-prone groups (i.e., whether it was an SC/ST or agricultural labor household) and five were randomly picked from the rest. For the control group (nonparticipants), households were randomly selected from the poverty-prone groups. In both cases, the households were chosen in such a way that both villages would be represented. As far as selection of SHGs and their control group was concerned, a few SHGs that functioned satisfactorily [SHGs (FS)], i.e., linked to CBs were selected. Also, a few additional groups of SHGs not linked to CBs—classified as SHGs (NFS)—were selected as a control group. A few individuals representing official agencies (NABARD; District Rural Development Agency [DRDA]; District Project Coordination Committee [DPCC], Project Steering Committee [PSC], MAVIM, MCED, and MITCON), the participating bank (Bank of Maharashtra [BOM]), an NGO, and local community organizations (VDAs and VDCs) were interviewed.

11. Given the unreliability of income estimates, account is taken of household size, caste affiliation (i.e., whether it belongs to a scheduled caste/scheduled tribe), whether it is headed by a widow and whether it owns land or any other asset. Although this introduces subjectivity, it is an improvement as it seeks to compensate for the generally poor quality of income data by taking into account various correlates of poverty. Even if the household income exceeds the poverty threshold of Rs 12,000, it is classified poor if other correlates of poverty are present (e.g., large household size—more than five persons, belongs to a SC/ST, and lacks assets).

12. These households earned extremely low incomes on a per capita basis and belonged to socially deprived groups.

13. For a rigorous test of targeting accuracy, a larger sample is necessary, based on a stratified random selection. Here the concern is confined to whether the participation of the affluent was excessive—in other words, whether it was more than considered unavoidable.

14. In general, the composition of an SHG is determined by the composition of the neighborhood to which it belongs. So, despite the fact that most of the respondents were relatively affluent, it was not surprising that the majority of the members were poor. Small loans, payable in bimonthly/monthly installments over a year, induced self-selection of the poor in SHGs through the initiative of NGOs/sahyoginis. Typically, the leaders are from the upper caste, relatively affluent, and better educated. Sahyoginis/field workers are essentially animators who interact closely with poor communities. Some individual beneficiaries admitted that they were discouraged from joining SHGs because of the inadequacy of loan amounts.

15. That there is no specific limit on loans to SHGs was confirmed by A. Dere, Assistant General Manager, NABARD, Pune, as the amount depends on the track record of the SHG, savings mobilization, proposal submitted, etc.

16. One problem is that income may be interpreted differently by different respondents (some may, for example, report incomes net of cost of labor services of household members while others may ignore them altogether). Another is attribution, i.e., how much of the extra income is attributable to the MRCP loan.

17. As these observations are equally pertinent for the returns on investments of SHG respondents, further comment is unnecessary.

18. Using an interest rate of 12 percent per annum and a maturity period of five years for a loan of Rs 20,000, a minimum return of 32 percent per annum is required.

19. In Tamil Nadu, for example, the loan recovery rate for non-SHGs is 38 percent as against 92 percent for SHGs (Karmarkar, 1999).

20. Based on a pilot project of NABARD for linking up SHGs in 1992, SHG intermediation led to reduction in the time spent by the bank staff on identification of borrowers, documentation, follow-up, and recoveries, implying a reduction of 40 percent in the transaction costs of banks. For the borrowers, on the other hand, elimination of cumbersome documentation procedures, time spent and costs incurred on repeated visits to banks, etc., meant a reduction of 85 percent in their transaction costs. However, there is some concern that the share of nonfarm activities financed by the loans is rather low. As argued in the following, this is due to small loans offered and high interest rates charged by SHGs (2 to 3 percent per month or 24 to 36

percent per annum as against 12 percent payable by individual borrowers from banks) among other reasons.

21. SHGs borrow at 12 percent per annum and lend it at substantially higher interest rates. The difference is credited to their accounts.

22. Repayment periods are short to allow high turnover of loans. This limits use of loans in agriculture, as the cycle of production is longer.

23. Sen (2000) distinguishes between constitutive and instrumental aspects of social exclusion. Not being able to participate in the mainstream of community activities constitutes deprivation. Besides, when social exclusion comes in the way of economic betterment, it has an instrumental role in deprivation.

24. The difficulties of documentation required and the potential for bribery can be gauged from the following list for individual beneficiaries:

    1. VDC's recommendation,

    2. land record of the site for the proposed activity,

    3. no objection certificate from village authorities (i.e., either the sarpanch [executive of the local panchayat] or the gram sevak),

    4. photographs,

    5. project feasibility report from MITCON,

    6. BPL card, and, if relevant,

    7. SC/ST certificate.

It may be recalled that Janardhan Ganpat Bhule—one of the poorest tribals in Sitewadi—spent Rs 1,000 on the documentation.

25. Although MCED concentrates on entrepreneurial skills and MITCON on project proposal, it is not clear who is responsible for project management advice.

26. No financial provisions were made for the formation of VDCs/VDAs. As a result, ad hocism in their formation was not surprising.

27. The exception is when it is not convenient for bank staff to attend it.

28. In an interview that the author conducted with the woman selling bangles in Karhati it was pointed out that her husband helped with domestic chores when she visited Baramati to procure bangles, but this may be unusual.

29. The fact that more than a few respondents shared this view suggests that the positive externalities were pervasive.

30. The sources of information were the block development officer (BDO), VDC members, bank officials, SHG members, sahyoginis, panchayat members, and friends and relatives.

31. Specifically, the frame maker and the vegetable/dried-fish seller were appreciative of this training, as the training helped them expand their business.

32. Each NGO is assigned six villages under the MRCP. Sahyoginis were the field workers of MAVIM or NGOs.

33. Section 11(2) states: "No company or association or partnership consisting of more than 20 persons shall be formed for carrying on business that has for its object the acquisition of gain by the company or by the individual members thereof, unless it is registered under the Act."

34. In the service area approach, a group of 20 to 25 villages is assigned to each bank branch for meeting the credit needs of potential borrowers.

35. Active participation of DCCBs is subject to amendments to the Cooperative Act to enable SHGs to borrow directly from the DCCBs (UNOPS, 1998).

Chapter 11

1. The government provides a special guarantee for the fund supplied by Agrani Bank.

2. Effective rate of interest is different from nominal rate when the latter is charged at a flat rate and not on the declining balance even if repayment is being done bimonthly/weekly.

3. The personnel of SEDP feel that the interest rates are very low and an interest subsidy to such an extent is not necessary. However, they consider the rate of interest charged by MFIs (the effective rate is in the range of 21 to 28 percent) too high and such a rate may not generate sufficient demand for the size of loans targeted by SEDP. In contrast, an impression has been created that even at the interest rates currently prevailing for commercial bank loans, there will be demand for SEDP loans. Therefore, in the next phase of SEDP proposals are being made to raise the interest rate in the highest category to 15 percent (for loans above taka 5 lakhs) and in the two lower categories, proposed interest rates are 12 and 14 percent.

4. All entrepreneurs in the present sample have been engaged in the same enterprise before receiving an SEDP loan.

5. It should be clarified that the results may suffer from selectivity bias since those who did not obtain loans are not included in the sample. However, this can still be a valid way to examine which factors influence the amount of credit sanctioned, once the borrowers are chosen. In fact, the choice of borrowers may depend on a different set of factors, influencing the initiative of the borrowers, while the amount sanctioned is determined mainly by the SEDP and Agrani Bank officials.

6. This may be due to some of the largest loans being sanctioned to new borrowers.

7. Only two types of activities within crop production are financed. These are nursery and selling the services of power tillers.

8. The official data on the purpose of the loan provide aggregates for noncrop agriculture and agro-processing. The present survey is small to provide a representative estimate of the distribution of loans. The observation on the predominance of loans for rice milling is based on general discussion with the borrowers and has been supported by the present data set.

9. The SEDP field officers closely monitor loan use thus making it possible to avoid using loans for consumption purposes as might be expected with such subsidized credit. Some cases of consumption use may not disclosed by the borrowers, however.

10. The probabilities of male and female unemployment have been based on data from the latest labor force survey report.

11. Project personnel provided a few examples as follows: (1) In the case of the service sector, demand originates in the locality. Most services have inelastic demand. For example, if a new hairdresser's shop is opened in addition to the existing shops, some of the existing buyers will go to the new shop with a very small expan-

sion in total demand. For the first few months, the new shop will find it difficult to make enough profit to survive, much less be able to repay loans. (2) For rice milling, if new enterprises are set up, each one may need to get raw materials (paddy) from places farther away, and thus the profit will rapidly fall.

For tailoring and trading, the market also lies within the locality; the price elasticity of demand is low (except during the months of festivals). Therefore, the new enterprises will find it difficult to attract clients.

Chapter 12

1. Census value added = gross margin. See Z. Bakht, "The Rural Nonfarm Sector in Bangladesh: Evolving Pattern and Growth Potential," *The Bangladesh Development Studies* (September-December), 1996, pp. 29-73.

2. This group of institutions will be referred to as the Grameen Bank Family.

3. Grameen Trust, Grameen Kallyan, and Grameen Fund do not directly engage with beneficiary groups. Rather, they serve the needs of the other sister concerns and promote research and development.

4. Professor Mohammad Yunus is the managing director of Grameen Bank.

5. It was gathered from interviews that the weaving community in the area had to depend on middlemen for sale of their produce, which often had to be on credit that ran into high figures (Tk 20,000 to 30,000). Over and above, a weaver had to bear a variable cost of Tk 20,000 per loom.

6. Under the system, the marketing agent provides raw materials to the producers and pays for the latter's labor on a piece-rate basis.

7. GU has been considering setting up a powerloom unit at Araihazar.

8. A GU unit office is an office setup, which coordinates production and delivery by weavers in about eight to ten villages. Details are presented later.

9. A total of 8,950 weavers have so far been engaged in the Sirajganj district, by organizing production through 2,095 weaver managers. The weaver-managers are loom owners who are in charge of organizing production in their factories/enterprises. They may often work on their looms as well.

10. Gano Sasthya Kendra (GSK) is a local NGO, specializing in health care services. It owns a pharmaceutical enterprise and runs a hospital for the poor in Dhaka city. Gano Sasthya Grameen Textile Mill is jointly owned by GSK and GU.

11. The statement contains unit-level information on the following aspects:

    1. volume of production in the previous month;

    2. number of member weavers;

    3. number of looms, including the nonoperative ones;

    4. volume of cumulative production; and

    5. volume of work (cloth in yard) yet to be produced by that unit.

12. GU hesitates to openly acknowledge the shift toward powerlooms. The unit in Sirajganj has 18 looms, employing 36 persons in two shifts. Although GU claims to be using the unit to train local weavers, in reality, the unit produced 20 percent of GU's gray cloth production during 1999-2000.

13. Export essentially means selling to the export-based readymade garment industry.

14. In addition to their own labor, these weaver managers employ 6,855 weavers. Thus, the GU reports mention of a total of 8,950 weavers.

15. Two units of Grameen Uddog were randomly selected from the six handloom units in Sirajganj: Chala-Belkuchi and Bawra-Belkuchi. A neighboring thana, Sirajganj Sadar, where GU is yet to open a handloom unit, was purposively selected as the control area. Information was collected through a structured questionnaire from 20 GU-weaver households from each of the handloom units. In addition, ten non-GU weaver households from the program area and ten weaver households from the control area were selected. Selection of the GU-weaver households was done randomly from the list made available by the local GU unit offices. Apart from the questionnaire-based survey, *focused group discussion* (FGD) was administered to incorporate the weaving community's perception.

16. A few leading garment entrepreneurs and GU officials were interviewed.

17. A few members of the yarn industry have also been interviewed.

18. After the success of Grameen Check, several other initiatives were taken following GU practices. The BRAC Check, whose production was organized through middlemen (from existing marketing agents), discontinued after being in operation for three years. The nonparticipants, who are unwilling to take GU orders, reported on nonavailability of work based on their past experience when the size of GU production was low. Moreover, these few respondents (three in total) have better market contacts than the rest.

19. GU sells about 5 percent of its production (either as fabric or as garments) through its retail outlets, which largely replaces imported fabrics. Traditional handloom products in the country include Saree, Lungi, and some specialized bedsheets.

20. As in the case of readymade garments, failure to deliver products on time will lead to cancellation of the work order.

21. As noted previously, the market for handloom products has declined, and the loom owners responded by varying the number of looms in operation. Along with it, the number of loom workers employed also changed. In cases of stable work orders, it is advantageous to engage in long-term contracts with the workers. If a sudden decline occurs in work orders from a monopsony, an entrepreneur (i.e., weaver-manager) will incur losses.

22. The bridge has been in operation since 1999, reducing the travel time from Sirajganj to Dhaka by almost two-thirds.

23. As noted, Sirajganj Check occupies part of the market. But its size is relatively small, even compared to GU's low scale of operation.

24. With a view to promote Grameen Check, GU has often organized fashion shows in European and U.S. cities. So far, product designs have mostly originated from the end market, mediated by the buying houses. However, GU has the choice to move into other products (non-Grameen Check), such as printed fabric.

25. We had previously noted that the seed money for GU was provided by the GBF (Grameen Bank Foundation) without any attachment. It is only after the first year that GU went into borrowing from commercial sources. The development of

corporate capital under the MFI/NGO umbrella is a new phenomenon in Bangladesh. Their comparative advantage lies in two areas: (1) the network of clients of the credit program is a resource, which may be fruitfully used for organization of production in a cost-effective way, and (2) command over financial resources (which includes donor grant and savings of group members) enables them to undertake investments.

26. BRAC identified individuals among marketing agents, who were assigned the responsibility of delivering a given quantity of products of prespecified design. These agents relied heavily on large loom owners, who had failed to deliver on time.

27. In the case of jute, market prices take account of quality differences and these prices are well integrated with the international prices, which the marketing agents receive.

28. NGOs, acting as conduits for indirect lending, are often termed "credit retailers."

29. The criterion of having "adequate workforce," from the bank's perspective, is to ensure minimum dislocation as a result of the project work.

30. There is no provision for supporting completely new entrepreneurs under direct lending. However, an established entrepreneur may venture into new products, thus diversifying his or her investment portfolio.

31. Palli Karma Sahayak Foundation (PKSF), a wholesaling outlet in credit, lends to NGOs at 3 to 5 percent per annum. In contrast, MEDU loans are available to the NGOs at an annual interest rate of 9.5 percent or more. There are, however, ceilings to PKSF's lending to individual NGOs.

32. MEDU loans amounting to less than Tk 50,000 are collateral free.

33. Previously, the MEDU interest rates were respectively 12 and 14 percent.

34. The subsidy is calculated on the basis of disbursement made under MEDU, and not on the basis of Agrani Bank's borrowings.

35. In operating such an account, the borrower is free to withdraw and repay money at any time during the period, and interests are calculated at the end of the year.

36. Current recovery rate of a MEDU loan is 88 percent, whereas recovery of a usual Agrani Bank loan is below 50 percent.

37. In contrast, the MIDAS operations (an NGO), confined to urban areas, support a wide variety of activities.

38. Noninterest cost (including expenses for paperwork) of borrowing from the two sources is roughly 2.5 percent of the loan amount. In borrowing under indirect lending, it is only 0.5 percent.

39. The discrepancy in average loan size with our survey findings is not as high as in indirect lending cases.

40. The concept of social mobilization often means different things to different people. In the context of the KST model, it may be interpreted in two different ways. First, a minimum degree of social mobilization (in the form of group formation) is involved for better banking. Second, once some success occurs in the credit operation, the group members may be mobilized for achieving other social and economic objectives.

41. During the early years, the project staff took the initiative to organize people in a neighborhood by seeking assistance from elderly/influential individuals who

called meetings. Even though the participants in the meeting were asked to raise their hands if they supported the candidacy of particular individuals for the positions of managers or presidents, very often such candidacy was sorted out prior to the meetings. Once KST credit operation became widely known in the area, it has often been the case that local youth had contacted the project office and assisted in doing the groundwork for formation of a VO in their area.

42. When the number of VOs supervised by the SOs was few, all weekly meetings were attended by the SOs. The frequency declined later. However, the VO managers regularly maintained contacts with their SOs at the project office, and/or in meetings at the ward/union levels.

43. Paras, if literally translated to English, would mean neighborhood. These are clusters of households that form a part of a village. There are currently about 1,000 VOs (covering at least 500 paras) in Kishoreganj Sadar Thana, although there are only 202 villages. Generally, the paras contain one male VO and one female VO in each.

44. A number of forms were to be filled in. We found that actual auditing was delayed by several months for a good number of VOs.

45. See Part I.

46. Some of the conventional MFIs have schemes on term deposit whereby they mobilize funds from the rural rich (nonbeneficiaries). The KST approach internalizes such activities.

47. One of the VO managers reported to us that he is planning to make explicit arrangements to share his remuneration with the leaders of the subgroups since it will enable him to coordinate a large group and thereby a large volume of credit transactions.

48. Even though some of the women's groups were found to be headed by male managers (due to nonavailability of qualified females in the group), there has been a search for their replacements with women managers. Thus, having a large number of women's groups with female managers is likely to create equal numbers of jobs (in the position of managers) for rural women as well.

49. Low figures for male VOs suggest lower interest among them.

50. Generally, a large part of such loans go to postharvest purchases of a paddy. Thus, working capital in an agro-processing industry is possibly no different than an investment on trade and services.

51. If a private exporter must rely on numerous small producers for its supply, it will be quite costly to organize this if the market is not sufficiently large. An example of the latter is the foodgrain and raw jute markets, where intermediaries have emerged over the years and market functions in a way to support the presence of private exporters. In the case of new products, an NGO/MFI does not have to make an additional investment for coordinating the producers.

52. The credit support was provided through the commercial banks that were allocated separate funds for lending to the agro-processing industry.

53. In one case, exporters of baby corn were linked with growers in the Kapasia region. In another, multinationals in the seed industry were linked with soybean growers through several NGOs. In the latter case, the NGOs were responsible for

providing the extension service and credit support to the growers and act as an agent of the seed company in return for a commission.

54. Given the design of the project, direct involvement by local institutions was peripheral. Even though it was, on paper, located within the Ministry of Agriculture, no attempt was made to transfer the ownership to any of the departments within the ministry.

Chapter 13

1. The EAS was introduced in 1,752 backward blocks in which the revamped Public Food Distribution System was in operation in 1992. The scheme aims at providing 100 days of unskilled manual work to the rural poor seeking employment. The Center and the States fund it in the 80:20 ratio, respectively. The scheme operates largely during lean agricultural periods.

2. Tehsil stands for a block of villages; a tahsildar is the officer in charge of a block.

3. However, a person who is between 15 and 18 years old can be given employment if there is no earning member in the family.

4. Following the high court directive, the EGS wage rate was hiked in conformity with the Minimum Wages Act. The piece-rates for different types of manual/unskilled work are so fixed that an average person working diligently for seven hours a day would earn a wage equal to the minimum wage prescribed for agricultural labor for the concerned zone, under the Minimum Wages Act.

5. The lack of coordination between technical and revenue departments often results in delays in execution of EGS projects, forcing the poor to seek alternative sources of employment.

6. This is down from 60:40. A few exceptions include canal works of medium and major irrigation projects, which involve rock cutting.

7. Often as a consequence of inflation of material costs some of these are deliberately included in labor costs in order to maintain this norm.

8. Often as a consequence of inflation of material costs some of these are deliberately included in labor costs in order to maintain this norm.

9. The definition of marginal and small farmers takes into account whether the land is irrigated or rainfed. Typically, the cutoff point for small farmers in rainfed areas is at least twice as much as in irrigated areas. In Baramati block (Pune district), for example, the cutoff point for small farmers is seven acres in irrigated areas and 18.25 acres in rainfed areas.

10. However, as there is a large subsidy (of Rs 45,000 in a total cost of Rs 70,000 to Rs 1,00,000), manipulation of these quotas is not implausible.

11. These estimates must be treated as notional.

12. About 50 percent of the reduction in EGS participation between 1988 and 1989 was a direct consequence of the lowering of the EGS expenditure and the hike in the wage rate. Taking this as an approximate measure of rationing, it follows that the extent of rationing was large.

13. Except for 1996 when there was a slight rise in EGS participation.

14. In 1995, for example, the share of females among the persons registered for the EGS was 57 percent while their share in total person-days of employment was about 37 percent.

15. Such estimates are no more than illustrative, as the registration numbers tend to be unreliable.

16. Even if employment created through additional crops is taken into account, this argument is likely to hold.

17. Since the cost of a well varies from Rs 70,000 to Rs 100,000, it is not too low to attract small contractors.

18. There may be elements of the poverty trap here. If there are a large number of assetless persons relative to the aggregate wealth in an agrarian economy, some of them may find employment at a wage equal to the energy intake at which efficient productivity is maintained while the rest are forced to eke out a bare subsistence. In the next period, the employed have a nutritional advantage over the rest, making it harder for the latter to improve their employment prospects and to break out of the poverty trap.

19. See, for example, Ranade (1998).

20. The top ten districts were Ahmadnagar, Auranagabad, Beed, Bhandara, Bhule, Nanded, Nasik, Osmanabad, Pune, and Solapur. For further details, see Ranade (1998).

21. This index was devised by the Centre for Monitoring the Indian Economy, taking into account social and physical infrastructure.

22. For details, see Bhende et al. (1990).

23. The poverty threshold is Rs 180 per capita per annum (at 1960-1961 prices). This threshold has been widely used in the Indian poverty literature.

24. Recall that there was a sharp hike in EGS wages in 1988.

25. The fact that EGS activity is concentrated in slack months implies that the greater the transfer benefits the greater would be the stabilization benefit as well. Net transfer benefit is gross earnings minus earnings in alternative non-EGS employment.

26. In the ICRISAT sample, the opportunity cost of time for EGS participants is non-negligible, implying that alternative employment opportunities exist. The opportunity cost is estimated to be 25 percent of gross earnings from EGS.

27. Until recently, it was stipulated that work would be provided within eight kilometers from the residence of a participant.

28. Available time allocation data suggest that women spend more time working than men. In Rajasthan villages, for example, in the age groups 19 to 34 and 34 to 44, the total work burden of females was substantially higher (142 percent and 111 percent, respectively). For other corroborative evidence, see Ravallion (1991).

29. Typically, those on daily wage contracts are closely supervised.

30. In the ICRISAT sample, an individual's long-term nutritional status, as proxied by his or her height, has a strong positive effect on participation, indicating that individuals with better nutritional status self-select themselves into the EGS.

31. The official explanation is slow disbursal of funds by the treasury. Irregular flow of funds is a recurring problem but it is not the only causal factor. Gross negligence and rampant corruption have more to do with such delays. An irksome re-

quirement is that wages must be paid on a particular day and if a participant fails to show up on that day, he or she must wait for a month to get paid. Some of the participants—especially the poor—do not complain for fear of abrupt dismissal.

32. Wages are by statute linked to the amount and kind of work done (and Byzantine formulae are posted to specify down to the size of rock broken, vertical and horizontal life of different types of loads, hardness of soil, etc.), and thus will vary significantly across individuals and projects. These complicated formulae are poorly understood by laborers and in themselves present a tailor-made opportunity for manipulation and corruption, an opportunity which is not passed over. On this issue, see also Echeverri-Gent (1993).

33. The estimating equation was of the following form:

$$WA_t = \alpha + \beta_1 WA_{t-1} + \beta_2 WA_{t-2} + \beta_3 WE_t + \beta_4 WE_{t-1} + \beta_5 WNF_t$$
$$+ \beta_6 WNF_{t-1} + \beta_7 + D_1 + \beta_8 D_2 + \varepsilon_t (3),$$

where *WA*, *WE*, and *WNF* denote farm, EGS, and nonfarm wages, respectively, with *t* as time subscript, and *D1* and *D2* as monthly dummies that took the value 1 for April and May, respectively, and 0 otherwise. In an alternative specification, a linear time trend (with *T* as an additional explanatory variable) was introduced. Using OLS, these equations were estimated with the ICRISAT data for the periods 1979 to 1984 and 1979 to 1989 (excluding the period 1985 to 1988 for lack of data). The results based on the latter are summarized here.

34. Specifically, if the wage series are random walks, a regression of one or the other can lead to spurious results. The null hypothesis of a random walk for each of the three wage series is rejected at the 5 percent level, using the Dickey-Fuller test.

35. The long-run effect of nonfarm wages is lower but significant. This presumably reflects that, while both the EGS and nonfarm activities tend to raise the reservation wage, it is only the former that has a further positive effect on agricultural wages through higher agricultural productivity. For details, see Gaiha (1997a,b).

36. Since this is an ongoing study (jointly with P. Scandizzo and K. Imai), a summary of the results is given.

37. (A) Participation in the EGS is specified in terms of whether an adult participated or not, duration of participation (number of days of participation in a year), and EGS earnings (in real terms). (B) Since monthly household income data were not available, coefficient of variation of labor earnings is used as a proxy.

38. The employment stabilization effect of EGS is shown from the fact that during the lean months (April-July) average monthly attendance on EGS sites was almost four times the average attendance during the peak season (October-January). See "Economic Liberalization, Targeted Programs and Household Food Security: A Case Study of India" by S. M. Dev et al., IFPRI, MTID Discussion Paper, May 2004.

39. In a few cases, even if the men succeeded in finding more rewarding work elsewhere, the wives joined EGS.

40. There were a few variations, i.e., either it was a joint decision of husband and wife or a family decision.

41. A few female participants attributed their awareness to their husbands.

42. The absence of this facility was a major concern of the female participants, as they were forced to divide their time between EGS and taking care of their children who often accompanied them to the work sites.

43. The absence of this facility was a major concern of the female participants, as they were forced to divide their time between EGS and taking care of their children who often accompanied them to the work sites.

44. Consider, for example, Alka Sudhakar Nipale with a total household income of Rs 12,000 per annum. Although she owned a few consumer durables (e.g., a tape recorder) that a poor household could not afford, the per capita income was only moderately higher than the poverty cutoff point (Rs 4,000 as against a poverty cut-off point of Rs 2,500). In other cases, per capita incomes were only slightly higher than the cutoff point.

45. Either the worker was paid the next day or the mukadam personally handed over the amount at the residence or friends/relatives/neighbors delivered it. In short, there were few cases of delays.

46. (A) Since there was some vagueness about whether the wages corresponded to slack or peak periods, it is appropriate to treat the figures cited here as the upper bounds for slack period wages. (B) Five poor tribals from Padoshi confirmed this. As against the daily agricultural wage of Rs 20, the EGS offered anywhere between Rs 30 and 50, depending on the volume and quality of work performed. A few respondents from Panodi, however, were somewhat ambiguous, as agricultural wages tended to be higher.

47. A few illustrations suffice here: (1) agricultural wages: Rs 60 per day (for males), Rs 40 per day (for females); (2) sugarcane sowing: either paid at the same rate as agricultural wages or Rs 1,000 per acre (on a contractual basis); (3) digging of wells: Rs 70 to 80 per day; and (4) watering of crops: Rs 50 per day.

48. Two of the poorest respondents—Kehar Punja Kadam in Padoshi and Ashok Karbhari Munde in Panodi—were somewhat unenthusiastic about the EGS, as an advance (either in cash or food grains) during a contingency was not permissible.

49. Another discordant note was struck by Manda Anil Pabal in Panodi. His reservation against EGS stemmed from the possibility of a reduction in EGS wages if the work performed did not meet certain specifications.

50. Another discordant note was struck by Manda Anil Pabal in Panodi. His reservation against EGS stemmed from the possibility of a reduction in EGS wages if the work performed did not meet certain specifications.

51. They could, for example, take time off to take care of their children or feed the cattle.

52. Eight out of the 12 poor respondents were emphatic that there were no malpractices in the payment of wages.

53. The response of one of the poorest EGS participants in Panodi is illustrative. He lamented, "We do not have access to the records maintained. We have to accept whatever the mukadam gives us. We do not complain because we are afraid that, if we complain, they may not give us the job in the future."

54. The case of Shankar Karbhari Dighe, an illiterate EGS participant in Panodi, is revealing. At the official rate of Rs 7 per square meter of digging and carrying mud, he and his wife (who worked as a team) should have been paid Rs 70 for dig-

ging and carrying ten square meters of mud per day, compared to an actual payment of Rs 35. As the alternatives were few and far between, they were forced to continue.

55. Shortly after his appointment as district collector of Dhule district in 1981, Arun Bhatia began an investigation into EGS corruption. He ordered his staff to verify 315 muster rolls randomly selected from the 44,500 that had been prepared in Dhule from September 1980 to August 1981. They detected 42 cases of misappropriation involving Rs 43,000. Bhatia estimated that total corruption in the district was at least Rs 8.6 lakhs—some 13.5 percent of total EGS expenditure in Dhule in 1980 to 1981 (Echeverri-Gent, 1993). Another estimate of corruption in the districts of Nasik, Dhule, and Jalgaon for 1993 to 1995 amounted to Rs 17 to 20 crores (Mhatre, 1997).

56. It was pointed out that technical departments/line agencies usually handed down projects ignoring the panchayat proposals. Rebuffed in this manner, the panchayats stopped taking any interest and the submission of proposals became a mere formality or ritual.

57. Seven out of the 12 poor participants, for example, denied the disincentive effects.

58. In fact, four control group respondents were emphatic that their dependence on antipoverty schemes would lessen substantially if access to credit improved. But this was contested by a few on the grounds of uncertainty of agricultural yields and returns.

59. Several respondents, including those in the control group, pointed out that the number of EGS sites had declined in recent years.

60. Kusum Nivruti Talpade (in Padoshi), for example, admitted candidly that the EGS discouraged her from seeking work outside the village. After completing work on one EGS site, she looked for work on another EGS site in the same village, rather than exploring farm/nonfarm opportunities in another village.

61. Person-days of employment declined from 188 million in 1986 to 1987 to 95 million in 1999 to 2000; see S. M. Dev et al. (2004). It is suggested that introduction of the countrywide program of JRY might have contributed to this.

Chapter 14

1. This report has relied on various mimeos, handouts, and project reports from both local and foreign sources, not separately listed here.

2. *External threat to the works:* The following illustrates the role of communities in resisting arbitrary moves of local politicians. The ARCDP projects in Ciabu were earlier threatened by the machinations of the congressional representative, who requested the national authorities to realign the projects to benefit her favored municipalities. In response, LGUs at the village and municipal levels sent petitions to the agrarian reform secretary to thwart the representative's request. The realignment request was turned down. In numerous other cases, however, the representatives were not as frustrated in their abuse of power (as in some instances reported in the CEDP). They could only be countervailed by communities who remained united

and vocal in their stand. *Internal threat to maintenance:* Communities, of course, were not themselves free from activities of its members. Some time after the completion of the construction phase, management problems arose. After a rash of complaints from the community regarding the lack of transparency in the use of the water fees, the move to constitute a Ciabu Potable Water System was approved in a general assembly. The decision remained an acrimonious affair within the village. The dispute reached a high point when the cooperative filed a case with the municipal council, where the matter remained unsettled. It was expected that the authorities would rule in favor of the larger majority of water users represented in the current water system management.

3. *Communal irrigation in the Philippines:* An irrigation facility with a coverage area of 1,000 hectares or less is classified as a communal system. Rivers with small catchment areas can be controlled with simple brush dams (which need to be rehabilitated after a significant flood). For centuries, local communities have built such dams to irrigate up to a few hundred hectares, with little assistance from any central authority. Most notable among these traditional associations are the *zanjeras*, which survive in the northern Philippines and are based on apportioning each farmer a parcel both at the upstream and downstream portions of the water source. These and other traditional associations are prototypes of the modern NIA.

4. Interestingly, the reason for their opposition lay in the land reform history of the area. In the 1970s, land reform covered only rice and corn lands. Landowners were then rumored to be planning conversion of the riceland into sugarland, as well as causing eviction of the tenants. To forestall this, a group of tenants requested an irrigation facility under the government irrigation program, thus establishing that rice was the major crop in the area. Henceforth, the irrigation facility was a source of animosity between landlords and some of the tenants.

5. The right to charge for use is given by the government's Water Board.

6. The project in Palanas (as in the other case studies) adopted a six-day workweek.

Bibliography

Chapter 1

Chadha, G. K. 1993. Nonfarm Sector in India's Rural Economy: Policy, Performance and Growth Prospects. Visiting Research Fellow Monograph Series No 220. Tokyo: Institute of Developing Economies.

Dev, S. M., and R. E. Evenson. 2003. Rural Development in India, Non-Farm and Migration (unpublished).

Haggblade, S., and D. C. Mead. 1998. An Overview of Policies and Programs for Promoting Growth of the Rural Nonfarm Economy. Paper presented at the IFPRI/World Bank Conference on Strategies for Stimulating Growth of the Rural Nonfarm Economy in Developing Countries, Airlie House, Warrenton, Virginia, May 17-21.

Hazell, P. B. R., and S. Haggblade. 1991. Rural-Urban Growth Linkages in India. *India Journal of Agricultural Economics,* 46(4): 515-529.

Hazell, P. B. R., and S. Haggblade. 1993. Farm-Nonfarm Growth Linkages and the Welfare of the Poor. In M. Lipton and J. van der Gaag (Eds.). *Including the Poor* (pp. 190-204). Washington, DC: World Bank.

Hazell, P. B. R., C. Ramasamy, and V. Rajagopalan. 1991. An Analysis of the Indirect Effects of Agricultural Growth on the Regional Economy. In P. Hazell and C. Ramasamy (Eds.), *The Green Revolution Reconsidered: The Impact of High-Yielding Rice Varieties in South India* (pp. 153-180). Baltimore: Johns Hopkins University Press.

Ho, S. P. S. 1986. The Asian Experience in Rural Nonagricultural Development and Its Relevance for China. World Bank Staff Working Paper, Number 757. Washington, DC: World Bank.

Hossain, M. 1988a. Credit for Alleviation of Rural Poverty: The Grameen Bank in Bangladesh. Research Report No. 65. Washington, DC: International Food Policy Research Institute.

Hossain, M. 1988b, Nature and impact of the Green Revolution in Bangladesh Research Report No. 67. Washington, DC: International Food Policy Research Institute.

Hossain, M. 2002. *Rural Nonfarm Economy in Bangladesh: A View from Household Survey.* Manila, Philippines: IRRI.

Leidholm, C. 1988. The Role of the Non-Farm Activities in the Rural Economies of the Asia-Pacific Region. Paper prepared for Conference on Directions and Strategies of Agricultural Development in the Asia-Pacific Region. Taipei, Taiwan, January 1988.

Otsuka, K. 1998. Rural Industrialization in East Asia. In Y. Hayami and M. Aoki (Eds.), *The Institutional Foundation of East Asian Economic Development* (pp. 447-475). London: Macmillan.

Reardon, T., K. Stamoulis, M. Elena Cruz, A. Balisacan, J. Berdugue, and K. Savadogo. 1998. Diversification of Households Incomes into Non-Farm Sources: Patterns, Determinants and Effects. Paper presented at the IFPRI/World Bank Conference on Strategies for Stimulating Growth of the Rural Nonfarm Economy in Developing Countries, Airlie House, Warrenton, Virginia, May 17-21.

Rosegrant, M. W., and P. Hazell. 2000. *Transforming the Rural Asian Economy: The Unfinished Revolution.* Hong Kong: Oxford University Press for the Asian Development Bank.

Shilpi, F. 2003. Understanding Rural Investment Climate in Bangladesh: Evidence from Enterprise and Household Surveys. World Bank (unpublished).

Start, D. 2001. The Rise and Fall of the Rural Non-Farm Economy: Poverty Impacts and Policy Opinions, *Development Policy Review* 4(19): 491-509.

UNDP. 2003. *Human Development Report 2003.* New York: UNDP.

World Bank. 2000. *World Development Report 2000.* World Bank.

World Bank. 2001. *World Development Report 2001.* World Bank.

Chapter 2

Asian Development Bank. 1997. *Microenterprise Development: Not by Credit Alone.* Manila: ADB.

Boomgaard, J. J., S. P. Davies, S. J. Haggblade, and D. C. Mead. 1992. A Subsector Approach to Small Enterprise Promotion and Research. *World Development* 20: 199-211.

Gibbons, D. S., and J. W. Meeham. 2000. The Microcredit Summit's Challenge: Working Toward Institutional Financial Self-Sufficiency While Maintaining a Commitment to Serving the Poorest Families (pp. 9-10, 22). Microcredit Summit Secretariat, June.

Goetz, A. M., and R. G. Gupta. 1996. Who Takes Credit? Gender, Power, and Control Over Loan Use in Rural Credit Programs in Bangladesh. *World Development* 24(1): 45-63.

Hashemi, S. M., S. R. Schuler, and A. P. Riley. 1996. Rural Credit Programs and Women's Empowerment in Bangladesh. *World Development* 24(4): 635-653.

Hayami Y., ed. 1998. *Toward the Rural-Based Development of Commerce and Industry: Selected Experiences from East Asia.* EDI Learning Resources Series. Washington, DC: Economic Development Institute, World Bank.

Hossain, M., and C. Diaz. 1997. Reaching the Poor with Effective Microcredit: Evaluation of a Grameen Bank Replication in the Philippines. *Journal of Philippine Development* 24(2): 275-308. Philippines: IRRI.

Islam, N. 1997. The Nonfarm Sector and Rural Development: Review of Issues and Evidence. 2020 Vision Discussion Paper 22. Washington, DC: International Food Policy Research Institute.

Kabeer, N. 1998. Money Can't Buy Me Love? Re-Evaluating Gender, Credit Improvement in Rural Bangladesh. IDS Discussion Paper 363. Sussex, U.K.

Kabeer, N. 2001. Conflicts Over Credit: Re-Evaluating the Empowerment Potential of Loans to Women in Rural Bangladesh. *World Development* 29(1): 63-84.

Khandker, S. R. 1998. *Fighting Poverty with Microcredit: Experience in Bangladesh.* New York: Oxford University Press.

Khandker, S. R., and O. H. Chowdhury. 1996. Targeted Credit Programs and Rural Poverty in Bangladesh. World Bank Discussion Papers No. 336, Washington, DC: World Bank.

Rahman, R. I. 1996. Microenterprise Development in Bangladesh: Country Report for the Project Review of Microenterprise Development in Selected DMCs. Dhaka: BIDS.

Rahman, R. 2000. Poverty Alleviation and Empowerment Through Micro Finance: Two Decades of Experience of Bangladesh. BIDS Research Monograph 30, Dhaka.

Ravallion, M., and Q. Wodon. 1997. Poor Areas or Only Poor People? Policy Research Working Paper No. 1798, World Bank, July.

Shilpi, F. 2003. Understanding Rural Investment Climate in Bangladesh: Evidence from Enterprise and Household Surveys, World Bank (unpublished).

Tendler, J., and M. A. Amorin. 1996. Small Firms and Their Helpers: Lessons on Demand. *World Development* 24(3): 407-426.

Zaman, H. 1999. Assessing the Impact of Microcredit on Poverty and Vulnerability in Bangladesh. Policy Research Working Paper 2145. Washington, DC: World Bank.

Chapter 3

Sen, B., and I. Ahmed. 1999. Social Mobilization in Kishwarganj Thana: A Survey of Issues with Grass-Roots Evidence, July 16.

Chapter 4

Asian Development Bank. 1997. *Micro Enterprise Development: Not by Credit Alone.* Manila: ADB.

Chapter 5

Ahmed, S. 2000. Creating Autonomous National and Sub-Regional Microcredit Funds. Dhaka: PKSF. Mimeo.

Bangladesh Institute of Development Studies. 1999. Benchmark Report, BIDS Study on PKSF's Monitoring and Evaluation System, BIDS, Dhaka.

World Bank, Private Sector Development and Finance, Bangladesh Dhaka Office. 1999. A Study on the Impacts of Microcredit on Borrowers of Partner Organization (POs) of Palli Karma-Sahayak Foundation (PKSF), World Bank.

Zohir, S., Mahmud, S., Sen, B., Asaduzzaman, M., Islam, J., Ahmed, N., Mamun, A. 2001. Monitoring and Evaluation of Microfinance Institutions.

Chapter 6

Ahmed, S. 2004. Microcredit: Giving the Poor a Chance. *The Daily Star,* February 13, pp. 12-14, Dhaka (Asia Pacific Region Micro Credit Summit Special).

Chowdhury, S. H. 2004. Managing Cost Effective Micro Finance Operations, *The Daily Star,* Dhaka, Micro Credit Special, February 14, pp. 13-14.

The Daily Star. 2004. Grameen Model Not for All, February 17, p. 16. Micro Credit Summit.

Gallardo, J. 1997. Leasing to Support Small Business and Micro Enterprises, Policy Research Working Paper, World Bank, December.

Matin, I., and Hulme, D. 2003. Programmes for the Poorest: Learning from the IGVGD Programme in Bangladesh, *World Development* 3(3), March: 647-665.

Palli Karma Sayak Foundation. 2000. Impact of Microcredit. PKSF: Dhaka.

Pretes, M. 2002. Micro Equity and Micro Finance, *World Development* 30(8), August: 1341-1353.

Schreiner, M., and Noller, G. 2003. Micro Enterprise Development in the U.S. and the Developing World, *World Development* 31(9), September: 1567-1580.

Sen, B., and S. Begum. 1998. Methodology of Identifying the Poorest at Local Level. WHO-ICO Division Technical Paper No. 27, Geneva, Switzerland.

Zaman, H. 1999. Assessing the Impact of Microcredit on Poverty and Vulnerability in Bangladesh. Policy Research Working Paper 2145. Washington, DC: World Bank.

Chapter 7

Ahmed, M. U., and Y. Shams. 1994. Nutritional Effects of Cash- versus Commodity-Based Public Work Programs. Bangladesh Food Policy Project, IFPRI, Washington, DC.

Bird, R. 1994. Decentralizing Infrastructure: For Good or for Ill? Policy Research Working Paper 1258. Washington, DC: World Bank.

Clay, E. 1986. Rural Public Works and Food-for-Work: A Survey. *World Development* 14:1237-1252.

Hirway, I., and P. Terhal. 1994. *Toward Employment Guarantee in India.* New Delhi: Sage.

Hossain, M., and M. M. Akash. 1993. Public Rural Works for Relief and Development: A Review of the Bangladesh Experience. International Food Policy Research Institute. Working Paper No. 7, June 1993, Washington, DC.

Islam, N. 1999. Poverty Alleviation Measures in India and Bangladesh: Lessons of Experiences (mimeo).

MRD. 2000. Auditor General's Report No. 3 of 2000, Ministry of Rural Development, Rural Employment Generation Program, December.

Neelakantan, M. 1994. Jawahar Rozgar Yojana: An Assessment through Concurrent Evaluation. *Economic and Political Weekly* 29(49): 3091-3097.

Osmani, S. R., and O. H. Chowdhury. 1993. Short-Run Impacts of Food-for-Work Program in Bangladesh. *The Bangladesh Development Studies* 11(1-2), March-June. Also, World Bank, Bangladesh from Counting the Poor to Making the Poor Count, PREMN, South Asia Region, April 29, 1998.

Parameswaran, I. 1994. Creating Rural Employment, *Economic and Political Weekly,* August 4, pp. 2065-2066.

Ravallion, M. 1989. On the Coverage of Public Employment Schemes for Poverty Alleviation, Washington, DC: World Bank (mimeo).

Ravallion, M. 1991a. Market Responses to Anti-Hunger Policies: Wages, Prices and Employment. In J. Dreze and A. Sen (Eds.) *The Political Economy of Hunger,* Vol. II. (pp. 241-278). Oxford: Clarendon Press.

Ravallion, M. 1991b. Reaching the Poor through Rural Public Works: Arguments, Evidence and Lessons from South Asia. *World Bank Research Observer* 6: 153-175.

Shanker, K. 1994. Jawahar Rozgar Yojana: An Assessment in UP, *Economic and Political Weekly,* July 16, pp. 1845-1848.

Thamarajakshi, R. 1997. *Micro Interventions for Poverty Alleviation in India.* New Delhi: ILO (SAAT).

WGTFI (Working Group on Targeted Food Interventions). 1993. Options for Targeting Food Interventions in Bangladesh. IFPRI in collaboration with the Government of Bangladesh.

World Bank. 1998. Bangladesh: From Counting the Poor to Making the Poor Count. PREMN, South Asia Region, April 29. World Bank.

World Bank. 2001. *World Development Report 2001.*

Chapter 9

Acharya, S., and A. Mitra. 2000. The Potential for Rural Industries and Trade to Provide Decent Work Conditions: A Data Reconnaissance in India. SAAT Working Paper. New Delhi: ILO.

Fan, S., P. Hazell, and S. Thorat. 1999. Linkages Between Government Spending, Growth and Poverty in Rural India. Research Report No. 110. Washington, DC: IFPRI.

Fisher, T. and V. Mahajan. 1997. *The Forgotten Sector: Non-Farm Employment and Enterprises in Rural India.* New Delhi: Oxford and IBH Publishing Co. Pvt. Ltd.

FWWB (Friends of Women's World Banking). 1998. India's Emerging Federation of Women's Savings and Credit Groups. Ahmedabad (mimeo).

Gaiha, R. 1999. Income and Expenditure Switching Among the Impoverished During Structural Adjustment in India. Faculty of Management Studies, University of Delhi (mimeo).

Gaiha, R. 2000. The Exclusion of the Poorest: Emerging Lessons from the Maharashtra Rural Credit Project, India. Rome: IFAD (PI).

Gaiha, R., P. D. Kaushik, and V. Kulkarni. 1998. Jawahar Rozgar Yojana, Panchayats, and Poor in Rural India. *Asian Survey* 38: 928-949.

Karmakar, K. G. 1999. *Rural Credit and Self-Help Groups: Micro Finance Needs and Concepts in India.* New Delhi: Sage.

Lanjouw, P., and A. Shariff. 2000. Rural Non-Farm Employment in India: Access, Incomes and Poverty Impact. New Delhi: National Council for Applied Economic Research (NCAER) (mimeo).

NABARD (National Bank for Agriculture and Rural Development). 1999. Report of the Task Force on Supportive Policy and Regulatory Framework for Micro Finance. NABARD, Mumbai (mimeo).

Rutherford, S. 2000. *The Poor and Their Money.* New Delhi: Oxford University Press.

Small Industries Development Bank of India (SIDBI). 2000. *SIDBI Report on Small Scale Industries Sector 2000.* India: Lucknow.

Study Group on the Rural Non-Farm Sector (SGRNFS). 1994. The Rural Non-Farm Sector in India (National Report). SGRNFS, New Delhi (mimeo).

Tirole, J. 1994. The International Organization of Government. *Oxford Economic Paper*, Vol. 46.

Chapter 10

Asian Development Bank (ADB). 1997. *Microenterprise Development: Not By Credit Alone.* Manila: ADB.

Gaiha, R. 2001. Micro Credit, Micro Enterprises and Rural Poor: A Review of SIDBI Foundation for Micro Credit (typescript).

Government of India (GOI). 1999. *Swarnjayanti Gram Swarozgar Yojana: Guidelines.* New Delhi: Ministry of Rural Development.

IFAD. 1997. India: Maharashtra Rural Credit Project—Mid-Term Review/Evaluation Report. Rome (mimeo).

IFAD. 1999. India: National Micro-Finance Support Project—Formulation Report. Rome: Asia and the Pacific Division, Program Management Department (mimeo).

Karmakar, K. G. 1999. *Rural Credit and Self-Help Groups: Micro Finance Needs and Concepts in India.* New Delhi: Sage.

Khandker, S. R. 1998. *Fighting Poverty with Microcredit: Experience in Bangladesh.* New York: Oxford University Press.

Mellor, J., ed. 1976. *Agriculture on the Road to Industrialization.* Baltimore: Johns Hopkins University Press.

Morduch, J. 1999. The Grameen Bank: A Financial Reckoning. Princeton, NJ: Princeton University (mimeo).

NABARD. 1999. Report of the Task Force on Supportive Policy and Regulatory Framework for Micro Finance (mimeo).

Sarap, K. 1986. Small Farmers' Demand for Credit with Special Reference to Sambalpur District, Western Orissa. PhD Dissertation, University of Delhi.

Sen, A. 2000. Social Exclusion: Concept, Application, and Scrutiny. Social Development Paper No. 1. Manila: Asian Development Bank.

Shankar, S. 2000. Partners in Empowerment: NGOs and Government in Maharashtra Rural Credit Project. To be published by IFAD (mimeo).

Srinivasan, T. N. 1994. Some Analytics of Borrowing and Lending. In P. Ghate et al., Eds., *Informal Finance: Some Findings from Asia* (pp. 207-217). New York: Oxford University Press.

UNOPS. 1997. IND/93/FOI—Maharashtra Rural Credit Project. Kuala Lumpur (mimeo).

UNOPS. 1998. IND/93/FOI—Maharashtra Rural Credit Project. Kuala Lumpur (mimeo).

Chapter 11

Kabeer, N. 1998. Money Can't Buy Me Love? Re-Evaluating Gender, Credit and Empowerment in Rural Bangladesh. Discussion Paper No. 363. Sussex: IDS.

Khandker, S. R. 1998. *Fighting Poverty with Microcredit: Experience in Bangladesh.* New York: Oxford University Press.

Mahmud, W. 1996. Employment Patterns and Income Formation in Rural Bangladesh: The Role of Rural Non-Farm Sector. *The Bangladesh Development Studies* 24 (3-4): 1-27.

MTR Scanteam. 2000. Agrani Bank Small Enterprises Development Project, Mid-Term Review Final Report, Dhaka.

Rahman, R. I. 2000. Tractor Use, Irrigation, and Agricultural Productivity in Bangladesh. In Abu Abdulah (Ed.). *Bangladesh Economy 2000: Selected Issues* (pp. 37-72). Dhaka: BIDS.

Scanteam, M. T. R. 2000. Agrani Bank Small Enterprises Development Project. Mid-Term Review Final Report, Dhaka.

SEDP. 1999-2000. *Annual Progress Report,* various years. Dhaka: Small Enterprises Development Project.

Chapter 12

Ahmed, S. 2000. Creating Autonomous National and Sub-Regional Microcredit Funds. Dhaka: PKSF (mimeo).

Bakht, Z. 1996. The Rural Nonfarm Sector in Bangladesh: Evolving Pattern and Growth Potential, in *The Bangladesh Development Studies,* September-December: 29-73. Dhaka.

Zohir, S. 1999. Strengthening Interlinkages between Rural Credit, Marketing and Agricultural Extension. Paper prepared for the UN-ESCAP, Bangladesh Institute of Development Studies, Dhaka, April.

Zohir, S. 2000. The KST Project: An Appraisal. Report prepared for the World Bank, final report, Dhaka, February.

Zohir, S., Mahmud, S., Sen, B., Asadazzaman, M., Islam, J., Ahmed, N., and Mahmun, A. 2000. Impact of Microfinance on Rural Households in Bangladesh: Findings from BIDS Surveys. Bangladesh Institute of Development Studies, Dhaka, December.

Chapter 13

Bhende, M. J., T. S. Walker, S. Lieberman, and J. Venkatraman. 1990. The EGS and the Poor: Evidence from Longitudinal Village Studies. ICRISAT, Patancheru (mimeo).

Dev, S. 1993. India's (Maharashtra) Employment Guarantee Scheme: Lessons from Long Experience. Indira Gandhi Institute of Development Research, Bombay (mimeo).

Dev, S. M., C. Ravi, B. Viswanathan, A. Gulati, and S. Ramachandar. 2004. Economic Liberalization, Targeted Programs, and Household Food Security: A Case Study of India. IFPRI-MTID Discussion Paper, May 2004.

Echeverri-Gent, J. 1993. *The State and the Poor.* Berkeley: University of California Press.

Gaiha, R. 1997a. Do Rural Public Works Influence Agricultural Wages? The Case of the Employment Guarantee Scheme in India. *Oxford Development Studies* 25: 315-344.

Gaiha, R. 1997b. Rural Public Works and the Poor: The Case of the Employment Guarantee Scheme in India. In S. Polachek (Ed.), *Research in Labor Economics,* vol. 16 (pp. 235-269). Greenwich, CT: JAI Press.

Mhatre, A. 1997. Twenty-Five Years of EGS: Strengths and Weaknesses (in Marathi), Mumbai: Y. B. Chavan Pratishthan Publications.

Ranade, A. 1998. Maharashtra's EGS: Regional Patterns, Scope for Reforms, and Replication. Indira Gandhi Institute of Development Research, Mumbai (mimeo).

Ravallion, M. 1991. Reaching the Poor through Rural Public Works: Arguments, Evidence and Lessons from South Asia. *World Bank Research Observer* 6: 153-175.

Chapter 14

Acharya, S., and V. G. Panwalkar. 1988. *The employment guarantee scheme in Maharashtra.* Bangkok: The Population Council.

Ahmed, M. U., and Y. Shams. 1994. Nutritional effects of cash- versus commodity-based public work programs. Bangladesh Food Policy Project, IFPRI, Washington, DC.

Ahmed, S. 2000. Creating autonomous national and subregional microcredit funds. PKSF, Dhaka. (Draft).

Apte, M. 1985. A study of poverty alleviation in rural India through employment program. ARTEP/ILO, New Delhi. (Draft).

Asian Development Bank (ADB). 1997. *Micro enterprise development: Not by credit alone.* Manila: ADB.

Balisacan, A. 1997. Getting the story right: Growth, redistribution, and poverty alleviation in the Philippines. *Philippine Review of Economics and Business* 34:1-37.

Bandyopadhayay, D. 1988. Direct intervention programs for poverty alleviation. *Economic and Political Weekly* (June 25).

Bangladesh Institute of Development Studies (BIDS). 1999. Benchmark report: BIDS study on PKSF's monitoring and evaluation system. BIDS, Dhaka.

Basu, K. 1981. Food for work program: Beyond roads that get washed away. *Economic and Political Weekly* 16: 37-40.

Besley, T., and S. Coate. 1992. Workfare versus welfare: Incentive arguments for work requirements in poverty alleviation programs. *The American Economic Review* 82: 249-261.

Bird, R. 1994. Decentralizing infrastructure: For good or for ill? Policy Research Working Paper 1258. Washington, DC: World Bank.

CDF. Various years. *CDF statistics*. Dhaka: Credit and Development Forum.

Chadha, G. K. 1993. Nonfarm sector in India's rural economy: Policy, performance, and growth prospects. Visiting Research Fellow Monograph Series No. 220. Tokyo: Institute of Developing Economies.

Clay, E. 1986. Rural public works and food for work: A survey. *World Development* 14: 1237-1252.

Dandekar, K., and M. Sathe. 1980. Employment guarantee scheme and food for work program. *Economic and Political Weekly* 15: 707-718.

Datt, G., and M. Ravallion. 1994. Transfer benefits from public works employment. *Economic Journal* 104: 1346-1369.

del Ninno, C. 2001. Improving the efficiency of targeted food programs in Bangladesh: An investigation of the VGD and RD programs. IFPRI, Washington, DC, May (unpublished).

Deolalikar, A. 1995. Special employment programs and poverty alleviation. *Asian Development Review* 13: 50-73.

Deolalikar, A., and R. Gaiha. 1993. What determines female participation in rural public works? The case of India's employment guarantee scheme. Paper presented at the Royal Economic Society Conference, Exeter. (Draft).

Dev, S. 1993. India's (Maharashtra) employment guarantee scheme: Lessons from long experience. Indira Ghandhi Institute of Development Research, Bombay. (Draft).

Development Academy of the Philippines. 1998. A study on the potential expansion of labor-based, equipment-supported infrastructure programs in the Philippines. Manila: International Labor Organization (Asia-Pacific) and the Development Academy of the Philippines.

Engkvist, R. 1995. Poverty alleviation and rural development through public works: The case of the employment guarantee scheme in Maharashtra. Department of Economics, University of Lund, Sweden. (Draft).

Estudillo, J., and K. Otsuka. 1999. Green revolution, human capital, and off-farm employment: Changing sources of income among farm households in Central Luzon, 1966-1994. *Economic Development and Cultural Change* 47: 497-523.

Gaiha, R. 1997a. Do rural public works influence agricultural wages? The case of the employment guarantee scheme in India. *Oxford Development Studies* 25.

Gaiha, R. 1997b. Rural public works and the poor: The case of the employment guarantee scheme in India. In S. Polachek (Ed.), *Research in Labor Economics*, vol. 16. Greenwich, CT: JAI Press.

Gaiha, R., P. D. Kaushik, and V. Kulkarni. 1998. Jawahar Rozgar Yojana, panchayats and poor in rural India. *Asian Survey* 38: 928-949.

Goetz, A. M., and R. G. Gupta. 1996. Who takes credit? Gender, power, and control
over loan use in rural credit programs in Bangladesh. *World Development* 24(1):
45-63.

Government of Bangladesh. 1999. Micro-credit programs by the public sector in
Bangladesh: A survey. Finance Division, Government of Bangladesh, Dhaka.

Government of India. 1999. Swarnjayanti Gram Swarozgar Yojana: Guidelines.
New Delhi: Ministry of Rural Development.

Government of India, Ministry of Rural Development. 1994. Report on concurrent
evaluation of Jawahar Rozgar Yojana (JRY), January-December 1992.

Hashemi, S. M., S. R. Schuler, and A. P. Riley. 1996. Rural credit programs and
women's empowerment in Bangladesh. *World Development* 24(4): 635-653.

Hazell, P. B. R., and S. Haggblade. 1991. Rural-urban growth linkages in India.
Indian Journal of Agricultural Economics 46(4): 515-529.

Hazell, P. B. R., C. Ramasamy, and V. Rajagopalan. 1991. An analysis of the indi-
rect effects of agricultural growth on the regional economy. In P. Hazell and C.
Ramsamy (Eds.) *The Green Revolution Reconsidered: The Impact of High-
Yielding Rice Varieties in South Asia* (pp. 153-180). Baltimore: Johns Hopkins
University Press.

Herring, R. J., and R. M. Edwards. 1983. Guaranteeing employment to the rural
poor: Social functions and class interests in the employment guarantee scheme in
Western India. *World Development* 11: 575-592.

Hirway, I., and P. Terhal. 1994. *Toward employment guarantee in India.* New
Delhi: Sage Publications.

Hossain, M., and M. M. Akash. 1993. Public rural works for relief and development:
A review of the Bangladesh experience. Working Papers No. 7, June, International Food Policy Research Institute, Washington, DC.

Hossain, M., and M. Asadazzaman. 1983. An evaluation of the special public works
program in Bangladesh. *Bangladesh Development Studies* 11(1-2, March-June):
191-226.

Husain, A. M. M. 1998. Poverty alleviation and empowerment. The Second IAS of
BRAC's Rural Development Programs, Dhaka.

IFPRI. 1994. Working group on targeted food interventions: Dhaka. IFPRI, Washington, DC.

International Labor Organization. 1983. *Target Wages, Actual Wages and Small
Contractors: The Pakyaw System in the Philippines,* vol. 1. *Main Report.* Manila: International Labor Organization.

Islam, N. 1997. The nonfarm sector and rural development: Review of issues and
evidence. 2020 Vision Discussion Paper No. 22. Washington, DC: International
Food Policy Research Institute.

Joshi, A. 1998. Mobilizing the poor? Activism and the employment guarantee
scheme, Maharashtra. IDS, Brighton. (Draft).

Khandker, S. R. 1998. *Fighting Poverty with Microcredit: Experience in Bangladesh.* New York: Oxford University Press.

Khandker, S. R., and O. H. Chowdhury. 1996. Targeted credit programs and rural poverty in Bangladesh. Discussion Paper No. 336. World Bank, Washington, DC.

Kochar, A. 1999. Smoothing consumption by smoothing income: Hours of work responses to idiosyncratic agricultural shocks in rural India. *Review of Economics and Statistics* 81: 50-61.

Kranti, S. 1997. Report of the baseline survey on micro-enterprises. Prepared for the MEDU, Agrani Bank, Dhaka.

Lanjouw, J. O., and P. Lanjouw. 1995. Rural nonfarm employment: A survey. Policy Research Working Paper No. 1463. World Bank, Washington, DC.

Lanjouw, P., and A. Shariff. 2000. Rural nonfarm employment in India: Access, incomes, and poverty impact. NCAER, New Delhi. (Draft).

Leidholm, C. 1988. The role of the nonfarm activities in the rural economies of the Asia-Pacific region. Paper prepared for Conference on Directions and Strategies of Agricultural Development in the Asia-Pacific Region, Taipei, Taiwan, January.

Mahmud, W. 1996. Employment patterns and income formation in rural Bangladesh: The role of rural nonfarm sector. *Bangladesh Development Studies* 24(3-4): 1-27.

Matin, I. 1998. Mis-targeting by the Grameen Bank: A possible explanation. *IDS Bulletin* 29(4): 51-58.

Mhatre, A. 1997. 25 years of EGS: Strengths and weaknesses (in Marathi). Y. B. Chavan Pratishthan Publication, Mumbai, India.

Morduch, J. 2000. The microfinance schism. *World Development* 28(4): 617-629.

National Bank for Agriculture and Rural Development. 1999. Report of the task force on supportive policy and regulatory framework for microfinance. NABARD, Mumbai. (Draft).

Neelakantan, M. 1994. Jawahar Rozgar Yojana: An assessment through concurrent evaluation. *Economic and Political Weekly* 29(49).

Osmani, S. R., and O. H. Chowdhury. 1993. Short-run impacts of food-for-work program in Bangladesh. *Bangladesh Development Studies* 11(1-2, March-June).

Palli Karma Sahayak Foundation (PKSF). 2000. Impact of microcredit. PKSF, Dhaka. (Draft).

Parameswaran, I. 1994. Creating rural employment. *Economic and Political Weekly* (August): 2065-2066.

Rahman, R. I. 2000. Poverty alleviation and empowerment through microfinance: Two decades of experience in Bangladesh. Research Monograph 20. Dhaka: BIDS, June.

Rahman, R. I., and S. R. Khandker. 1995. Role of targeted credit programs in promoting employment and productivity of the poor in Bangladesh. *Bangladesh Development Studies* 22(2-3): 49-52.

Ravallion, M. 1989. On the coverage of public employment schemes for poverty alleviation. World Bank, Washington, DC. (Draft).

Ravallion, M. 1991. Reaching the poor through rural public works: Arguments, evidence, and lessons from South Asia. *World Bank Research Observer* 6: 153-175.

Rosegrant, M. W., and P. B. R. Hazell. 2000. *Tranforming the Rural Asian Economy: The Unfinished Revolution.* Hong Kong: Oxford University Press for the Asian Development Bank.

Scanteam, MTR. 2000. Agrani Bank small enterprises development project, Mid-Term Review. Final Report, Dhaka.

SEDP. 1999 and 2000. Annual progress report. Dhaka: Small Enterprises Development Project.

Sen, B., and I. Ahmed. 1999. Social mobilization in Kishwarganj Thana: A survey of issues with grass-roots evidence. July 16 (Draft).

Sen, B., and S. Begum. 1998. Methodology of identifying the poorest at the local level. Technical Paper (November 27). WHO, Geneva.

Shanker, K. 1994. Jawahar Rozgar Yojana: An assessment in UP. *Economic and Political Weekly* (July 16): 1845-1848.

Shuler, S. R., S. M. Hasheim, A. P. Riley, and S. Akhter. 1996. Credit programs, patriarchy, and men's violence against women in rural Bangladesh. *Social Science and Medicine* 43(12): 1729-1742.

Tendler, J., and M. Alves Amorin. 1996. Small firms and their helpers: Lessons on demand. *World Development* 24(3): 407-426.

Thamarajakshi, R. 1997. *Micro interventions for poverty alleviation in India.* New Delhi: ILO (SAAT).

UNOPS. 1997 and 1998. IND/93/FOI-Maharashtra rural credit project. Kuala Lumpur. (Draft).

WFP. 1997. Report on work norms and wage rates for food-assisted works in Bangladesh, Part I. A study undertaken for the Government of Bangladesh, May, Dhaka.

Zohir, S. 2000. The KST project: An appraisal. Report prepared for the World Bank. Final report, February, Dhaka.

Zohir, S., Mahmud, S., Sen, B., Asaduzzaman, M., Islam, J., Ahmed, N., and Mamun, A. 2000. Impact of microfinance on rural households in Bangladesh: Findings from BIDS surveys. Bangladesh Institute of Development Studies, December, Dhaka. (Draft).

Index

Page numbers followed by the letter "t" indicate tables.

Training *(continued)*
  poor excluded from, 155
  screening through, 69
  specialized, 34, 38-39, 44, 45, 56,
    73, 88-89, 99, 100, 105,
    155-156, 168-169 (*see also*
    Job training)
  technology-related and technical,
    31, 34, 38-39, 105
  types of, 38-39
Transaction costs, 102
Transport
  employment in, 5t-7t, 9, 10t-11t
  gender composition of, 9, 10t-11t
  livelihood (household) and
    microenterprises, 20t
  support for, 55, 186
  urban goods distribution through
    rural, 19
Transportation
  availability of, 257
  industry activities, increasing, 12
  nonfarm sector, impact on, 13, 76,
    108, 112-113
Tribal groups, 142-143, 260
Trishal thana (Mymensingh district,
    India), 173
Tumbaga Access Road Project (ARP),
    281, 282-284, 290, 293-296,
    294t, 295t, 298, 302, 303t
Typhoons, 290

Underemployment, 296
Underpayment, 116, 269, 272, 274, 291
Unemployment
  combating, 111, 123, 124, 131, 134,
    247
  seasonal, 126, 134
  targeting of, 212
  in unskilled labor market, 190
United Nations Development Programme
    (UNDP), 43, 44, 60, 195, 234,
    236, 238, 240, 251
Unskilled labor, 123, 190, 268, 296
Urban enterprises, 36, 106
Urban sector, 5t-7t, 9, 124
Urbanization, increasing, 9
Urban/rural links, 19, 106-107,
    108-109, 112, 250

USAID (United States Agency for
    International Development),
    58, 250, 252
User associations, 288-289, 290, 290t,
    291, 306-307, 308

Village assistants (extension agents),
    248
Village committees (ayojan samitis),
    144, 147, 153, 154
Village development assemblies
    (VDAs), 61, 70, 157,
    165, 170
Village development councils (VDCs),
    61, 67, 70, 71t-72t, 80t, 82,
    87, 157-158, 160, 161, 163,
    164, 165, 170
Village organizations (VOs)
  borrower screening, involvement
    in, 87
  credit incentives of, 234
  credit program monitoring by,
    244-245
  financial viability of, 236
  lending activities, role in, 80t
  managers, 58-59, 60, 82, 86,
    235-236, 238, 238t, 243-245,
    243t, 248, 251
  marketing, role in, 54, 247
  members, 43, 44, 241-242, 241t,
    242t, 246t, 247, 249t, 251
  poverty targeting through, 240-242,
    241t, 242t
  presidents, 235, 238, 238t
  purpose of, 43, 239, 240
  rural development, models
    of, 238
  size of, 237
  training through, 89
Village-level extension workers
    (volunteers), 247
Village-level workers (VLWs), 144,
    163, 255
Villages (barangays), 276, 277, 279,
    283, 292
Violence, credit impact on, 30
Voluntary organizations of poor, 122